MCQs in Clinical Imaging

Asif Saifuddin

Senior Registrar in Diagnostic Radiology
Leeds General Infirmary, St James's University Hospital,
and Bradford Royal Infirmary
UK

D0103195

CHAPMAN & HALL MEDICAL
London · Glasgow · New York · Tokyo · Melbourne · Madras

Published by Chapman & Hall, 2–6 Boundary Row, London SE1 8HN

Chapman & Hall, 2–6 Boundary Row, London SE1 8HN, UK

Blackie Academic & Professional, Wester Cleddens Road, Bishopbriggs, Glasgow G64 2NZ, UK

Chapman & Hall Inc., 29 West 35th Street, New York NY10001, USA

Chapman & Hall Japan, Thomson Publishing Japan, Hirakawacho Nemoto Building, 6F, 1-7-11 Hirakawa-cho, Chiyoda-ku, Tokyo 102, Japan

Chapman & Hall Australia, Thomas Nelson Australia, 102 Dodds Street, South Melbourne, Victoria 3205, Australia

Chapman & Hall India, R. Seshadri, 32 Second Main Road, CIT East, Madras 600 035, India

First edition 1993

© 1993 Asif Saifuddin

Typeset in Great Britain by Columns Design and Production Services Limited
Printed in Great Britain by St Edmundsbury Press, Bury St Edmunds, Suffolk

ISBN 0 412 47970 2

A catalogue record for this book is available from the British Library

Library of Congress Cataloging-in-Publication data
Saifuddin, Asif.
    MCQs in clinical imaging / Asif Saifuddin. — 1st ed.
        p.   cm.
    Includes bibliographical references and index.
    ISBN 0–412–47970–2 (alk. paper)
    1. Diagnostic imaging—Examinations, questions, etc.   I. Title.
    [DNLM: 1. Diagnostic Imaging—examination questions.   WN 18 S1325m]
    RC78.7.D53S27   1993
    616.07'54'076—dc20
    DNLM/DLC
    for Library of Congress   92–49566
                                                                           CIP

To my parents, for all the interest they have taken in my education throughout the years, and to my wife, Saira, and my son, Hashim, for their remarkable patience during the preparation of this work

Contents

Foreword

In producing this book of *MCQs in Clinical Imaging*, Dr Asif Saifuddin has provided much more than just another crammer for examination candidates. Certainly it provides useful practice for those approaching the Final Examination for the Fellowship of the Royal College of Radiologists, but one of the book's main strengths is the answer section, which provides concise, accurate accounts of the subject matter on which each question is based. Thus, for examination candidates, the book has a dual role, providing both practice and knowledge.

The information has been obtained from the most up-to-date and authoritative sources, mainly North American and British. Dr Saifuddin has ensured further refinement and immediacy by having questions and answers verified by a consultant radiologist experienced in the particular subspeciality. Thus Fellowship candidates can rely on both the format of the questions and the accuracy of the answers. They should note, however, that at the time of writing, the Fellowship MCQ papers have more emphasis on traditional X-ray radiology and other subjects within the new syllabus than does this book, and that Dr Saifuddin has no intention of trying to replicate the balance of a Fellowship paper. By concentrating on the newer methods of imaging, Dr Saifuddin has also avoided replicating material that is already available in other books of MCQ, which this book therefore complements. Those studying radiology in other countries will also find the contents useful, whatever the format of their examinations.

There is much interest at present, and certainly there will be more in the future, in continuing education and maintenance of competence of practising radiologists. I see this book playing a useful role in this process by providing the established radiologist with up-to-date information on clinical imaging. Such knowledge acquired at the fireside or in the library will complement attendance at courses and conferences and prevent the reader being labelled a 'poor performer'.

All radiologists can benefit by owning a copy.

Keith Simpkins FRCP FRCPE FRCR FRACR (Hon.)
Consultant Radiologist,
Leeds General Infirmary,
and Senior Clinical Lecturer,
University of Leeds

Introduction

Multiple-choice questions form an integral part of the Final Examination for the Fellowship of the Royal College of Radiologists. Over the last five years there have been major advances in imaging techniques, which have resulted in many new imaging features of various disease processes being described. Although there are several MCQ books already published, none has concentrated heavily on imaging other than plain radiography and conventional contrast studies. The aim of this book is to provide the candidate with questions on imaging procedures such as angiography, scintigraphy, ultrasound, computed tomography (CT) and magnetic resonance imaging (MRI). The book is divided into nine chapters covering the major systems and sub-specialities, with a final chapter that contains questions on skeletal trauma and extracranial head and neck imaging.

The questions and answers have been prepared from various texts and articles, as listed at the end of the book. In an attempt to make the information contained as clear and accurate as possible, the questions and answers have been assessed by several consultant radiologists which may have resulted, on occasion, in slight changes to some answers compared to the source of the information.

It is important for candidates answering multiple-choice questions to understand what the examiner means. In accordance with accepted conventions, in this book words have been used in the questions that can be taken to mean the following. A feature is considered 'typical' if that feature should be present in the vast majority of cases, and if absence of that feature would make the diagnosis very unlikely. A feature is considered 'characteristic' if its presence is very suggestive of a particular diagnosis. This feature, however, would not have to be present in a high proportion of cases. The words 'majority', 'most commonly' or 'usually' simply mean more than 50% of cases. Words such as 'commonly', 'frequently' or 'likely' have not been used in the questions.

Several abbreviations have been used and their meanings should be self-evident. In relation to MRI, T1-WSE and T2-WSE scans refer to T1- or T2-weighted spin echo sequences. PDW and GE refer to proton density weighted and gradient echo sequences respectively.

I would like to acknowledge all those colleagues who gave up their time to look at questions. They include Drs R. J. Arthur, R. F. Bury, B. M. Carey, A. G. Chalmers, A. Coral, H. C. Irving, D. Lintott,

M. Nelson, G. J. S. Parkin, P. J. Robinson, R. J. H. Robertson, J. Straiton, U. M. Sivananthan, K. C. Simpkins and S. E. W. Smith. Their contribution was very significant. Finally, I would again like to thank Dr K. C. Simpkins for his advice and encouragement throughout this project.

Asif Saifuddin 1992

1 Cardiovascular system

1.1. Concerning the cardiac first-pass study following an intravenous bolus of ^{99m}Tc-labelled red cells
A right ventricular ejection fraction can be assessed.
B dilution of activity in the right atrium is a feature of ventricular septal defect (VSD).
C any residual activity in the lungs as the bolus enters the left ventricle is abnormal.
D activity should enter the lungs before the thoracic aorta.
E atrial septal defect (ASD) and VSD cannot be differentiated.

1.2. Concerning the multiple uptake gated acquisition (MUGA) scan
A a left ventricular aneurysm can be demonstrated.
B hypertrophic obstructive cardiomyopathy (HOCM) can be diagnosed.
C a left bundle branch block (LBBB) can be demonstrated.
D the left ventricular ejection fraction should be calculated from the left anterior oblique (LAO) view.
E a myocardial infarction produces a wall motion abnormality at rest.

1.3. Concerning myocardial perfusion imaging with thallium-201
A approximately 90% of injected thallium is taken up by myocardial cells.
B stress images should be acquired immediately after injection of the radionuclide.
C the right and left ventricles are normally equally well visualized.
D balanced triple vessel disease is a recognized cause of a false negative study.
E a focal defect on stress images is diagnostic of myocardial ischaemia.

1.4. Concerning the radionuclide assessment of myocardial infarction
A infarct imaging with ^{99m}Tc pyrophosphate is best performed 12–18 hours after the suspected infarct.
B the sensitivity of pyrophosphate infarct imaging is independent of infarct site.
C the 'doughnut' sign is a feature of transmural infarction.
D single-photon emission computed tomography (SPECT) is more sensitive than planar imaging.
E chemotherapy is a recognized cause of a false positive study.

1.5. Concerning CT in the assessment of masses affecting the heart and pericardium
A   the majority of left atrial myxomas are seen adjacent to the atrial septum.
B   left atrial thrombus usually appears as a filling defect in the atrial appendage on postcontrast CT.
C   angiosarcoma is a recognized cause of a mass in the right atrium.
D   metastatic disease is a recognized cause of diffuse thickening of the pericardium.
E   primary mesothelioma is the commonest cause of a solid mass involving pericardium.

1.6. Concerning MRI in the assessment of the heart
A   the pericardium should not measure more than 1 mm in thickness.
B   areas of high signal intensity in non-haemorrhagic pericardial effusions are typical features of T1-WSE scans.
C   areas of signal void in the pericardium are a recognized finding in constrictive pericarditis.
D   amyloid heart disease is a recognized cause of uniform myocardial thickening.
E   apical myocardial thickening is a recognized feature of hypertrophic obstructive cardiomyopathy (HOCM).

1.7. Concerning SE and GE (cine) MRI of the heart
A   the left ventricular ejection fraction (LVEF) can be measured on GE images.
B   valvular regurgitation can be demonstrated on GE images.
C   left atrial myxomas are typically hypointense to myocardium on SE images.
D   intracavitary thrombus and slowly moving blood cannot be differentiated.
E   the normal left ventricular end-diastolic wall thickness is 20 mm.

1.8. Concerning SE and GE (cine) MRI in the assessment of ischaemic heart disease
A   segmental diastolic myocardial thinning is a feature of old infarction.
B   focally reduced signal intensity on T2-WSE scans is a feature of old infarction.
C   focally increased signal intensity on T2-WSE scans is a feature of acute infarction.
D   increased intracavity signal intensity adjacent to an infarct is a recognized finding.
E   the end-diastolic wall thickness in acute infarction is usually normal.

1.9. Concerning CT of the pericardium
A   normal thickness on non-ECG gated studies should not exceed 4 mm.
B   absence of the left hemipericardium can be demonstrated.
C   benign and malignant pericardial effusions can be distinguished by the attenuation values of the fluid.
D   small effusions are typically seen as curvilinear collections anterior to the right ventricle.
E   constrictive pericarditis characteristically causes diffuse pericardial thickening.

1.10. Which of the following are true concerning thoracic aortic dissections?
A On CXR, displacement of the calcified intima more than 1 cm from the aortic outline is a specific sign.
B Widening of the mediastinum is the commonest finding on CXR.
C At aortography, inability to pass the catheter beyond the aortic arch is a feature of a De Bakey type III dissection.
D At aortography, an intimal flap is seen in less than 50% of cases.
E At aortography, the false lumen usually lies anterolaterally to the true lumen in the ascending aorta.

1.11. Concerning CT in the assessment of thoracic aortic dissections
A complete thrombosis of the false lumen is typical.
B pulmonary atelectasis adjacent to the descending aorta is a feature.
C a round shape to the aortic lumen indicates an atheromatous aneurysm rather than aortic dissection.
D the diagnosis can only be made on postcontrast scans.
E the aorta is invariably dilated.

1.12. Concerning MRI of the thoracic aorta and pulmonary arteries
A the entire thoracic aorta can be demonstrated on a single slice.
B in aortic dissections, flow void may be seen in both the true and false channels on SE images.
C in aortic dissection, the origin of the intimal flap cannot be detected.
D persisting signal throughout the cardiac cycle is a finding in pulmonary hypertension.
E pulmonary embolism can be demonstrated in the central vessels.

1.13. Concerning pulmonary angiography in the assessment of pulmonary embolus
A angiography is best performed within 24 hours of the suspected event.
B an intraluminal filling defect is the most specific abnormality.
C segmental hypervascularity is a recognized finding.
D segmental hypovascularity is a recognized finding.
E a region of hypoperfusion in the right middle lung field is a normal finding.

1.14. Which of the following are true of pulmonary angiography?
A With arteriovenous malformation (AVM), a fusiform venous aneurysm is a recognized finding.
B With AVM, the arterial supply is always via a pulmonary artery.
C Multiple feeding arteries are a recognized feature of AVM.
D Peripheral pulmonary artery aneurysms are a typical finding in Marfan's syndrome.
E A hypoplastic left pulmonary artery is demonstrated in the 'scimitar' syndrome.

1.15. Concerning CT of mediastinal vascular abnormalities
A an aberrant right subclavian artery appears as a tubular structure that passes between the oesophagus and trachea.
B there is an association between a right aortic arch and a retro-oesophageal mass.
C an aberrant left brachiocephalic artery appears as a tubular structure passing behind the oesophagus.
D enhancing vessels on the chest wall are a feature of superior vena cava (SVC) obstruction.
E a persistent left SVC appears as a tubular structure behind the left pulmonary artery.

1.16. Concerning Doppler ultrasound of the carotid arteries
A the normal internal carotid artery waveform shows continuous forward flow during diastole.
B the normal internal carotid demonstrates a laminar flow pattern.
C the common carotid artery waveform can distinguish between proximal and distal obstruction.
D a peak systolic internal carotid velocity of 100 cm/s indicates a 60% stenosis.
E 'spectral broadening' in the Doppler waveform is seen distal to a stenosis.

1.17. Which of the following are true of intracranial aneurysms?
A The commonest aneurysm to bleed is located at the origin of the anterior communicating artery.
B Aneurysm rupture is a recognized cause of subdural haematoma.
C Giant intracranial aneurysms are usually atherosclerotic in origin.
D Giant intracranial aneurysms typically present with subarachnoid haemorrhage.
E Peripheral enhancement on postcontrast CT is a finding with completely thrombosed giant intracranial aneurysms.

1.18. Concerning CT in the assessment of intracerebral and subarachnoid haemorrhage (SAH) due to ruptured 'berry' aneurysms
A the presence of blood in the suprasellar cistern indicates a ruptured posterior communicating aneurysm.
B medial temporal lobe haematomas are typically due to ruptured posterior communicating aneurysms.
C midbrain haemorrhages are associated with basilar tip aneurysms.
D regions of parenchymal low density are a recognized finding.
E abnormal sulcal enhancement on postcontrast scans is a recognized finding.

1.19. Which of the following are true of intracranial AVM?
A They are most commonly located in the posterior cranial fossa.
B They most commonly present with subarachnoid haemorrhage.
C Precontrast CT is usually normal.
D On postcontrast CT, enhancement is seen in most cases.
E An avascular mass on cerebral angiography is a recognized finding.

1.20. Concerning MRI in the assessment of intracranial vascular abnormalities
A serpiginous areas of signal void are a recognized feature of AVM on SE sequences.
B on T2-WSE scans, hypointensity of the basal cisterns is a recognized finding with recurrent SAH.
C signal void is a recognized feature of occult vascular malformations.
D on SE scans, congenital berry aneurysms do not show signal void.
E intraluminal signal within a giant aneurysm indicates thrombosis.

1.21. Which of the following are true of arteriovenous malformation of the spinal cord?
A they most commonly present in adults with subarachnoid haemorrhage.
B myelography is usually normal.
C intramedullary lesions may be supplied by the artery of Adamkiewicz.
D arteriography may demonstrate an associated intraspinal aneurysm.
E a 'glomus' type lesion is typically supplied by a single artery.

1.22. Concerning CT in the assessment of the abdominal aorta
A enlargement of the psoas muscle is a recognized finding in rupture of an aortic aneurysm.
B demonstration of gas around an aortic graft indicates graft infection.
C gas within the aortic wall is a feature of mycotic aneurysm.
D the lumen of a chronic pseudoaneurysm (false aneurysm) does not enhance on postcontrast scans.
E hyperdensity of the aortic wall is a recognized finding in acute aortic dissection.

1.23. Concerning CT in the evaluation of the inferior vena cava (IVC)
A in azygos continuation of the IVC, the intrahepatic cava is not seen.
B in transposition of the IVC, the suprarenal cava is seen to the right of the aorta.
C in duplication of the IVC, two venae cavae are seen above the renal veins.
D tumoural thrombosis can be differentiated from non-tumoural thrombosis.
E inhomogeneous enhancement following bolus intravenous contrast injection indicates partial thrombosis.

1.24. Which of the following are true of hepatic venography?
A In free hepatic venography, sinusoidal opacification is always abnormal.
B In wedge hepatic venography, opacification of the portal veins may be a normal finding.
C Absence of sinusoidal filling is a feature of cirrhosis.
D A reduction in the degree of hepatic vein branching is a feature of cirrhosis.
E Failure of opacification of a main hepatic vein is diagnostic of the Budd–Chiari syndrome.

1.25. Which of the following are recognized angiographic findings in hepatic cirrhosis?
A   Stretching of the intrahepatic branches of the hepatic arteries.
B   Failure of opacification of the gastroduodenal artery following a common hepatic artery injection.
C   A dense hepatogram on hepatic arteriography.
D   Opacification of the portal vein following a hepatic artery injection.
E   An absent hepatogram on arterioportography.

1.26. Concerning angiography in the assessment of benign liver lesions
A   hepatic adenomas are typically supplied by a centrally located artery.
B   arterioportal shunting is a typical feature of focal nodular hyperplasia.
C   cavernous haemangiomas usually opacify via the portal veins.
D   neovascularity is a typical feature of regenerating hepatic nodules.
E   vascular lakes are a recognized finding in haemangioendothelioma.

1.27. Concerning angiography in the assessment of malignant liver tumours
A   neovascularity is a recognized feature of cholangiocarcinoma.
B   arterioportal shunting is a feature of hepatoma.
C   hepatoma is a recognized cause of a hypovascular mass.
D   enlargement of the supracoeliac aorta is a feature of hepatoblastoma.
E   metastases from thyroid carcinomas are typically hypovascular.

1.28. Concerning ultrasound in the assessment of portal hypertension
A   demonstration of a hypoechoic tubular structure within the falciform ligament is a specific finding.
B   failure of an increase in SMV diameter during inspiration is a finding.
C   a coronary vein diameter greater than 5 mm is a recognized feature.
D   demonstration of Doppler signal in the portal vein excludes thrombosis.
E   prehepatic portal hypertension due to portal vein compression typically shows no abnormality on Doppler ultrasound.

1.29. Concerning ultrasound in the assessment of collateral vessels in portal hypertension
A   in portal hypertension due to isolated splenic vein occlusion, oesophageal varices are invariably identified.
B   cavernous transformation of the portal vein is seen in 100% of patients with complete portal vein thrombosis.
C   short gastric veins are the commonest demonstrated collaterals draining blood to the SVC.
D   demonstration of hepatofugal flow in the coronary vein excludes the presence of oesophageal varices.
E   paraumbilical veins are most commonly seen to originate from the left portal vein.

1.30. Concerning angiography of the pancreas
A   microcystic adenoma is a recognized cause of a hypervascular mass.
B   vascular encasement is a recognized feature of mucinous adenoma.
C   the majority of islet cell tumours are hypervascular.
D   demonstration of enlarged draining veins from an islet cell tumour indicates that it is malignant.
E   arterial narrowing by a mass differentiates adenocarcinoma from chronic pancreatitis.

1.31. Concerning angiography in the assessment of gastrointestinal bleeding
A   bleeding from a duodenal ulcer is best demonstrated by injection into the gastroduodenal artery.
B   early opacification of the ileocolic vein is a feature of caecal angiodysplasia.
C   punctate contrast extravasation is a recognized finding.
D   hyperaemia of the bowel wall is a recognized feature.
E   multiple saccular aneurysms of the medium and small mesenteric arteries are a recognized finding.

1.32. Concerning radionuclide scanning in the assessment of gastrointestinal bleeding
A   following injection of labelled colloid, a positive result only occurs if the patient is actively bleeding.
B   using labelled RBCs, bleeding may be detected several hours after the injection.
C   a serpiginous area of abnormally located activity is typical of small bowel bleeding.
D   a positive labelled colloid test is more accurate at determining the site of bleeding than a late positive labelled RBC test.
E   a bleeding site in the duodenum is best detected using the labelled colloid technique.

1.33. Concerning the angiographic features of fibromuscular dysplasia of the renal arteries
A   the proximal one-third of the renal artery is typically involved.
B   isolated lesions of the renal artery branches are not found.
C   alternating areas of stenosis and dilatation is the commonest finding.
D   a long, smooth stenosis is a recognized feature.
E   enlargement of the capsular artery may be demonstrated.

1.34. Which of the following are true concerning renal angiography?
A   Intrarenal aneurysms are a feature of Wegener's granulomatosis.
B   Distortion of the renal contour on the nephrogram phase is a feature of Page kidney.
C   An arteriocaliceal fistula is a recognized finding following renal trauma.
D   An angiomyolipoma is a recognized cause of a hypovascular mass.
E   Parasitization of adjacent vessels is a feature of renal adenocarcinoma.

1.35. Concerning Doppler ultrasound of the native kidney
A   the arcuate arteries normally show no flow during diastole.
B   a resistivity index (RI) of 0.5 is abnormal.
C   acute hydronephrosis is a recognized cause of an increase in the RI.
D   a ratio of peak systolic velocity in the renal artery to that in the aorta of 4.0 is abnormal.
E   hypernephromas are characteristically associated with increased flow in diastole.

1.36. Concerning the radionuclide assessment of renovascular hypertension due to renal artery stenosis (RAS)
A   with DTPA scintigraphy, deterioration of the baseline renal function following ACE inhibition is a feature.
B   on a MAG3 renogram, a reduction in cortical retention following ACE inhibition is a typical feature.
C   bilateral renal artery stenosis cannot be identified.
D   renal artery stenosis causing a 40% reduction in arterial lumen can be identified.
E   DTPA scintigraphy without ACE inhibition is usually abnormal when there is a 90% reduction in arterial lumen.

1.37. Which of the following are true of thromboangiitis obliterans (Buerger's disease)?
A   It is commoner in women.
B   The femoral artery is most commonly affected.
C   There is an association with hyperlipidaemia.
D   Multiple segmental occlusions in an otherwise normal-calibre artery is a typical angiographic finding.
E   Arterial spasm is a recognized angiographic feature.

1.38. Concerning Doppler ultrasound in the assessment of peripheral vascular disease
A   reversal of flow in early diastole is a normal feature of the Doppler waveform in the iliac artery.
B   an increase in peak systolic velocity is typically seen distal to a stenosis.
C   continuous forward flow in diastole is a recognized finding distal to a stenosis.
D   stenoses and occlusions cannot be distinguished.
E   stenoses in the tibial arteries can be detected with a sensitivity of 90%.

1.39. Concerning real-time and Doppler ultrasound in the diagnosis of deep vein thrombosis (DVT)
A   enlargement of the vein is a feature of acute thrombosis.
B   partial compressibility of a vein excludes the presence of thrombus in that segment.
C   non-occlusive thrombi can be demonstrated by colour flow Doppler ultrasound.
D   augmentation of the Doppler signal in the superficial femoral vein after calf compression excludes the presence of thrombus in intervening veins.
E   a continuous uniform Doppler signal in the common femoral vein is normal.

1.40. Concerning the MRI feature of thrombus and flowing blood
A   tumour thrombus and blood clot cannot be differentiated.
B   acute thrombus may appear hyperintense on both T1- and T2-WSE scans.
C   old thrombus typically produces a signal void within the vessel lumen.
D   signal from slow-flowing blood can be differentiated from thrombus using GE techniques.
E   flow signal can be differentiated from thrombus by using a presaturation pulse.

Cardiovascular system: Answers

1.1. A True
Following a bolus intravenous injection of ^{99m}Tc-labelled RBCs, both the right ventricular and left ventricular ejection fractions can be determined if ECG gating is performed. The ejection fraction is calculated by subtracting the end-systolic counts from the end-diastolic counts, dividing by the end-diastolic counts and multiplying by 100 to give a percentage. Typical values are 55% (range 45–65%) for the right ventricle and 65% (range 55–75%) for the left ventricle.

1.1. B False
Dilution of right atrial activity is seen with an atrial septal defect since, as the bolus enters the right atrium, it is diluted by blood entering through the defect from the left atrium.

1.1. C False
A little activity persists in the lungs in normal patients. Significant persistence of activity in the lungs after the bolus has entered the left ventricle is seen with pulmonary recirculation, as occurs with a left-to-right shunt. However, any cause of decreased pulmonary transit time of blood can cause this appearance, mimicking a shunt.

1.1. D True
If activity is seen in the left ventricle and aorta at the same time as it enters the lungs, a right-to-left shunt is indicated.

1.1. E False
If the shunt is greater than 2 : 1, ASD and VSD can be differentiated. Features indicating an ASD include dilution of right atrial activity, a large right ventricle and small left ventricle. VSD is suggested by a large left ventricle and no dilution of right atrial activity. In both cases there will be persistent lung activity with consequent poor visualization of the aorta.

(Reference 5)

1.2. A True
A left ventricular aneurysm demonstrates paradoxical movement during ventricular systole and appears as a focal outward bulge. The phase and amplitude images will show that the abnormal portion of ventricle moves but the movement is in time with atrial contraction. The left ventricular phase histogram may show a bimodal distribution representing normal myocardium and aneurysm.

1.2. B True
In the case of hypertrophic cardiomyopathy, the MUGA scan may show an increase in the left ventricular ejection fraction. During systole, constriction of the mid-left ventricular cavity is typically seen and isolation of the distal

ventricular cavity may occur. A thickened ventricular septum is best demonstrated on oblique views.

1.2.   C True
LBBB results in delayed contraction of the left ventricle after right ventricular systole. This is demonstrated on the phase images as delayed contraction of the whole of the left ventricle. LBBB may be seen in association with hypertrophic cardiomyopathy.

1.2.   D True
On the LAO view, the left ventricle is demonstrated without any superimposition of the right ventricle, as occurs on the anterior view. The interventricular septum is also best demonstrated on this view.

1.2.   E True
A myocardial infarction will result in focal hypokinesis or akinesis at rest, whereas myocardial ischaemia may only appear as a wall motion abnormality under conditions of haemodynamic stress.

(Reference 5)

1.3.   A False
Following the intravenous injection of thallium-201, approximately 4% of the injected dose is extracted by myocardial cells. Myocardial uptake of thallium is dependent upon myocardial blood flow and the integrity of the ATPase-dependent sodium–potassium pump. Therefore, although thallium images generally reflect myocardial perfusion, they also reflect myocardial cell function.

1.3.   B False
Peak uptake of radionuclide occurs approximately 10 minutes after injection, and stress imaging should therefore begin after a suitable delay but within 30 minutes of injection. If patients cannot exercise adequately, the test may be performed after intravenous injection of dipyridamole, which causes coronary vasodilatation and increases myocardial uptake of thallium. Rest images are typically performed about 3 hours after stress images.

1.3.   C False
Clear visualization of the right ventricle on thallium imaging occurs when there is right ventricular hypertrophy, pulmonary arterial hypertension or global decrease in left ventricular uptake.

1.3.   D True
There are several causes for false negative thallium studies, a rare cause being balanced triple vessel disease. In this case there is a global reduction in thallium uptake throughout the myocardium equally in each vessel distribution. The images may appear normal although there is an absolute reduction in uptake which may be detected by quantitative techniques. Other causes of a false negative test include inadequate exercise, imaging continued for more than 30 minutes after injection and the presence of adequate collateral blood supply.

1.3.   E False
A defect on stress images may be due to ischaemia, in which case there will be a degree of reversibility manifest by 'filling in' of the defect on delayed redistribution images. A defect on stress images that does not reverse on redistribution images indicates an infarct. However, because myocardial uptake of thallium is partly dependent upon cell function, defects may occur due to other conditions such as myocarditis.

(Reference 5)

1.4.   A False
In patients who have had a recent myocardial infarct, there is a delay between the onset of infarction and the development of a positive pyrophosphate study. Imaging performed within 12–18 hours is usually negative. Maximal intensity of uptake of pyrophosphate by infarcted tissue is typically 24–72 hours after the infarct. Therefore, imaging should be performed ideally 1–3 days after the suspected event. Images usually revert to normal 7–10 days after the infarct.

1.4.   B False
The sensitivity of the pyrophosphate study is related to infarct site and the thickness of the wall involved. The study is virtually 100% sensitive for full-thickness anterior infarcts and approximately 20% sensitive for subendocardial inferior infarcts.

1.4.   C True
A 'doughnut' appearance is indicative of a large transmural infarct. It is produced by the fact that with such an infarct there will be no blood flow to the central necrotic myocardium, resulting in absence of pyrophosphate uptake, while the surrounding ischaemic myocardium, which still has some blood supply, will take up the radionuclide.

1.4.   D True
The use of ^{99m}Tc-SPECT has been advocated for detection of infarction and also for measuring the mass of infarcted tissue. SPECT imaging has the advantage over planar imaging in that it reduces the effects of uptake in overlying tissues such as the sternum and ribs. However, one limitation of the technique is that it cannot be used when the patient is confined to the intensive care unit.

1.4.   E True
Any condition that leads to the presence of cellular necrosis, or the abnormal accumulation of calcium, will give rise to a positive study. Causes of false positive pyrophosphate studies include unstable angina, myocardial trauma (e.g. post-cardioversion), radiotherapy, myocardial tumours, valvular calcification, myocarditis, pericarditis and ventricular aneurysm.

(References A, B, D, E, 18; C, 5)

1.5.   A True
Myxomas are the commonest primary cardiac tumours. Approximately 75% are located in the left atrium, 20% in the right atrium and the rest in the ventricles. Of left atrial myxomas, 90% are attached to the atrial septum in the

region of the fossa ovalis, which is an important feature in distinguishing them from left atrial thrombus.

1.5.   B True
Left atrial thrombus usually appears as a filling defect in the left atrial appendage or in a more dependent portion of the atrium. Another feature that helps differentiate them from atrial myxomas is their shape, which is characteristically more angulated.

1.5.   C True
Angiosarcoma is the commonest primary malignant tumour of the heart. Although they can occur anywhere in the heart, 80% are found either in the right side of the heart or in the pericardium. Rarely, metastases may occur to the lungs, lymph nodes and liver. Metastatic tumours to the heart are much more common than primary tumours, the commonest being lung, breast, leukaemia and lymphoma. Melanoma has the highest percentage of metastases to the heart. Tumour can also extend into the heart via the SVC, IVC and pulmonary veins.

1.5.   D True
Metastases involve the pericardium more commonly than primary tumours and are frequently associated with pericardial effusions. Small tumour implants may be missed at CT but larger mass lesions also occur.

1.5.   E False
Solid pericardial masses are more likely due to metastatic disease than primary pericardial tumours, which are rare. Primary tumours of the pericardium may be benign or malignant, the commonest malignant tumour being mesothelioma. Pleural mesothelioma may also invade the pericardium. Benign tumours include the intrapericardial teratoma and lipoma.

(Reference 9)

1.6.   A False
On MRI, the normal pericardium appears as a 1–3 mm thick curved line of low signal intensity around the heart. It is surrounded by high-signal-intensity pericardial fat and limited internally by either high-signal-intensity epicardial fat or intermediate-signal-intensity myocardium. The anterior pericardium may appear artificially thickened due to motion-related signal loss. With imaging at high field strength the pericardium may be obscured by chemical shift artefact.

1.6.   B False
On T1-WSE scans, pericardial effusions cause an increase in the width of the pericardial stripe. Non-haemorrhagic effusions typically have uniform low signal intensity, whereas effusions due to trauma, uraemia or tuberculosis may show regions of high signal intensity.

1.6.   C True
Constrictive pericarditis results in irregular thickening of the pericardium which may demonstrate intermediate signal intensity. Calcifications are demonstrated as irregular areas of signal loss which typically occupy the atrioventricular groove.

1.6. D True
Amyloid is one of the commonest causes of restrictive cardiomyopathy, which may clinically resemble constrictive pericarditis. MRI shows uniform thickening of the myocardium with normal left ventricular size and contraction. Small pleural effusions are common.

1.6. E True
Hypertrophic cardiomyopathy is associated with marked thickening of the left ventricular myocardium, which typically affects the septum to the greatest degree. However, myocardial hypertrophy may occasionally be concentric or apical. Systolic anterior motion of the anterior leaflet of the mitral valve is a feature and can be demonstrated by GE MRI.

(Reference 3)

1.7. A True
Left and right ventricular function can be assessed both qualitatively by visual inspection of the GE studies, and also quantitatively by measurement of end-diastolic and end-systolic ventricular volumes, from which ejection fractions and cardiac output can be calculated. In addition, direct measurement of systolic wall thickening provides an excellent index of myocardial contractility, which correlates with myocardial perfusion.

1.7. B True
Using fast gradient echo (GE) techniques, blood appears bright and the signal generated by flowing blood increases with increased blood flow velocity. Signal intensity is lost in GE images when there is intravoxel spin-phase cancellation as occurs with flow disturbances due to regurgitant jets, shunt jets and flow distal to stenoses. Consequently, a regurgitant jet is seen as an area of diminished signal intensity which is typically flame-shaped.

1.7. C False
Left atrial myxomas are typically isointense or slightly hyperintense compared to myocardium on SE imaging. GE imaging can demonstrate not only the site of attachment of the tumour, but also any associated prolapse of the mass into the mitral valve during diastole.

1.7. D False
The appearance of intracavitary thrombus may be variable. Typically, it has slightly increased signal intensity on the first echo image and a slightly dark appearance on the second echo image of a double echo T2-WSE sequence. In contrast, slow-moving blood may appear similar on the first echo but should get considerably brighter on the second echo image due to even echo rephasing.

1.7. E False
The left ventricular wall thickness at end-diastole is normally about 10 ± 2 mm and the thickness increases by 35–70% at end-systole. Wall thickening is generally best assessed on the short axis view, where the perceived thickness is less exaggerated due to through-plane wall curvature.

(Reference 12)

1.8. A True
In cases of myocardial infarction, MRI demonstrates both segmental diastolic wall thinning and decreased segmental systolic wall thickening, and motion of that area of myocardium. However, rarer diseases such as sarcoidosis can produce similar features.

1.8. B True
Another MRI feature of old myocardial infarction is a reduction in the signal intensity of the infarcted tissue on T2-WSE scans. This is due to a decrease in the water content of scar tissue as compared to normal myocardium.

1.8. C True
When myocardial infarctions are less than 2–3 weeks old, MRI typically demonstrates an area of increased signal intensity of the infarcted tissue due to acute intracellular oedema. However, this finding is not always present. Also, the sensitivity of MRI for acute infarction may vary significantly with TE and the degree of T2 weighting.

1.8. D True
Increased intracavity signal intensity on SE sequences adjacent to an area of decreased wall motion is due to slow blood flow. This may be seen both with acute and old infarction but is not a specific finding, since it may be observed in normal individuals, particularly along the inferolateral portion of the left ventricle, between the papillary muscles.

1.8. E True
Segmental decrease in systolic wall thickening is possibly the most reliable indicator of myocardial infarction. End-diastolic wall thickness is usually within the normal range in acute infarction, since fibrosis and scar formation typically develop approximately 2 weeks or more after the acute event. GE imaging may demonstrate the area of infarction as a relatively hypointense region compared to normal myocardium, possibly due to haemorrhage and resultant iron deposition.

(Reference 12)

1.9. A True
The pericardium can be identified on CT because of the presence of fat in the epicardial space and mediastinum. It can be identified in 95% of adults, usually anterior to the ventricles and less commonly inferolaterally. The normal pericardium may appear thickened if imaged tangentially, especially around the anterior sternopericardial ligaments. Also, the pre- and retroaortic pericardial recesses may be seen at the level of the carina since they commonly contain a small amount of fluid.

1.9. B True
Congenital pericardial defects may be classified as partial (almost always on the left), absence of the left hemipericardium (the commonest type) or total absence of the pericardium. The characteristic CT signs of absent left hemipericardium include inability to identify the fibrous layer of the parietal pericardium along the left heart border, displacement of the main pulmonary artery toward the left lung and direct contact of the lung with the heart.

1.9. C False
Most pericardial effusions appear as near-water-density collections between the mediastinal and epicardial fat. Exudative or haemorrhagic effusions may have soft tissue densities but benign and malignant effusions cannot reliably be differentiated by their CT numbers. Small soft-tissue-density effusions can be differentiated from pericardial thickening since they change shape with position and do not enhance.

1.9. D False
Pericardial thickening tends to occur over the anterolateral surface of the heart, whereas small effusions typically collect behind the left ventricle and to the left of the left atrium. Pericardial effusions may occasionally be loculated due to adhesions, usually as a result of surgery or pericarditis. These loculated effusions are commoner in a posterior or right anterolateral location.

1.9. E True
Pericardial thickening in constrictive pericarditis is typically diffuse but not necessarily regular. The pericardium may measure from 0.5 cm to 2 cm in thickness. CT will also demonstrate associated dilatation of the SVC and IVC, ascites and pleural effusions and dilatation of the atria with small ventricles. Radiation, pericarditis and trauma are other causes of pericardial thickening.

(Reference 2)

1.10. A False
Displacement of the calcified intima by more than 1 cm is seen in about 7% of patients with thoracic aortic dissection. However, as the only finding it is an unreliable sign of aortic dissection, since it may be seen in tortuous, atherosclerotic aortas without dissection. A double contour to the aortic outline is seen in up to 40% of cases, but overlap of the ascending and descending aorta can simulate this finding.

1.10. B True
CXR demonstrates a widened mediastinum in up to 81% of cases. This may be due to the dissection itself or to displacement of the aorta by a large false channel or mediastinal haematoma. The aortic contour may also be irregular. Pleural effusion is seen in approximately 27% of patients.

1.10. C True
Thoracic aortic dissections were originally classified into De Bakey types I, II and III. Type I dissections begin in the ascending aorta and extend distally for a variable distance. Type II dissections are localized to the ascending aorta, while type III dissections begin at the isthmus and extend distally for a variable distance. According to the Stanford classification, dissections are classified into two types. Type A dissections include any that involve the ascending aorta and type B dissections include the remainder.

1.10. D False
Aortography can demonstrate an entry site and intimal/medial flap with opacification of a second channel in approximately 87% of cases. If the false channel is not demonstrated it may be completely thrombosed or the entry site may be proximal to the injection site. Aortography also demonstrates

compression of the true lumen, aortic valvular regurgitation, increased aortic wall thickness and obstruction of aortic branches.

1.10.   E True
Also, in the aortic arch, the false lumen lies superoposteriorly to the true lumen and continues posterolaterally in the descending thoracic and abdominal aorta. These features are also evident at CT.

(Reference 10)

1.11.   A False
The false lumen may be partially or rarely completely filled with clot. A more typical appearance on dynamic sequential postcontrast CT is of delayed filling and washout of contrast in the false lumen compared to the true lumen, due to slower blood flow in the former. The demonstration of an intimal flap with contrast enhancement on each side of it is the most important finding in the CT diagnosis of aortic dissection.

1.11.   B True
Atelectasis in the left lower lobe may mimic a false channel, as may the innominate veins, the SVC, the left superior intercostal vein and the left pulmonary veins. Also, streak artefacts can mimic an intimal flap. However, intimal flaps are typically gently curved structures of uniform thickness conforming to the configuration of the aorta, whereas streak artefacts are straight and vary in thickness.

1.11.   C False
In atherosclerotic aneurysm the lumen is almost always round, whereas in aortic dissections the true lumen is flattened in over half the cases. However, dissections can extend around the circumference of the aorta, resulting in a round shape to the true lumen. Also, in cases where the true lumen is completely thrombosed, the CT differentiation from an atherosclerotic aneurysm can be difficult. Calcification may line the inner surface of thrombus within an atherosclerotic aneurysm, mimicking the calcified, displaced intimal flap.

1.11.   D False
Findings on precontrast CT that allow a diagnosis of aortic dissection to be made include demonstration of displaced, calcified atheromatous plaques and a moderate increase in density of the false lumen due to recently clotted blood. Also, a non-calcified intimal flap may be seen on precontrast scans in anaemic patients.

1.11.   E False
In a recent large series, almost 60% of patients with dissections showed no aortic dilatation. Other CT findings in cases of dissection include demonstration of pericardial and pleural fluid, mediastinal haematoma and extension of haemorrhage into the lung parenchyma.

(Reference 4)

1.12. A True

It is possible with MRI to obtain a long-axis view of the thoracic aorta showing the whole curve of the aortic arch, and if the aorta is not tortuous the whole of the thoracic aorta can be imaged on a single slice. This is achieved by obtaining multiple oblique sagittal scans parallel to a plane projected through the centres of the ascending and descending aorta.

1.12. B True

Flow void will occur if there is rapid flow in both the true and false channels. However, particularly in diastole, slower flow in the false lumen may give rise to signal and the true lumen may become flattened.

1.12. C False

MRI is able to locate the site of the intimal tear precisely, which is another advantage over CT in the assessment of aortic dissection. MRI is also successful in demonstrating the extent of dissection and involvement of major aortic branches, including the brachiocephalic, coeliac, superior mesenteric and renal arteries. The intimal flap is often best seen using GE techniques.

1.12. D True

The persisting pulmonary artery signal is due to the slow flow of blood in patients with pulmonary hypertension. Other MRI features include right ventricular wall thickening and convex bulging of the interventricular septum into the left ventricle. The septum is usually either straight or bulges slightly into the right ventricle.

1.12. E True

Pulmonary emboli in the central pulmonary arteries can be demonstrated on both ECG gated and non-gated studies. Emboli typically produce an intense signal within the vessel lumen, which should normally appear devoid of signal on SE images due to rapidly flowing blood.

(Reference 3)

1.13. A True

There is evidence to suggest that in almost half the cases of PE the emboli affect the smaller vessels (microemboli) and that fragmentation, partial lysis and resolution of larger emboli occur after 24 hours. Consequently, a standard non-magnification pulmonary angiogram obtained more than 24 hours after the acute episode may be falsely normal.

1.13. B True

An intraluminal filling defect was observed in 94% of pulmonary angiograms in one series. This finding, together with demonstration of an abrupt termination to a pulmonary artery branch, are the two major signs in the diagnosis of pulmonary embolus. The latter sign, however, is not specific.

1.13. C True

Other angiographic signs are less reliable for diagnosing PE. Segmental hypervascularity may be seen in association with pulmonary infarction, although atelectasis may contribute to this finding. Tortuous arterial collaterals may also be seen, especially with pulmonary infarction.

1.13. D True
Segmental hypovascularity (oligaemia) is often wedge-shaped, especially when viewed in lateral or oblique projection. Other angiographic findings in PE include absence of a draining vein from the affected segment(s), organization of thrombus manifest as eccentric, concentric plaque, web or stenosis, recanalization with opacification of a new channel and pruning and attenuation of pulmonary artery branches.

1.13. E True
On the parenchymal phase of the arteriogram, a relatively hypovascular region is frequently observed in the mid-portion of the right lung field. This is unrelated to embolic disease and is a consequence of the anatomy of the right lung. Also, vascular crowding due to collapse may occur in the absence of embolus.

(References A, C, D, E, 10; B, 4)

1.14. A True
The angiographic appearance allows pulmonary AVMs to be divided into two types. The simple type is the commonest, accounting for 80% of cases, and is characterized by a direct artery-to-vein communication. The venous limb is dilated, resulting in a fusiform or saccular venous aneurysm.

1.14. B False
In 95% of simple AVMs the blood supply is via pulmonary arteries and in the remainder it is via a systemic artery. The latter type can be associated with paradoxical embolization. Pulmonary AVMs are multiple in approximately one-third of patients, and 70% of lesions are found in the lower lobes, often near the surface of the lungs.

1.14. C True
The second type of pulmonary AVM is referred to as the complex type and accounts for 20% of cases. Angiography demonstrates one or more feeding arteries and prominent draining veins with an interposed network of tortuous, dilated vascular channels. Acquired pulmonary AV fistulae may be due to schistosomiasis, trauma, infection or the pulmonary angiodysplasia associated with hepatic cirrhosis.

1.14. D False
In connective tissue disorders such as Marfan's syndrome, there is dilatation of the main pulmonary artery and proximal right and left pulmonary arteries. Peripheral pulmonary artery aneurysms are not characteristic of this disorder. Most aneurysms of the pulmonary arteries are acquired (e.g. mycotic, traumatic or due to an arteritis).

1.14. E False
Hypoplasia of the right pulmonary artery is seen with the hypogenetic lung (scimitar) syndrome. Other features of this condition include hypoplasia of the right lung with systemic arterial supply and partial or total anomalous pulmonary venous drainage from the right lung into systemic veins, usually to the IVC but also to the left atrium.

(Reference 10)

1.15. A False
An aberrant right subclavian artery is the commonest anomaly of the aortic arch, occurring in about 0.5% of the population. It arises distal to the left subclavian artery, as the fourth branch of the arch, and extends obliquely upward towards the right shoulder, passing behind the oesophagus into the posterior mediastinum. The condition is almost always asymptomatic but may be associated with congenital heart diseases, especially Fallot's tetralogy and coarctation.

1.15. B True
Usually, a right aortic arch lies to the right of the oesophagus and trachea to continue caudally on the right side of the spine. However, it may cross the midline posterior to the oesophagus to descend to the left of the spine. CT will then demonstrate a retro-oesophageal mass, the vascular nature of which will be confirmed on postcontrast scans.

1.15. C True
An aberrant left brachiocephalic artery is an uncommon anomaly of a right aortic arch. CT will demonstrate the right-sided arch, which has three branches, the third of which is the left brachiocephalic artery. This vessel crosses the midline between the oesophagus and spine. The findings are similar to those of an aberrant left subclavian artery, but differ in that four arch branches occur in this case.

1.15. D True
Causes of SVC obstruction include thrombosis due to a central venous catheter or compression by a mediastinal tumour or mediastinal fibrosis. CT may demonstrate a mass compressing the cava but other signs will indicate the haemodynamic significance of this. These include the demonstration of venous collaterals in the axilla or anterior and lateral chest wall, and more intense enhancement of the azygos vein if the obstruction is between the right atrium and the entrance of the azygos vein into the SVC.

1.15. E False
A persistent left SVC appears on postcontrast CT as an enhancing tubular structure that lies anterolateral or lateral to the aortic arch. It passes lateral to the left pulmonary artery and anterior to the left hilum to drain into the coronary sinus. A right SVC is usually present and the two may be joined by the left brachiocephalic vein. In the absence of the right SVC, the right brachiocephalic vein will be seen anterior to the branches of the aortic arch.

(Reference 9)

1.16. A True
The internal carotid artery waveform is characterized by broad systolic peaks and continuous forward flow during diastole. The external carotid artery waveform typically shows sharp systolic peaks with little or no diastolic flow, or in some cases reversed flow in early diastole.

1.16. B True
A laminar flow pattern is characterized by a thin Doppler shift envelope that encompasses a clear window. This indicates that the majority of the blood cells are travelling at almost the same speed. However, Doppler flow disturbances

may also be encountered in the normal carotid artery system at areas of tortuosity or branching and especially at the carotid bulb.

1.16. C True
If there is obstruction to the distal common carotid or internal carotid arteries, the common carotid artery waveform will demonstrate high pulsatility features compared to the low pulsatility normally seen. Conversely, obstruction in the brachiocephalic or proximal common carotid artery produces a low-amplitude or damped common carotid artery waveform.

1.16. D False
Internal carotid peak systolic velocities of less than 125 cm/s are considered normal. Velocity increases typically do not occur until there is at least a 50% stenosis present. Peak systolic velocities of 125–250 cm/s are associated with a 50–80% stenosis, and velocities greater than 250 cm/s are found with greater than 80% reduction in vessel lumen.

1.16. E True
'Spectral broadening' implies that there is a greater difference in the range of velocities between the fastest and slowest-moving blood cells which characterizes turbulent blood flow. The Doppler waveform may show loss of the clear spectral window that is a feature of the laminar flow.

(Reference 11)

1.17. A True
The most common sites for congenital 'berry' aneurysms are the origin of the posterior communicating artery, anterior communicating artery, and middle cerebral bifurcation. Approximately 80% of aneurysms are supratentorial and 22% are multiple. Bleeding occurs most commonly in the sixth decade of life.

1.17. B True
An aneurysm can rupture into the subarachnoid space, ventricles or brain parenchyma. Rarely a superficial aneurysm can rupture into the subdural space. Blood in the subarachnoid space can alter the normal CSF reabsorption, resulting in acute hydrocephalus and ventricular dilatation. This can be transient, with a return to normal CSF flow following reabsorption of the subarachnoid blood, or may be permanent if adhesions and basal arachnoiditis develop.

1.17. C True
Aneurysms greater than 2.5 cm in diameter are referred to as giant aneurysms and may be seen on both pre- and postcontrast CT. They most commonly arise from the cavernous portion or bifurcation of the internal carotid artery, middle cerebral artery, basilar artery and vertebrobasilar junction. They account for approximately 25% of mass lesions in the posterior fossa.

1.17. D False
Bleeding is uncommon from giant aneurysms and symptoms are usually due to their mass effect. Non-thrombosed aneurysms appear as well-defined spherical, slightly hyperdense masses on precontrast CT which show marked, uniform enhancement on postcontrast scans. Partial thrombosis is seen in most giant

aneurysms. The thrombus is slightly hyperdense compared to the vessel lumen on precontrast CT, and mural calcification is also common.

1.17. E True
Completely thrombosed giant aneurysms appear on precontrast CT as well-defined, round or oval masses of heterogeneous increased density. Calcification is common. They may have a slightly hyperdense rim that enhances on postcontrast scans. Central enhancement or surrounding oedema is not a feature.

(Reference 9)

1.18. A False
The location of a ruptured aneurysm can often be predicted by the pattern of extravasated blood if CT is performed within 5 days of the bleed. Blood in the anterior interhemispheric fissure suggests an anterior communicating or anterior cerebral artery aneurysm. Blood in the septum pallucidum is indicative of a ruptured anterior communicating aneurysm. Blood in the suprasellar cistern is usually not a helpful localizing sign.

1.18. B True
Haematomas in the medial aspect of the temporal lobe are virtually pathognomonic of a ruptured posterior communicating aneurysm. The combination of temporal haematoma and pericallosal or interhemispheric blood can be due to an anterior communicating aneurysm.

1.18. C True
Basilar tip aneurysms may also result in perimesencephalic and/or interpeduncular cistern haemorrhages. Third and fourth ventricular blood with little or no blood in the lateral ventricles is suggestive of a vertebral artery aneurysm. Haemorrhages involving the basal ganglia are seen with internal carotid or middle cerebral artery aneurysms and CT may not be able to distinguish these haematomas from hypertensive haemorrhages.

1.18. D True
Parenchymal hypodensities that develop after aneurysm rupture usually represent areas of infarction. These may result secondary to vessel spasm or vessel compression by intraparenchymal haematomas. Other less common causes of infarction in association with aneurysms would be due to embolization from intra-aneurysmal clot or impairment of cerebral blood flow from hydrocephalus.

1.18. E True
On postcontrast CT, abnormal enhancement of the basal cisterns and cortical sulci is seen in up to 50% of patients with SAH. Enhancement is probably the result of an inflammatory response to subarachnoid blood and is associated with an increased incidence of hydrocephalus.

(Reference 9)

1.19. A False
Intracranial AVMs are most commonly located in the cerebral hemisphere, followed by the basal ganglia. The posterior fossa is the least common site. The lesions are usually single but multiple AVMs have been reported. The lesion is composed of a vascular nidus, usually located in the white matter, and multiple, tortuous, enlarged vessels which represent feeding arteries and draining veins.

1.19. B True
Also, once an SAH has occurred, the probability of a second bleed increases. Perventricular rupture of AVMs is also common. Those AVMs that involve the midbrain can cause compression of the aqueduct or result in periaqueductal gliosis, causing hydrocephalus.

1.19. C False
Precontrast CT may show abnormalities in up to 81% of patients. The commonest finding is an area of heterogeneous increased and decreased density. The hypodense regions are likely due to associated brain atrophy, whereas the regions of increased density are probably due to a combination of microscopic haemorrhage, thrombosis and calcification. The latter occurs in 30% of cases and may be extensive. Cystic changes can also be seen.

1.19. D True
More than 95% of AVMs will show enhancement on postcontrast scans. The enhancement is usually heterogeneous with ill-defined margins. Dilated arteries and veins are occasionally demonstrated as serpiginous structures. Postcontrast CT is essential to make the diagnosis of AVM in the 20% of patients that have normal precontrast scans.

1.19. E True
In a very small percentage of cases, the AVM is not detected with angiography and is termed occult or cryptic. Angiography will be either normal or demonstrate an avascular mass and the negative result may be due to thrombosis or slow flow through the lesion. Precontrast CT demonstrates occult AVMs as an area of increased density which may show patchy enhancement following contrast.

(Reference 9)

1.20. A True
Signal voids on SE sequences are the result of rapid flow in the vascular components of the AVM. Calcification within the lesion may also result in signal loss. Gradient echo (GE) techniques can be useful in distinguishing between signal void due to haemorrhage or calcium from that due to flow as seen on SE sequences, as the latter will demonstrate flow-related enhancement on GE sequences.

1.20. B True
The ability of MRI to detect acute SAH has been very limited, with the exception of focal clots, and at present precontrast CT is the imaging method of choice. However, chronic recurrent SAH may result in haemosiderin and

ferritin deposition in the meninges, which appears as low signal intensity lining the cisternal spaces on T2-WSE scans and GE sequences.

1.20. C True
Occult or cryptic vascular malformations are defined as those that are not demonstrated by angiography. MRI shows a characteristic appearance consisting of an outer rim of signal void, most prominent on T2-WSE scans, due to haemosiderin, and central areas of higher but differing signal intensities representing multiple haemorrhages at different stages of evolution.

1.20. D False
Flow void is usually identified in congenital berry aneurysms but the smaller lesions may be difficult to distinguish from normal vessels. Flow-related enhancement may be seen on GE sequences.

1.20. E False
The classical MRI appearance of a giant aneurysm is of a region of flow void surrounded by thrombus, the signal intensity of which will depend upon the pulse sequence and the age of the clot. However, intraluminal signal may be seen with non-thrombosed aneurysms due to complex flow patterns.

(Reference 3)

1.21. A False
Spinal AVMs account for approximately 5% of all spinal 'tumours'. Most patients are in the fourth to sixth decades and usually present with progressive myelopathy. Rarely, presentation may be with an acute neurological deficit or subarachnoid haemorrhage. These forms of presentation are commoner in children.

1.21. B False
Myelography may demonstrate characteristic serpiginous intradural filling defects in at least 50–70% of cases. Lesions are predominantly retromedullary and commonest in the thoracic or thoracolumbar regions. The malformation consists of large 'arterialized' veins that drain a relatively small nidus which is usually supplied by posterior retromedullary arteries or dural vessels. The venous component may commonly appear as a single coiled vessel.

1.21. C True
Intramedullary AVMs are supplied by multiple radiculomedullary arteries, including the artery of Adamkiewicz. These are usually high flow lesions that produce symptoms due to a 'steal' phenomenon or due to actual cord compression by the angiomatous tissue. They are more commonly located in the cervical cord, especially in children.

1.21. D True
In adults, presentation with subarachnoid haemorrhage occurs in less than 30% of cases and almost always indicates an associated spinal aneurysm. Spinal AVMs may be associated with port wine stains, Weber–Osler–Rendu

disease, vertebral haemangiomas and the Klippel–Trenaunay–Weber syndrome (cutaneous angioma of an extremity with hypertrophy of the limb).

1.21.   E True
A 'glomus' type lesion is another type of spinal AVM that appears as a small plexus or tuft of vessels that is usually supplied by a single artery and does not have predominant draining veins. Less commonly, AVMs can be entirely extradural and may present with radiculopathy.

(Reference 10)

1.22.   A True
CT features of ruptured abdominal aortic aneurysm include obscuration or anterior displacement of the aneurysm by an irregular high-density mass (typically around 70 HU) which may extend into the perirenal spaces or less commonly into the pararenal space. The kidney may be displaced anteriorly by the haematoma which may also obscure or enlarge the psoas muscles.

1.22.   B False
Gas may normally be seen around a graft in the immediate postoperative period and tends to appear as a single collection anterior to the graft. The CT features of graft infection include the presence of irregular, septated fluid collections around the graft, occasionally associated with pockets of gas, which tend to be multiple and in a posterior location. These gas collections usually form more than 10 days after the surgery.

1.22.   C True
A mycotic aneurysm usually has a saccular appearance and can be diagnosed by CT if gas is seen in its wall. CT may also demonstrate associated splenic infarcts, and other vessels typically show lack of atheromatous change. Infected atheromatous aneurysms may also have gas in their walls.

1.22.   D False
A chronic aortic pseudoaneurysm (false aneurysm) usually appears on CT as a well-defined, round mass with an attenuation value similar to the native aorta on precontrast scans. On postcontrast images, the lumina of both aneurysms as well as their communication may enhance.

1.22.   E True
Aortic dissections usually arise in the thoracic aorta and extend into the abdominal aorta. Hyperdensity in the aortic wall may occur due to the presence of fresh intramural haematoma. However, increased density of the wall will also be seen in atherosclerosis. In patients under 60 years of age, visibility of the aortic wall on precontrast CT is suggestive of anaemia, due to reduction in the CT numbers of blood in anaemic patients.

(References A, B, C, D, 2; E, 16)

1.23.   A True
In azygos continuation of the IVC, a normal IVC is seen from the confluence of the common iliac veins to the level of the renal veins. The intrahepatic segment of the cava is absent. The azygos and commonly the hemiazygos veins

are dilated and are seen bilaterally in the retrocrural spaces, either side of the aorta. This anomaly may be isolated or associated with the asplenia/polysplenia syndromes.

1.23. B True
With transposition of the IVC, a single vessel ascends to the left of the vertebral column and crosses either anterior or posterior to the aorta at the level of the renal veins. It then continues cephalad to enter the right atrium.

1.23. C False
With duplication of the IVC, two vessels are seen, one on either side of the aorta, below the level of the renal veins. They may be of equal or unequal sizes. At the level of the renal veins, the left-sided cava joins the right-sided IVC via a vessel that crosses either anterior or posterior to the aorta. Caudally, the left-sided IVC terminates at the left common iliac vein, differentiating it from a dilated left gonadal vein.

1.23. D True
Tumoural and non-tumoural thrombosis can be differentiated in cases where CT demonstrates hypervascularity in the former on dynamic postcontrast scans. In the absence of this finding, the two appear identical. CT may also demonstrate focal enlargement of the vein and, in cases of complete occlusion, absent enhancement on postcontrast scans and collateral circulation via paravertebral, ascending lumbar, azygos/hemiazygos, gonadal, periureteric and other retroperitoneal veins.

1.23. E False
Inhomogeneity of caval enhancement on postcontrast CT is common and a 'pseudothrombus' artefact may occur in the suprarenal cava due to poor mixing between the densely opacified renal venous blood and the less densely opacified infrarenal caval blood. This artefact is most prominent at and just above the origin of the renal veins but may be seen in the intrahepatic cava also.

(Reference 2)

1.24. A False
In the normal liver, the free hepatic venogram shows symmetrical arborization of the hepatic vein branches, and up to the fifth order branches are visualized. Some sinusoidal opacification is seen under normal circumstances, if the injection pressure is great enough.

1.24. B True
With wedge hepatic venography in the normal case, several hepatic veins are opacified, and if the injection is forceful portal vein branches may also be seen. The flow in the portal veins is always hepatopetal and washout is rapid. Depending upon the force of the injection, the sinusoids may not be discernible due to an intense parenchymal stain (with forceful injection), or a granular, confluent and homogeneous sinusoidal pattern may be seen.

1.24.   C True
The degree of sinusoidal filling is related to the severity of liver disease. With early cirrhosis, sinusoidal filling is seen but is irregular. With advancing disease, the sinusoidal filling becomes more inhomogeneous and may eventually be absent.

1.24.   D True
With mild cirrhosis, only the third or fourth order hepatic vein branches are identified, with a gradual reduction in the degree of branching as the cirrhosis progresses. The hepatic veins also appear narrowed and irregular.

1.24.   E False
Hepatic venography shows characteristic features in the Budd–Chiari syndrome. There is typically absence of a main hepatic vein, which is replaced by a 'spider-web'-like appearance due to the formation of collateral veins and recanalization of the hepatic veins. However, it should be noted that inability to catheterize hepatic veins and absent hepatic venous opacification may occur with advanced cirrhosis.

(Reference 10)

1.25.   A True
Early cirrhosis is characterized by either a normal appearance to the intrahepatic arteries or stretching due to hepatomegaly associated with diffuse fatty infiltration. At this stage of the disease, the hepatogram is normal and there may be a slight delay in arterial filling and emptying. Portal venous flow is also normal.

1.25.   B True
In moderately severe cirrhosis, there is increased intrahepatic resistance resulting in some reversal of portal flow. This must be compensated for by an increase in hepatic artery flow. Arteriography will demonstrate enlargement of the common hepatic artery and the increased flow in this vessel results in reversal of flow in the gastroduodenal artery, which is consequently not opacified following a common hepatic artery injection.

1.25.   C True
A dense, inhomogeneous hepatogram is seen following hepatic arteriography in severe cirrhosis. The liver is reduced in size, resulting in a 'corkscrew' appearance to the hepatic arteries. The intrahepatic arteries do not enlarge in uncomplicated cirrhosis. This finding would suggest the development of a hepatoma. Both tumour and regenerating nodules will result in inhomogeneity of the hepatogram.

1.25.   D True
In advanced cirrhosis, arterioportal shunting may occur, resulting in opacification of the portal vein on hepatic arteriography. There will also be reversal of portal vein flow associated with the formation of portosystemic collaterals.

1.25.   E True
In early cirrhosis, the hepatogram during arterioportography is prompt and homogeneous and the portal veins appear normal. With moderate disease, the hepatogram becomes less homogeneous and delayed and the portal veins show

some tortuosity. In advanced disease, the portal veins appear distorted and show calibre changes, and the hepatogram may be absent peripherally or totally absent.

(Reference 10)

1.26. A False
Hepatic adenomas are supplied by peripherally located, enlarged hepatic arteries. The majority of tumours are hypervascular and well-defined, round masses. Some tumours are hypovascular but demonstrate fine neovascularity. Arteriovenous shunting is occasionally seen. Areas of haemorrhage or focal necrosis may be present, appearing as relatively hypovascular or avascular regions.

1.26. B False
Arteriovenous shunting is rarely seen with FNH. The tumours are well defined and hypervascular. Large lesions are supplied by enlarged, peripherally located hepatic arteries, whereas smaller lesions have a centrally located artery which often divides in a spoke-wheel pattern.

1.26. C False
Cavernous haemangiomas do not opacify via the portal veins. The feeding arteries are usually not enlarged unless the lesion is very large. In the late arterial and parenchymal phases, there is dense initially peripheral opacification of well-defined, nodular vascular spaces. This opacification persists late into the venous phase.

1.26. D False
Neovascularity is not a feature of regenerating hepatic nodules. These lesions usually appear as isodense or hypovascular masses that cause displacement of the portal and hepatic veins. The hepatic artery branches are of normal calibre and are centrally located. They may be stretched if the nodules are large.

1.26. E True
Haemangioendotheliomas are tumours of childhood which may present with hepatomegaly and congestive cardiac failure. On angiography, the feeding arteries are enlarged, tortuous and do not taper. The tumour is hypervascular and vascular lakes are seen. However, in contrast to cavernous haemangiomas, contrast clears rapidly from these spaces. Arteriovenous shunting is present but neovascularity is not a feature.

(Reference 10)

1.27. A True
The angiographic features of cholangiocarcinoma are similar to those of anaplastic hepatoma. It is typically an avascular/hypovascular tumour that demonstrates vascular encasement, occlusion and displacement of primarily the intrahepatic vessels (both hepatic arteries and portal veins). About 50% of cases show neovascularity.

1.27. B True
The typical angiographic features of well-differentiated hepatomas are enlarged arterial feeders, coarse neovascularity, vascular lakes, dense tumour stain, and arterioportal shunting. Portal venous invasion or occlusion is seen in 73–100% of hepatomas. Growth into the hepatic veins and IVC are not uncommon. Intravenous growth of the tumour is seen on the late arterial and venous phases of the hepatic angiogram as parallel vascular streaks.

1.27. C True
Anaplastic hepatomas may have identical angiographic features to cholangiocarcinoma (see above). Occasionally the tumour may be totally avascular. In these cases, enlarged feeding arteries and shunting are not usually identified.

1.27. D True
Hepatoblastoma is the third commonest intra-abdominal tumour in children. Angiography demonstrates enlargement of the supracoeliac aorta and the hepatic artery. The tumour is hypervascular, and marked neovascularity, vascular lakes and avascular areas due to necrosis are typical findings. Arteriovenous shunting is not characteristic.

1.27. E False
Metastases from oesophageal, bronchial and pancreatic primaries are typically hypovascular, whereas those from carcinoid, choriocarcinoma, cystadenocarcinoma, islet call carcinomas, renal cell carcinomas, thyroid and uterine carcinomas are typically hypervascular.

(Reference 10)

1.28. A False
The demonstration of a hypoechoic structure in the centre of the falciform ligament, resulting in a 'target' sign, may be a normal finding. However, the demonstration of hepatofugal flow within such a structure is a specific finding, representing recanalized umbilical or paraumbilical veins.

1.28. B True
In normal individuals, the diameter of the splenic and superior mesenteric veins increases markedly in inspiration. In portal hypertension, this finding is either diminished or absent. Respiratory variation of less than 20% in SMV diameter has been found to be 85% sensitive and 100% specific for portal hypertension.

1.28. C True
The finding of a dilated coronary vein is said to be specific for portal hypertension. A diameter of 7 mm indicates severe portal hypertension with the risk of oesophageal variceal bleeding.

1.28. D False
The demonstration of Doppler signal in the portal vein can occur with either a completely patent vein, partial thrombosis, or signal arising in periportal venous collaterals when the vein is completely occluded. Therefore the Doppler findings must be correlated with the real-time appearances.

1.28. E False
In prehepatic portal hypertension due to portal vein compression, Doppler ultrasound will demonstrate increased velocities in the narrowed, compressed segment of the portal vein and possibly also turbulent flow.

(References A, 11; B, C, D, E, 17)

1.29. A False
With prehepatic portal hypertension, shunting can occur via both portohepatic and portosystemic collaterals. With left-sided portal hypertension due to splenic vein thrombosis, one of the portohepatic collateral pathways is via the short gastric veins, perigastric venous plexus, and coronary vein to the portal vein. Oesophageal varices will develop if there is obstruction to the coronary vein.

1.29. B False
Cavernous transformation is seen in approximately 50% of patients with complete occlusion of the portal vein and may be seen as soon as 4 weeks following thrombosis. It is due to the development of a network of collaterals in the porta hepatis and hepatoduodenal ligament. The majority of patients will also have oesophageal varices.

1.29. C False
Short gastric veins are visualized as tubular structures between the spleen and stomach with a reported sensitivity of only 10% due to their small size. The coronary vein has been demonstrated in 64-90% of patients as it enters the splenic or proximal portal vein. Venous collaterals in the lesser omentum can also be seen and result in thickening of that structure.

1.29. D False
Oesophageal varices can also be seen if there is hepatofugal flow in the coronary vein. The reported sensitivity of ultrasound in visualizing gastro-oesophageal varices varies from 18% to 83% depending on their size. Ultrasound cannot differentiate between submucosal varices (which bleed) and para-oesophageal varices (which do not bleed).

1.29. E True
Paraumbilical veins less commonly arise from branches of the left portal vein, intrahepatic portal radicles or rarely the right portal vein. They extend through the falciform ligament and connect with inferior epigastric veins in the anterior abdominal wall.

(Reference 17)

1.30. A True
On arteriography, microcystic adenoma appears as a hypervascular lesion with dilated feeding arteries, dense tumour blush and densely opacified veins. Neovascularity is present but vascular encasement is not a feature. The splenic vein may be occluded by compression and arteriovenous shunting is occasionally seen.

1.30.   B True
Mucinous (macrocystic) adenoma is usually hypovascular with sparse neovascularity. Vascular encasement and splenic vein occlusion may both be present. Angiographically, the benign tumour cannot be differentiated from its malignant counterpart (in the absence of demonstrable metastases).

1.30.   C True
With good technique, angiography can detect 90% of islet cell tumours. Approximately 90% of glucagonomas and 66% of insulinomas are hypervascular. However, the angiographic appearance of the different types of tumours is similar and does not allow a specific diagnosis to be made.

1.30.   D False
The only angiographic feature that allows a diagnosis of malignancy is the demonstration of hypervascular hepatic metastases. Typically, islet cell tumours are round and well defined. Irregular tumour vessels may be seen in larger lesions, which may also cause displacement of enlarged pancreatic arteries. The tumour stain varies with the vascularity of the lesion. Some tumours have dilated vascular spaces, and enlarged draining veins may be observed.

1.30.   E False
Arterial and venous occlusion (especially the splenic vein) and occasionally smooth arterial narrowing due to fibrosis may be seen in chronic pancreatitis. There are no specific angiographic features for pancreatic adenocarcinoma, and similar findings can occur in carcinomas of the ampulla, duodenum, distal CBD, stomach and with metastases.

(Reference 10)

1.31.   A True
A gastroduodenal arteriogram is best for demonstration of bleeding from the pyloroduodenal region. If the gastroduodenal artery cannot be catheterized, a common hepatic or coeliac arteriogram should be performed. Small intestinal and colonic bleeding (up to the splenic flexure) requires a superior mesenteric arteriogram, and distal large bowel haemorrhage is assessed by injection of the inferior mesenteric artery.

1.31.   B True
The angiographic features of caecal angiodysplasia include early and dense opacification of the ileocolic vein that persists late into the venous phase, and demonstration of one or more vascular 'tufts' along the antimesenteric border of the caecum and ascending colon. Caecal angiodysplasia may account for chronic gastrointestinal bleeding in up to 50% of individuals over 55 years of age. However, it has also been seen as an incidental finding at arteriography in 15% of non-bleeding patients.

1.31.   C True
There are various angiographic features that are seen in patients with gastrointestinal haemorrhage. Punctate contrast extravasation is seen from mucosal erosions or shallow ulcers. Massive contrast extravasation is seen from ulcers, aneurysm rupture, diverticula or trauma. This appears as dense

extravascular contrast accumulation that persists past the venous phase. Intraluminal contrast is seen to outline mucosal folds.

1.31. D True
Hypervascularity of the bowel mucosa may occur during the parenchymal phase of the arteriogram and is usually associated with dilated arteries and early, dense opacification of the veins. It is a feature of haemorrhagic gastritis and inflammatory diseases. Focal areas of hyperaemia may be the only sign of an intermittently bleeding duodenal ulcer.

1.31. E True
Polyarteritis nodosa affects the mesenteric vessels in 50% of cases and may present with GI bleeding. Arteriography demonstrates multiple saccular aneurysms of the small and medium-sized branches of the mesenteric arteries. The main vessels (e.g. SMA trunk) are spared, in contrast to the necrotizing angiitis of Wegener's granulomatosis and mycotic aneurysms. Lumen irregularity of the medium and small arteries may also be seen.

(Reference 10)

1.32. A True
Intravenously injected labelled colloid is rapidly taken up by the liver and spleen, and consequently colloid remains in the blood compartment for only a brief period. Therefore, for this technique to be successful, it is essential that bleeding should be actually taking place during the first few minutes after injection of the radiopharmaceutical. If this is the case, then it is possible to detect very small amounts of blood in the bowel lumen.

1.32. B True
Technetium-labelled RBCs will remain in the vascular compartment for several hours following intravenous injection. The demonstration of gastrointestinal bleeding then depends upon the accumulation within the bowel of a sufficient concentration of blood to produce a spot of radioactivity which is visible against the background of normal blood-pool activity. Due to the prolonged retention of radiopharmaceutical in the blood compartment, intermittent bleeding may be detected on scans obtained several hours after injection.

1.32. C True
Extravasated blood tends to move along the bowel lumen fairly rapidly and therefore sequential images usually show a change in the position of the abnormal focus of activity. This may help in determining the anatomical location of the bleeding. A serpiginous appearance to the abnormal activity is typical of small bowel bleeding whereas colonic blood tends to move in a linear fashion around the margins of the abdomen.

1.32. D True
Extravasated radioactivity seen soon after the injection of either labelled colloid or a blood-pool marker indicates bleeding at or near the point where the activity was first seen. However, deduction of the site of bleeding from a late positive blood-pool image is more difficult since the first appearance of an abnormal focus of activity on a late image may indicate bleeding from that site, accumulation of blood from a more proximal bleeding point in the bowel, or

rarely, from a more distal bleeding point (due to reflux or retrograde movement of the activity).

1.32.   E False
Due to the high level of hepatic and splenic activity following use of labelled colloid, bleeding from upper abdominal sites may be more difficult to detect than with use of a blood-pool marker. Other sources of difficulty in interpretation include activity in the renal pelvis and bladder activity. In the latter case, imaging should be performed after the patient has emptied his or her bladder.

(Reference 14)

1.33.   A False
Fibromuscular dysplasia (FMD) of the renal arteries accounts for about one-third of the cases of renovascular hypertension. The majority of patients are women (ratio of 3 : 1) and all age groups may be affected, although patients are generally younger than those affected by atherosclerosis. The lesions typically occur in the distal two-thirds of the renal arteries, unlike atheromatous disease. Occasionally the whole of the artery may be affected, and rarely lesions are confined to the proximal third of the vessel.

1.33.   B False
In approximately 79% of cases, the main renal artery is affected and in 17% the branch vessels are also involved. Isolated involvement of the branch vessels is seen in about 4% of cases. The disease is bilateral in two-thirds of cases, and with unilateral disease the right renal artery is more commonly affected. FMD is the commonest cause of renovascular hypertension in children and the branch vessels are most commonly affected.

1.33.   C True
Alternating areas of stenosis and aneurysm formation result in the 'string of beads' sign, which is seen in approximately 60–70% of cases. Irregular stenoses without aneurysm formation is the second commonest appearance, seen in 15–25% of cases.

1.33.   D True
Other rarer angiographic appearances include a long smooth stenosis due to smooth muscle hyperplasia (5–15% of cases), a focal stenosis of the main renal artery (1–2% of cases) and a long segmental stenosis of the main renal artery (<1% of cases). Arterial stenoses are also seen in neurofibromatosis, and arteriographic findings in these patients may resemble those of FMD. However, the proximal renal artery or orifice is typically affected in neurofibromatosis.

1.33.   E True
Significant obstruction to the renal artery by stenoses results in formation of intrarenal collateral vessels, and an enlarged capsular artery may be seen anastomosing with intrarenal branches. Angiography may also demonstrate dissection of the renal artery, and FMD is the commonest cause of spontaneous renal artery dissection.

(Reference 10)

1.34. A True
Intrarenal aneurysms are also seen on polyarteritis nodosa and after trauma. The kidney is affected in 85% of patients with polyarteritis nodosa, and the interlobar, arcuate, interlobular and occasionally the segmental branches are affected. In contrast, aneurysms due to atherosclerosis affect the main renal artery or its major branches and are commonly located at a bifurcation.

1.34. B True
Page kidney refers to the renin–angiotensin-mediated hypertension due to renal compression by a process (haematoma, cyst, tumour) in the perinephric or subcapsular space. Haematoma following blunt trauma is the commonest cause. Arteriography shows stretching and splaying of the intrarenal vessels, slow arterial washout and distortion of the renal contour with parenchymal thinning on the nephrogram phase.

1.34. C True
Arteriocaliceal fistula is a rare complication of penetrating renal injury. Other arteriographic findings following blunt or penetrating renal trauma include arterial occlusion, intimal disruption or dissection, arteriovenous fistula, aneurysm, renal fracture, perirenal haematoma manifest as a lateral, concave compression of the renal parenchyma, and retroperitoneal haematoma.

1.34. D True
The typical arteriographic features of renal angiomyolipoma are seen in less than 50% of cases. Most lesions are hypervascular, but if the fatty component predominates the mass may be hypovascular. Supplying arteries are enlarged and tortuous. Neovascularity and aneurysms are seen but arteriovenous shunting is not a feature. The angiographic appearance may simulate renal cell carcinoma.

1.34. E True
Other angiographic findings in renal cell carcinoma include dilated, tortuous, poorly tapering supplying arteries, coarse neovascularity and aneurysm formation, dense, inhomogeneous tumour stain in lesions that are hypervascular, arteriovenous shunting and tumour growth into the renal vein and IVC. Parasitization of surrounding vessels occurs from branches of the lumbar, adrenal, subcostal, superior and inferior mesenteric arteries. Hypervascular lymph node metastases may be seen.

(Reference 10)

1.35. A False
The main renal artery, interlobar, arcuate and interlobular arteries all demonstrate a low impedance Doppler flow pattern characterized by continuous forward flow during diastole. Peak systolic velocities are usually less than 100 cm/s with a gradual decrease in velocities as the vessels are traced distally.

1.35. B False
The resistivity index (RI) is a measure of renovascular impedance and is equal to the peak systolic velocity minus the peak diastolic velocity divided by the peak systolic velocity. Values above 0.9 are considered abnormal. Absence of

diastolic flow results in an RI of 1.0, whereas reversed diastolic flow results in an RI of greater than 1.0.

1.35. C True
Acute hydronephrosis can cause abnormal impedance as a result of back pressure from the distended high-pressure calices that impede flow during diastole in the surrounding vessels, leading to diminished, absent or even reversed diastolic flow. Many chronic renal diseases resulting in diffuse scarring can also cause an elevated RI.

1.35. D True
Renal artery stenosis (RAS) is a cause of increased peak systolic velocity. One criterion for this diagnosis is a peak systolic velocity of over 180 cm/s. Another criterion for RAS is a renal artery peak systolic velocity to aortic peak systolic velocity of greater than 3.5, which has correlated with a stenosis of greater than 50%.

1.35. E True
Tumour vascularity can be demonstrated by pulsed Doppler ultrasound. Tumour vessels tend to be abnormal, showing frequent arteriovenous anastomoses resulting in high velocities due to the pressure gradient. Tumour vessels also tend to have little muscle in their walls, resulting in low-impedance flow manifest as a high diastolic component.

(Reference 11)

1.36. A True
In the assessment of renal artery stenosis (RAS) with radionuclides, a baseline study is typically performed and followed by a repeat study following administration of an angiotensin-converting enzyme (ACE) inhibitor (e.g. captopril or enalapril). ACE inhibition decreases the glomerular filtration rate (GFR) in the kidney with RAS, and a decrease in the baseline renal function on a DTPA scan is characteristic.

1.36. B False
If the study is performed using MAG3 (^{99m}Tc-mercaptoacetyltriglycine), ACE inhibition results in an increase in cortical retention in the kidney with RAS.

1.36. C False
Bilateral RAS can be diagnosed if there is a substantial bilateral decrease in the baseline renal function on DTPA scintigraphy following ACE inhibition. Branch RAS causing hypertension can also be diagnosed.

1.36. D False
Approximately 60–75% of the renal arterial lumen must be occluded before clinical renovascular hypertension develops. Subcritical (less than 60–75%) asymptomatic RAS is associated with normal baseline renal radionuclide studies which are unchanged with ACE inhibition. With a stenosis obliterating 60–90% of the arterial lumen, recent onset of hypertension and an otherwise normal kidney, baseline renal studies may be normal, but there is typically a dramatic decrease in renal function following ACE inhibition.

1.36. E True
Severe RAS that occludes 90% or more of the renal artery lumen is nearly always associated with a significant decrease in renal function and a decrease in size of the affected kidney. Despite activation of the renin–angiotensin system, its effectiveness in maintaining GFR is limited and baseline studies are typically abnormal, but non-specific.

(Reference 15)

1.37. A False
Thromboangiitis obliterans is an inflammatory disease of unknown aetiology involving the medium and small arteries of the extremities and leading to segmental occlusions. The disease occurs predominantly in young men less than 40 years of age.

1.37. B False
The foot and calf vessels are affected most frequently, with popliteal and superficial femoral artery occlusions being less common. Consequently, intermittent claudication of the foot and lower calf is the commonest presenting symptom. There is also a high frequency of upper limb involvement. Superficial thrombophlebitis is also a recognized part of the disease.

1.37. C False
Thromboangiitis obliterans may be differentiated clinically from atherosclerosis by the absence of predisposing conditions such as diabetes mellitus and hyperlipidaemia. Also, patients are typically younger (almost invariably under 50 years of age). There is, however, a definite association with cigarette smoking.

1.37. D True
The angiographic features of thromboangiitis obliterans are also fairly typical and allow differentiation from atheroma. Multiple segmental occlusions are seen, with an abrupt transition from a normal-calibre vessel to an occlusion, or rarely to a thready, partially recanalized vessel. The occlusion ends as abruptly as it begins, with continuation of a normal-calibre vessel.

1.37. E True
Areas of arterial spasm are a feature and may mask the presence of an extensive collateral network of arteries, unless vasodilators are used. Other findings include multiple segmental stenoses or occlusions of the metatarsal and digital arteries.

(Reference 10)

1.38. A True
The peripheral circulation perfuses a high-resistance bed of skin and muscle. Consequently, the normal Doppler waveform in the resting state is triphasic. The three components are rapid forward flow during systole, followed by reversed flow in early diastole and then slow forward flow in late diastole. By the end of diastole there may be no flow. Following exercise, there is reduction in peripheral resistance which may be manifest as continuous antegrade flow.

1.38.   B False
The abnormal Doppler examination shows abnormalities of both waveform and velocity. There is frequently a decrease in peak systolic velocity above and below a diseased segment and increased velocity within a short-segment stenosis. Turbulence may be demonstrated as the jet of blood exits from the stenosis. Long-segment stenoses or stenoses distal to an occlusion may lack increased velocity.

1.38.   C True
Characteristically, there is a non-specific loss of triphasic flow distal to a stenosis or occlusion, attributed to a decrease in peripheral resistance due to ischaemia. However, loss of triphasic flow may occur proximal to a stenosis also, and triphasic flow may persist distal to a stenosis, usually in patients who have a palpable pulse distal to the stenosis.

1.38.   D False
Vessel occlusion is detected by noting a complete lack of flow, usually with extremely damped flow in the segment immediately above the occlusion. Occasionally, elevated velocities may be detected in the areas of reconstitution of flow, where collateral vessels enter below the occluded segment. The ability to differentiate stenosis from occlusion is good, with reported sensitivities ranging from 93% to 100% in large series.

1.38.   E False
The technique of Doppler ultrasound in the evaluation of peripheral vascular disease is limited by a number of factors, one of which is the difficulty in assessment of the tibial vessels due to their small calibre. Other problems include poor visualization of the iliac artery due to overlying bowel gas and difficulty in assessing the superficial femoral artery at the adductor canal, where the vessel lies deep to muscle.

(Reference 13)

1.39.   A True
Acute thrombus usually expands the vein and, being soft, will change shape in response to compression by the transducer. Fresh thrombus may also be free floating and with real-time ultrasound can be seen to move within the vein with respiration or compression. The echogenicity of the thrombus is unreliable in assessing its age, and both fresh and old thrombus can appear hyperechoic or hypoechoic relative to muscle.

1.39.   B False
Partial compressibility of the vein may occur with fresh thrombosis (see above), and only if the vessel completely collapses with pressure can thrombosis be reliably excluded. The inability to demonstrate a vein next to its accompanying artery is another sign that should suggest the possibility of thrombosis, since thrombus may be isoechoic with surrounding muscle.

1.39.   C True
Colour Doppler diagnosis of a normal vein is made by identifying complete filling of the lumen by colour. Thrombus is diagnosed by the absence of colour or the outlining of non-occlusive thrombus by colour signal from flowing blood around it. Sometimes colour does not fill the vein because the machine is

insensitive to slow flow. In these cases, distal compression can augment the colour signal by increasing the venous velocity and so filling the vessel with colour.

1.39. D False
Augmentation of Doppler signal following distal compression excludes the presence of occlusive thrombosis between the site of compression and the point from which the signal is being obtained. However, this sign does not exclude the presence of non-occlusive thrombosis.

1.39. E False
The normal venous Doppler waveform varies with respiration (so-called phasic change). The absence of phasic variation is the most important abnormal Doppler signal. A continuous, non-changing signal suggests obstruction proximal to the vessel being examined. The cause may be due to thrombosis or a pelvic mass compressing the vein. However, phasic variation may be lost because of prior DVT. Loss of phasic variation in the common femoral vein suggests thrombosis in the iliac vein.

(Reference 11)

1.40. A False
The characteristics of blood clot (bland thrombus) depends upon the age of the thrombus and the pulse sequence used. Tumour thrombus may show similar signal characteristics to the primary tumour and may enhance following Gd-DTPA, allowing differentiation from bland thrombus.

1.40. B True
Fresh thrombus may have a variable *in vivo* appearance depending upon the formation of paramagnetic haemoglobin breakdown products and the degree of clot retraction. Hyperintensity on both T1- and T2-WSE scans is a recognized feature.

1.40. C False
As thrombus matures and becomes incorporated into the vessel wall, there is a gradual reduction in signal intensity, and old thrombus usually has low to intermediate signal on both T1- and T2-WSE scans. Unlike intracerebral haematoma, haemosiderin is not commonly demonstrated and therefore signal void is not a typical feature.

1.40. D True
Using fast GE techniques (e.g. FLASH), flowing blood appears hyperintense whereas thrombus typically has low to moderate signal intensity. However, if the flip angle is too small or the TR too long, thrombus may appear bright and difficult to differentiate from flowing blood.

1.40. E True
Presaturation results in reduction or elimination of the signal due to flow, whereas the signal from thrombus is unaffected. Signal from flowing blood can also be differentiated from thrombus by the presence of ghost artefact due to pulsatile flow. Also, flow signal will have different intensities on different planes of section, whereas the signal from thrombus will not be affected.

(Reference 3)

2   Chest

2.1. Concerning the pulmonary manifestations of AIDS
A   consolidation is typically the earliest CXR abnormality in *Pneumocystis carinii* pneumonia (PCP).
B   cavitation is not a feature of PCP.
C   Kaposi's sarcoma (KS) typically presents as a solitary lung nodule.
D   lymphocytic interstitial pneumonitis (LIP) is a recognized cause of bilateral interstitial shadowing.
E   the gallium scan is usually negative in patients with KS.

2.2. Which of the following radiographic features suggest a subpulmonary effusion rather than a raised hemidiaphragm?
A   The 'hemidiaphragm' peaks more laterally than usual.
B   A sharp posterior costophrenic angle.
C   Lack of visibility of vessels behind the 'hemidiaphragm'.
D   An associated triangular, retrocardiac shadow.
E   A smoothly convex upper contour to the 'hemidiaphragm' on the lateral view.

2.3. Which of the following are typical CT features of a pleural effusion?
A   Anterolateral displacement of the diaphragmatic crus.
B   A sharp interface between the fluid and liver or spleen.
C   Fluid lying behind the posteromedial aspect of the right lobe of the liver.
D   The attenuation coefficient of the fluid can differentiate benign from malignant effusions.
E   The attenuation coefficient of the fluid can indicate haemothorax.

2.4. Concerning ventilation/perfusion (V/Q) scanning
A   pneumonia is a cause of V/Q mismatch affecting ventilation to a greater extent.
B   perfusion defects due to pulmonary emboli may be subsegmental.
C   matched non-segmental V/Q defects are typical of parenchymal lung disease.
D   a single unmatched segmental perfusion defect is diagnostic of pulmonary embolus.
E   a matched V/Q defect with a corresponding CXR abnormality excludes pulmonary embolus.

2.5. Concerning ventilation/perfusion (V/Q) scanning
A   pulmonary vasculitis is a cause of unmatched segmental perfusion defects.
B   pulmonary venous hypertension can be suggested on the perfusion scan.
C   emphysematous bullae are a cause of matched V/Q defects.
D   acute bronchospasm is a cause of bilateral segmental perfusion defects.
E   radiotherapy to the lung typically causes a greater defect of perfusion than ventilation.

2.6. Concerning solitary pulmonary nodules in adults
A calcification excludes the diagnosis of bronchial carcinoma.
B a doubling time of less than 28 days is typical of bronchial carcinoma.
C a doubling time greater than 18 months occurs with granuloma.
D most lesions greater than 4 cm diameter are bronchial carcinomas.
E a lesion with a lobulated margin is invariably malignant.

2.7. Which of the following are true of lung involvement by *Aspergillus fumigatus*?
A Mycetoma formation occurs in ankylosing spondylitis.
B Pleural thickening adjacent to a mycetoma cavity is a recognized CT feature.
C Irregular air-spaces within the mycetoma are a feature on CT.
D Central bronchiectasis is a characteristic CT finding in allergic bronchopulmonary aspergillosis (ABPA).
E Tubular densities are a recognized CT finding in ABPA.

2.8. In the differentiation of a peripheral lung abscess from empyema, which of the following suggest empyema?
A An air–fluid level of similar length on frontal and lateral CXR.
B Displacement of the hilum away from the lesion.
C If CT demonstrates an obtuse angle between the lesion and chest wall.
D If CT demonstrates air densities within the wall.
E If CT demonstrates compression of lung adjacent to the mass.

2.9. Which of the following are recognized CXR findings in bronchial carcinoma?
A An air crescent sign.
B Lobar consolidation.
C Satellite nodules.
D Multiple well-defined pulmonary nodules.
E Diffuse nodular pleural thickening.

2.10. Concerning the CT assessment of bronchial carcinoma
A demonstration of mediastinal lymphadenopathy indicates that the tumour is not surgically curable.
B contact between the tumour and the mediastinum indicates mediastinal invasion.
C pleural thickening adjacent to the mass indicates chest wall invasion.
D the accuracy of CT in detecting nodal disease is independent of the site of involved nodes.
E the incidence of nodal involvement at presentation is independent of primary tumour size.

2.11. Concerning MRI in the evaluation of bronchial carcinoma
A   atelectatic lung is best differentiated from central mass on T1-WSE images.
B   on T2-WSE scans, increased chest wall signal intensity adjacent to a mass is a feature of chest wall invasion.
C   MRI is more accurate than CT in the evaluation of superior sulcus (Pancoast) tumours.
D   pericardial invasion can only be diagnosed if effusion is present.
E   benign and malignant lymphadenopathy can be distinguished by differences in relaxation times in the majority of cases.

2.12. Concerning radiation-induced lung damage
A   acute radiation pneumonitis usually develops within days of treatment.
B   pulmonary shadowing is typically confined by fissures.
C   reduction of vessels distal to a central radiation field is a recognized CT finding.
D   bronchiectasis is a recognized finding at CT.
E   MRI can distinguish radiation fibrosis from recurrent tumour.

2.13. Concerning bronchiectasis
A   the V/Q scan is usually normal.
B   on CT, a bronchial diameter greater than its accompanying artery indicates bronchial dilatation.
C   CT cannot distinguish bullae from cystic bronchiectasis.
D   flame-shaped opacities are a recognized CT finding.
E   there is an association with male infertility.

2.14. Concerning emphysema
A   CT demonstrates areas of reduced parenchymal attenuation.
B   CT demonstrates vessel branching angles of greater than 80°.
C   CT can differentiate centriacinar from panacinar emphysema.
D   subpleural bullae are a feature of paraseptal emphysema.
E   following infection a bulla may disappear.

2.15. Which of the following are true of malignant mesothelioma?
A   It usually presents within 10 years of asbestos exposure.
B   CT typically demonstrates nodular pleural thickening.
C   Associated large pleural effusions typically cause contralateral mediastinal shift on CXR.
D   Identical CT features occur with metastatic adenocarcinoma.
E   The presence of effusion differentiates malignant mesothelioma from a benign pleural plaque.

2.16. Concerning progressive massive fibrosis (PMF)
A   it is defined as nodules with diameter greater than 3 cm.
B   in silicosis it is due to conglomeration of nodules.
C   CT demonstrates a smooth lateral border paralleling the chest wall.
D   CT demonstrates the lesions as spherical masses.
E   a calcified rim is a recognized CT finding.

2.17. Concerning respiratory tract involvement with amyloid
A   the CXR in systemic pulmonary amyloid is usually normal.
B   calcification of hilar nodes is a feature.
C   airway amyloid is typically focal.
D   a cartilage-containing nodule is a recognized feature.
E   pleural effusions are a recognized finding.

2.18. In the CT assessment of mediastinal fat abnormalities
A   mediastinal lipomatosis is typically seen in a paraspinal location.
B   mediastinal lipomatosis has homogeneous fat density.
C   mediastinal lipomatosis typically causes tracheal deviation.
D   omental herniation cannot be differentiated from a pericardial fat pad.
E   fat herniation through the oesophageal hiatus appears as a paraspinal fatty mass.

2.19. Recognized CT features of rounded atelectasis include which of the following?
A   An air bronchogram within the mass.
B   Increasing size of the mass on follow-up scans.
C   Associated pulmonary fibrosis.
D   Increased density at the centre of the mass.
E   An acute angle between the mass and adjacent pleura.

2.20. Concerning CT in the assessment of pulmonary collapse
A   an obstructing mass cannot be differentiated from collapsed lung.
B   with left upper lobe (LUL) collapse there is foreshortening of the aorto-pulmonary window.
C   a collapsed right upper lobe (RUL) is bordered posteriorly by the minor fissure.
D   a collapsed right middle lobe (RML) loses contact with the anterior chest wall.
E   a collapsed lower lobe appears as a wedge-shaped paraspinal density.

2.21. Which of the following are recognized causes of cavitating pulmonary masses?
A   Behçet's disease.
B   Systemic temporal arteritis.
C   Churg–Strauss syndrome.
D   Necrotizing sarcoid angiitis.
E   Pulmonary alveolar proteinosis.

2.22. Concerning the CT features of intrathoracic lymphoma
A   isolated enlargement of posterior mediastinal nodes is characteristic of nodular sclerosing Hodgkin's disease.
B   enlargement of paracardiac nodes is typical at presentation.
C   nodal masses characteristically enhance on postcontrast CT.
D   diffuse reticulonodular shadowing is a recognized feature.
E   chest wall invasion is a recognized finding.

2.23. Concerning the technique and anatomical features of high-resolution CT (HRCT) of the lungs
A   slice thicknesses are typically of the order of 1–2 mm.
B   reconstruction using a high-resolution (bone) algorithm increases the visualization of noise in the image.
C   motion artefact is most commonly seen in the RUL.
D   the majority of interlobular septa are visualized.
E   the intralobular pulmonary artery branch is typically seen within the pulmonary lobule.

2.24. Concerning HRCT of the lungs
A   lymphangitis carcinomatosa is a cause of smooth thickening of the interlobular septa.
B   cryptogenic fibrosing alveolitis (CFA) results in irregular thickening of the interlobular septa.
C   irregular linear opacities in CFA are typically subpleural.
D   cystic air spaces in lymphangioleiomyomatosis typically involve the upper lobes.
E   cystic air spaces in CFA are typically surrounded by abnormal lung parenchyma.

2.25. Concerning HRCT of the lungs
A   nodular thickening of bronchovascular bundles is a feature of sarcoid.
B   nodules in silicosis tend to affect the upper lobes.
C   poorly defined areas of increased parenchymal attenuation are specific for the active phase of CFA.
D   subpleural consolidation is characteristic of eosinophilic pneumonia.
E   the combination of consolidation and nodules is diagnostic of pulmonary lymphoma.

2.26. Concerning CT in the assessment of intrathoracic neoplasms
A   a nodule with an average CT number of 100 HU is always benign.
B   mediastinal invasion is a feature of tracheal carcinoma.
C   hilar lymphadenopathy is a feature of bronchial carcinoid.
D   mediastinal lymphadenopathy is a typical feature of small cell carcinomas at presentation.
E   absence of calcification excludes a diagnosis of hamartoma.

2.27. Concerning CT of asbestos-related lung disease
A   subpleural cysts are a recognized HRCT finding in asbestosis.
B   on HRCT, curvilinear subpleural lines in dependent lung are diagnostic of asbestosis.
C   postcontrast CT typically shows no enhancement in round atelectasis.
D   pleural plaques are not seen in a paraspinal location.
E   diffuse benign pleural thickening typically spares the mediastinal pleura.

2.28. Concerning HRCT of focal lung disease
A   the presence of an air-bronchogram within atelectatic lung excludes tumour as the cause of obstruction.
B   identification of fat within an area of consolidation is diagnostic of lipoid pneumonia.
C   low attenuation surrounding an area of consolidation is a feature of invasive pulmonary aspergillosis.
D   enhancement is a feature of bronchial carcinoid.
E   demonstration of a vascular connection to a nodule is diagnostic of AVM.

2.29. Concerning CT of the postpneumonectomy space
A   the space should contain homogeneous soft-tissue attenuation material by 1 year.
B   pleural calcification indicates chronic infection.
C   tumour recurrence usually manifests as rib destruction.
D   the degree of obliteration of the space is related to the degree of overinflation of the contralateral lung.
E   herniation of the right lung posterior to the mediastinum may occur after left pneumonectomy.

2.30. Concerning CT in the assessment of the lung
A   in PCP cystic changes are typically initially associated with areas of consolidation.
B   peripheral enhancement is a feature of pulmonary infarct on postcontrast scans.
C   pulmonary metastases are usually demonstrated in the outer third of the lungs.
D   cavitation is demonstrated in 25% of metastases.
E   nodular retrocrural masses are a recognized feature of lymphangioleiomyomatosis.

2.31. Concerning lymphangioleiomyomatosis
A   women are affected in 85% of cases.
B   the disease process is limited to the chest.
C   reticulonodular shadowing with increased lung volumes is characteristic.
D   chylous pleural effusions occur in the majority of cases.
E   the majority of patients die of respiratory failure.

2.32. Which of the following are true of chylothorax?
A   It is invariably unilateral.
B   It is typically painful.
C   CT numbers of the effusion are usually negative.
D   Mediastinal malignancy is the commonest cause.
E   Trauma to the thoracic duct is most commonly due to cardiac surgery.

2.33. Which of the following are recognized features of thoracic sarcoid?
A   Pleural effusion.
B   Middle lobe collapse.
C   Bronchial stenosis.
D   Superior vena cava syndrome.
E   Pulmonary arterial hypertension.

2.34. Concerning Wegener's granulomatosis
A   respiratory tract disease occurs in all patients at some stage of the disease.
B   pulmonary nodules only resolve with treatment.
C   nodules are usually single.
D   tracheal stenosis is a recognized complication.
E   cavitation within nodules indicates secondary infection.

2.35. Concerning the intrathoracic extension of thyroid masses
A   anterior displacement of the trachea is a recognized finding.
B   calcification differentiates benign from malignant lesions.
C   precontrast CT typically demonstrates areas of greater-than-muscle attenuation.
D   lateral displacement of the brachiocephalic veins is typical.
E   areas of high signal are a feature on T2-WSE MRI.

2.36. Which of the following are true of thymoma?
A   It is the commonest cause of a thymic mass in adults.
B   There is an association wth hypogammaglobulinaemia.
C   Calcification is not a recognized CT finding.
D   It typically spreads by blood-borne metastases to the lungs.
E   Early invasive thymoma is best demonstrated by T2-WSE MRI.

2.37. Which of the following are true of anterior mediastinal masses?
A   Thymic cysts are most commonly associated with treated Hodgkin's lymphoma.
B   Thymic hyperplasia usually appears on CT as a focal mass in the gland.
C   CT cannot differentiate rebound thymic hyperplasia from recurrent lymphoma.
D   Absence of fat density on CT excludes the diagnosis of benign cystic teratoma.
E   Malignant germ cell tumours usually show calcification on CT.

2.38. Which of the following are true of cystic mediastinal masses?
A   Attenuation values of over 100 HU are a recognized feature of bronchogenic cysts on precontrast CT.
B   Bronchogenic cysts may arise in the posterior mediastinum.
C   Pericardial cysts are most common in the left cardiophrenic angle.
D   Neurenteric cysts are usually located between the oesophagus and the trachea.
E   Mediastinal pancreatic pseudocysts in adults usually follow traumatic pancreatitis.

2.39. Which of the following mediastinal masses may demonstrate enhancement on postcontrast CT?
A   Castleman's disease (giant lymph node hyperplasia).
B   Aggressive fibromatosis.
C   Lymphangioma.
D   Nerve sheath tumours.
E   Paraganglionomas.

2.40. Concerning CT in the assessment of the trachea

A the average coronal measurement of the adult trachea is 12 mm.

B there is an association between diffuse tracheal dilatation and bronchiectasis.

C a reduction in sagittal tracheal diameter is typical of sabre-sheath trachea.

D thickening of the tracheal wall is a feature of relapsing polychondritis.

E nodular thickening of the tracheal wall is a feature of laryngeal papillomatosis.

Chest: Answers

2.1. A False

PCP is the commonest fatal infection in AIDS, affecting about 60% of patients at least once in the course of their disease, with one-quarter of these initial episodes being fatal. The earliest CXR abnormality is usually bilateral perihilar interstitial shadowing which may progress over 3–5 days to confluent consolidation involving the entire lung.

2.1. B False

Atypical CXR features in PCP include cavitation with abscess formation, unilateral disease, upper lobe predominance, lobar involvement and cystic parenchymal disease leading to an increased risk of pneumothorax. Pulmonary fibrosis is also a recognized complication although the exact incidence is not known.

2.1. C False

Pulmonary KS occurs in approximately 20% of patients with epidemic KS and is almost always preceded by cutaneous or visceral KS. The CXR may show diffuse interstitial shadowing indistinguishable from PCP. The presence of lymphadenopathy, pleural effusions and nodules is more suggestive of KS. CT characteristically shows the nodules to be predominantly peribronchial.

2.1. D True

LIP is associated with many autoimmune disorders and is considered diagnostic of AIDS if histologically proven in a seropositive patient. The CXR features are non-specific.

2.1. E True

However, PCP is commonly associated with diffuse uptake of gallium with a sensitivity of 95–100%. A normal gallium scan has a predictive value of over 90% in excluding PCP. The specificity of a positive scan is low since uptake may occur in other opportunistic infections.

(Reference 7)

2.2. A True

Subpulmonary effusions may be unilateral or bilateral, and when unilateral are more commonly right-sided. The apex of the 'hemidiaphragm' is characteristically more lateral and also the contour on either side of the peak is straighter. The medial slope tends to be gradual and the lateral slope steep. These features tend to be accentuated on expiration, a phenomenon that is attributed to the pulmonary ligament.

2.2. B False

With subpulmonary effusion both the lateral and posterior costophrenic sulci are usually blunted, ill-defined or shallow. However, occasionally both the lateral and posterior angles may appear sharp.

2.2. C True
With a subpulmonary effusion, the posterior basal aspect of the lower lobe is pushed out of the posterior costophrenic sulcus by fluid. Therefore its vessels will not be visible through the 'hemidiaphragm' on the frontal CXR.

2.2. D True
This triangular retrocardiac shadow effaces the medial aspect of the 'diaphragmatic' contour and the left paravertebral interface with lung and is due to paramediastinal extension of fluid. On the left side, there will also be a greater distance between the contour and the stomach fundus.

2.2. E False
On the lateral CXR, subpulmonary effusion typically has a flat upper border beneath the lower lobe and may be limited anteriorly by the major fissure. A small amount of fluid may enter the fissure. The diagnosis of subpulmonary effusion can be made by obtaining a decubitus radiograph, since they are rarely loculated. Alternatively, ultrasound can be performed.

(Reference 4)

2.3. A True
On CT, pleural effusion typically appears as a homogeneous crescentic density in the dependent part of the chest. Since pleural fluid accumulates between the crus and the vertebral column it pushes the crus anterolaterally. Ascitic fluid accumulates anterolateral to the crus, pushing it posteromedially against the spine.

2.3. B False
A sharp interface between the fluid and liver or spleen is a feature of ascites. The interface between pleural fluid and the liver or spleen is typically hazy, probably due to partial volume effect from the obliquely sectioned diaphragm. This sign must be assessed away from the domes of the diaphragm and is less useful with the use of thin sections.

2.3. C True
This constitutes the bare area of the liver which is between the upper and lower layers of the coronary ligament and is extraperitoneal. Therefore ascitic fluid cannot accumulate at this site. However, ascites can lie above and below the coronary ligament.

2.3. D False
Similarly CT numbers cannot differentiate transudative from exudative effusions, but the low attenuation of the fluid will allow demonstration of any associated pleural thickening or mass which may not be possible on the CXR.

2.3. E True
CT may demonstrate haemothorax as a non-homogeneous collection with CT attenuation coefficients significantly higher than those of water. Haemothorax may eventually organize and cause massive pleural thickening (fibrothorax).

(Reference 4)

2.4.   A True
Alveolar consolidation is the main pathology and blood flow is affected to a lesser extent. However, usually the V/Q scan shows perfusion to be reduced to a similar degree. A central bronchial carcinoma may cause an unmatched perfusion defect if it involves the pulmonary artery also, but is more commonly associated with a defect of ventilation also.

2.4.   B True
However, pulmonary emboli usually result in defects that are segmental. Parenchymal lung disease, fat emboli and tumour emboli may also cause subsegmental perfusion defects.

2.4.   C True
Parenchymal lung disease is not primarily vascular and therefore typically results in non-segmental defects.

2.4.   D False
A picture of multiple, bilateral, segmental, unmatched perfusion defects is considered diagnostic of pulmonary embolus. Fat and tumour emboli can also result in segmental unmatched perfusion defects, with a normal CXR in the latter case. A single unmatched segmental perfusion defect may be due to pulmonary embolus but is considered to have a low probability.

2.4.   E False
The combination of a matched V/Q defect and a CXR abnormality may be associated with pulmonary embolus since an area of consolidation on the CXR may be due to pulmonary infarction. The defects of pulmonary embolus may persist indefinitely or may clear within 24 hours.

(Reference 5)

2.5.   A True
The vasculitis caused by polyarteritis nodosa, systemic lupus erythematosus and other collagen vascular diseases may cause scan appearances identical to pulmonary embolus. The defects may regress following steroid therapy.

2.5.   B True
With pulmonary venous hypertension, there will be a reversal of the normal perfusion gradient with increased perfusion to the upper zones and non-segmental perfusion defects at the bases. Basal segmental defects may therefore be less well demonstrated. The cardiac outline may also be enlarged.

2.5.   C True
The defects due to emphysematous bullae may also show air trapping on the washout phase of a xenon-133 ventilation scan.

2.5.   D True
The perfusion defects of acute bronchospasm are matched by ventilation defects. The scan may return to normal if repeated following bronchodilators. Temporary ventilation defects are a feature of chronic obstructive airways

disease due to mucus plugging of bronchi. They may resolve with physio-therapy.

2.5. E True
Radiotherapy affects the pulmonary vasculature. The resulting perfusion defects may have straight boundaries and rarely correspond to bronchopulmonary segments.

(Reference 5)

2.6. A False
Bronchial carcinoma may engulf a calcified granuloma but in these cases the calcification is almost invariably eccentric in location. Concentric (laminated) calcifications are typical of tuberculous or fungal granulomas, whereas a popcorn type is seen in cartilage-containing nodules. Punctate calcifications may be seen in hamartomas, granulomas and metastases, usually from osteosarcoma and chondrosarcoma.

2.6. B False
This rate of growth is too rapid for carcinoma and occurs with infection, infarction, histiocytic lymphoma and metastases from tumours such as germ cell neoplasms and some sarcomas. Average doubling time for bronchial carcinoma is 4.2–7.3 months.

2.6. C True
Other conditions with a doubling time greater than 18 months include hamartoma, bronchial carcinoid and round atelectasis. Rarely, bronchial carcinoma may take as long as 24 months to double in size.

2.6. D True
Other possible causes of solitary pulmonary nodules greater than 4 cm are less common and include solitary metastasis, lung abscess, Wegener's granulomatosis, round pneumonia, round atelectasis and hydatid cyst. Conversely, carcinomas less than 9 mm in diameter are virtually invisible on the CXR, and therefore lesions of this size that are clearly seen are likely to be calcified and therefore unlikely to be primary carcinomas.

2.6. E False
Lobulation and notching can be seen with almost all kinds of nodules. However, the more pronounced these features are, the greater the chance the lesion is a bronchial carcinoma. The presence of spiculation is almost specific for carcinoma. Conversely, a lesion with a well-defined, non-lobulated edge is rarely due to bronchial carcinoma.

(Reference 4)

2.7. A True
Other chronic fibrotic diseases that may be complicated by mycetoma are tuberculosis (commonest), sarcoid, interstitial pulmonary fibrosis, bronchiectasis and lung abscess or infarct. Mycetoma may become invasive if host immunity is compromised by steroids or in diabetes mellitus.

2.7. B True
The pleural thickening adjacent to a mycetoma cavity may be up to 2 cm in thickness. The cavity wall also appears thick since it is lined by fungal hyphae. Typically there is a mobile intracavitary mass. In some cases the mycetoma will fill the cavity and no longer be mobile. An air–fluid level may also be seen within the cavity.

2.7. C True
CT can also diagnose mycetoma formation earlier than the CXR by demonstrating fungal strands either lining the cavity wall or within the cavity before a fungus ball has been produced. In the semi-invasive form of pulmonary aspergillosis, the CXR demonstrates a focal chronic consolidation which may progress to cavitation and mycetoma formation. Aspergillosis is also a rare cause of a solitary pulmonary nodule with appearances indistinguishable from a neoplasm.

2.7. D True
The bronchiectasis of ABPA affects lobar and segmental bronchi with normal distal airways. CT is twice as sensitive as the CXR in demonstrating this complication, which is considered specific to ABPA. The bronchiectasis typically has a perihilar and upper lobe distribution and is bilateral. Upper lobe volume loss is a secondary effect but overall lung volume is usually increased.

2.7. E True
Tubular opacities are due to bronchocele formation, which occurs in up to 30% of cases and is the result of mucoid impaction of central bronchi. The CXR demonstrates well-defined, branching band-like opacities that point towards the hilum.

(Reference 4)

2.8. A False
An empyema is typically lens-shaped and an air–fluid level will therefore appear longer on one projection than the other. Abscesses are usually spherical, resulting in a similar length to the air–fluid level regardless of the projection. Also the air–fluid level in an empyema may reach the chest wall, whereas that in an abscess is unlikely to do so since an abscess is often surrounded by lung parenchyma.

2.8. B True
Since an empyema compresses lung it will cause displacement of normal structures. Abscess causes destruction of lung tissue rather than compression and is therefore less likely to result in displacement of normal structures.

2.8. C True
CT typically demonstrates a lens-shaped fluid collection with tapering margins adjacent to the chest wall. Abscesses are usually spherical, forming acute angles with the chest wall. However, a large empyema may also have an acute angle at its point of contact with the chest wall.

2.8. D False

Air densities in the wall are typical of an abscess. The wall of an empyema is usually smooth and of uniform thickness and may show enhancement of postcontrast CT scans. The wall of an abscess is typically thick and irregular and may also show distorted air bronchograms within it.

2.8. E True

Compression of adjacent lung is seen with empyema for the same reasons as outlined in (B). Consolidation adjacent to the lesion is of little help in differentiation since it can occur with both abscess and empyema, which are usually both complications of pneumonia.

(Reference 4)

2.9. A True

Rarely a meniscus or air crescent sign may be seen with a cavitating bronchogenic carcinoma. The sign may be due to an intracavitary tumour mass, aspergilloma, or other formed debris within the cavity. Cavitation in lung carcinoma occurs in approximately 10% of cases. The cavity wall is typically thick and irregular but may rarely be thin and smooth.

2.9. B True

Lobar consolidation is particularly seen with bronchoalveolar cell carcinoma. Other presentations include focal segmental or non-segmental consolidations, or as a diffuse air-space filling process that may extensively involve both lungs. CT may demonstrate stretching, spreading and narrowing of the involved bronchi. Other forms of adenocarcinoma may also rarely produce a pattern of consolidation.

2.9. C True

Satellite nodules are a feature of inflammatory disease, particularly tuberculosis, and have been reported in about 10% of cases of tuberculoma and 8% of cases of tuberculosis. However, they also occur in approximately 1% of bronchial carcinomas.

2.9. D True

Metastases are by far the commonest cause of such a CXR finding. However, bronchial carcinoma may cause this finding in three situations. Firstly, 25% of bronchoalveolar cell carcinomas will have this appearance. Secondly, haematogenous lung-to-lung metastases can occur with all types of bronchial carcinomas, usually in the presence of distant metastases. Thirdly, multiple primary tumours of the lung may occur with an incidence of 0.72–3.5%.

2.9. E True

Pleural involvement by primary lung cancer occurs in 8–15%, usually manifest by pleural effusion. Occasionally, a peripheral carcinoma, typically an adenocarcinoma, will directly invade the pleura, causing diffuse pleural thickening similar to mesothelioma. However, a visible parenchymal mass is usually evident.

(Reference 6)

2.10. A False
Enlargement of mediastinal lymph nodes in patients with bronchial carcinoma may be due to coexistent benign causes such as reactive hyperplasia, previous tuberculosis, sarcoid or pneumoconiosis. Nodes less than 10 mm in their short axis are unlikely to be involved by tumour. Also, enlarged nodes that contain extensive calcification are most likely benign.

2.10. B False
The CT and MRI signs which indicate mediastinal invasion are visible tumour surrounding the vessels, oesophagus or mainstem bronchi. Mere contact with the mediastinum is not enough to diagnose invasion and these tumours are considered indeterminate. Also, apparent interdigitation of tumour with mediastinal fat is an unreliable sign.

2.10. C False
Pleural thickening adjacent to a tumour may be secondary to invasion but this sign is not diagnostic. The greater the degree of pleural thickening or contact of the mass with the pleura, the greater the chance of parietal pleural invasion. A chest wall mass (in the absence of other possible causes) or adjacent rib destruction allow a diagnosis of chest wall invasion to be made.

2.10. D False
Left hilar and aortopulmonary window node involvement is more difficult to detect accurately than nodal disease at other sites. However, left-sided tumours may commonly involve the subcarinal and right-sided nodes, making the poor sensitivity of CT in detecting left-sided nodes less important in tumour staging.

2.10. E False
Hilar and mediastinal nodal metastases are often present at the time of initial diagnosis, especially with adenocarcinoma and small cell carcinoma. Primary tumours greater than 3 cm in diameter have a greater incidence of nodal metastases than smaller tumours. Also, the likelihood of nodal metastases is greater with centrally located tumours.

(Reference 4)

2.11. A False
Atelectatic lung is better differentiated from a central mass on T2-WSE images, in which case the central tumour typically appears hypointense and the distal lung may produce very high signal. However, one study has shown contrast-enhanced CT to be more sensitive than MRI in making this distinction.

2.11. B True
Chest wall invasion is better demonstrated on T2-WSE scans than on T1-WSE scans since tissue contrast between tumour, fat, chest wall and muscle is best using the former scanning sequence. However, plain radiographs and CT demonstrate rib destruction better than MRI.

2.11. C True
MRI has proved more sensitive than CT in assessing the extent of spread of Pancoast tumours. Structures such as vessels and the brachial plexus are separated from the apical pleura by a thin rim of fat, and preservation of this excludes invasion, whereas obliteration or impingement suggests invasion. Sagittal and coronal images are of greatest value.

2.11. D False
The normal pericardium is seen on MRI as a low-signal-intensity line not exceeding 2.6 mm in thickness. Pericardial invasion may be indicated by interruption of this line. Similarly, interruption of the low-signal-intensity line of the diaphragm can indicate invasion.

2.11. E False
Studies have shown considerable overlap in relaxation times of benign and malignant nodes, making MRI unreliable in the distinction between the two. Furthermore, MRI cannot easily demonstrate calcification in benign nodes.

(Reference 6)

2.12. A False
Acute radiation pneumonitis develops between 1 and 6 months after treatment. Symptoms include dyspnoea, productive cough, fever and night sweats. However, patients are commonly without symptoms. In the fibrotic (chronic) phase the patient is usually asymptomatic. However, if fibrosis is extensive, cough, dyspnoea, haemoptysis and recurrent infections may occur, as may finger clubbing.

2.12. B False
Typically, the pulmonary shadowing corresponds to the field of radiation and bears no relationship to normal anatomical structures. The initial changes consist of a diffuse haze in the irradiated region which obscures the vessels. This is followed by development of pulmonary consolidation which progresses to the fibrotic stage.

2.12. C True
This reduction in vascularity distal to a central radiation portal is thought to be due to radiation-induced vascular stenoses. Increased transradiancy of the lung on the CXR is a less common manifestation of radiation therapy and presumably results from diminished pulmonary perfusion. Pleural effusion may complicate the acute phase of radiation pneumonitis, and spontaneous pneumothorax is also a recognized feature.

2.12. D True
Fibrosis and volume loss occur 12–18 months after therapy, and when particularly severe may be associated with the presence of bronchiectasis. This is usually not detectable on the CXR but may be demonstrated by CT.

2.12. E True
Radiation fibrosis has low signal on both T1 and T2-WSE MRI scans, whereas recurrent tumour may show increased signal on T2-WSE scans. However, acute radiation pneumonitis may have similar signal to recurrent tumour, as

may secondary infective pneumonitis and haemorrhage, although recurrent tumour would not be likely to occur within 6 months of treatment.

(Reference 4)

2.13.   A False
In bronchiectasis, there is impairment of both ventilation and perfusion but typically affecting ventilation to a greater extent. A normal V/Q scan with a normal CXR virtually excludes bronchiectasis.

2.13.   B True
This may be manifest on CT as the 'signet ring' sign, which comprises a ring shadow due to a dilated bronchus and a small nodule due to its accompanying artery. Bronchi are also visible more peripherally than usual due to their thickened walls. CT may also demonstrate air–fluid levels in dilated bronchi.

2.13.   C False
CT features indicating cystic bronchiectasis include a linear arrangement of cysts (string of beads sign), continuity of cysts with proximal airways, generally thicker cyst walls and identification of an accompanying artery. Additional non-specific signs of bronchiectasis include atelectasis, pleural thickening, consolidation and hyperinflation.

2.13.   D True
Flame-shaped opacities are produced by dilated, fluid-filled airways imaged in long axis. When imaged in transverse plane they appear as rounded opacities. They can be traced through adjacent CT sections to unfilled airways.

2.13.   E True
Bronchiectasis and infertility occur as part of the immotile cilia syndrome. This has autosomal recessive inheritance and presents with bronchitis, rhinitis and sinusitis. Approximately 50% of cases are associated with dextrocardia (Kartagener's syndrome). Female fertility is usually not affected. Patients with cystic fibrosis may also be infertile.

(Reference 4)

2.14.   A True
CT is more sensitive in detecting emphysema than the CXR. Parenchymal changes include generalized or focal areas of reduced attenuation, which may be either poorly or well-defined. Bullae are also seen and appear as well-defined areas of lower attenuation with hairline walls.

2.14.   B True
Other vascular abnormalities demonstrated by CT include pruning of small vessels, vascular distortion around areas of reduced parenchymal density and vascular attenuation. The reduction in number and size of pulmonary vessels and their branches is also seen on the CXR and is most marked in the middle and outer part of the lung. Vessels may be unduly straightened or curved. Bullae are avascular.

2.14. C True
Centriacinar emphysema is characterized by punctate areas of low attenuation in a homogeneous background, upper lobe predilection and vascular pruning. Panacinar emphysema shows widespread areas of low density, predominantly in the lower zones, attenuated vessels and increased vascular branching angles.

2.14. D True
CT in paraseptal emphysema also demonstrates areas of low attenuation distributed along septa, vessels and airways. However, bulla formation also occurs in other forms of emphysema and in isolation.

2.14. E True
A bulla may also disappear after haemorrhage. Infection, haemorrhage and pneumothorax are the major complications of bullae. CT may demonstrate wall thickening and an air–fluid level in the case of an infected bulla.

(Reference 4)

2.15. A False
The interval between exposure and mesothelioma is typically between 20 and 40 years. Crocidolite is the most carcinogenic form of asbestos but most cases of mesothelioma are caused by chrysotile since this was the commonest used asbestos. A history of asbestos exposure in patients with mesothelioma is obtained in about 50% of cases.

2.15. B True
The pleural thickening of mesothelioma characteristically extends to encase the lung. Pleural effusion is also common. CT may also demonstrate invasion of adjacent lung, mediastinum and chest wall, which are all recognized features of mesothelioma but usually occur relatively late. Lymphatic and haematogenous metastases may be seen in 50% of cases at post-mortem but are typically clinically silent.

2.15. C False
The pleural effusion accompanying mesothelioma may be large enough to obscure the tumour and then cannot be distinguished from other causes of large effusions. However, mesothelioma typically encases the lung and fixes the mediastinum. For this reason, contralateral mediastinal shift that usually occurs with massive effusions is not as commonly seen with mesothelioma.

2.15. D True
Other conditions which may mimic malignant mesothelioma are pleural lymphoma and metastatic thymoma. Pleural spread from bronchial and breast primaries are the commonest sites of adenocarcinoma.

2.15. E False
Pleural effusions can also occur in benign asbestos-related pleural disease. Features that suggest malignant mesothelioma are nodular involvement of the pleural fissures and ipsilateral volume loss with fixation of the mediastinum. Also, benign pleural plaques can be seen to enlarge on serial examinations.

(Reference 4)

2.16. A False
Nodules in silicosis or coal-worker's pneumoconiosis that have a width greater than 1 cm are defined as progressive massive fibrosis (PMF). Typically, PMF starts as a mass near the periphery of the lung. It usually occurs following at least 20 years of dust exposure. The CXR frequently demonstrates background nodulation in the lungs.

2.16. B True
Consequently, PMF nodules in silicosis may be calcified as the individual nodules of silicosis may calcify. The masses of PMF are usually round or oval on the frontal CXR and have lobular or irregular edges.

2.16. C True
The medial border is typically irregular. Characteristically, emphysema occurs between the lateral border of the nodule and the chest wall. The mass may also parallel a major fissure.

2.16. D False
The depth of the mass may be significantly less than its lateral diameter and consequently the nodules of PMF are typically lens-shaped, a feature best appreciated on lateral or oblique CXR or CT. The masses are usually bilateral. Very slow growth and migration toward the hilum are observed over many years. This migration is accompanied by emphysema.

2.16. E True
Calcification may also be seen within the mass (see B). These features help to distinguish the lesion from a carcinoma, especially if the mass is unilateral (unusually). CT may also demonstrate cavitation within the mass.

(Reference 4)

2.17. A True
This is the case even when lung involvement is severe enough to cause major abnormalities of pulmonary function. When the CXR is abnormal it usually demonstrates reticulonodular shadowing, which is usually diffuse and symmetrical but may be segmental. With disease progression, the nodular shadowing may become conglomerate.

2.17. B True
In amyloid, hilar and mediastinal lymphadenopathy is often massive and may occur with or without parenchymal shadowing. Nodal calcification is common and may occasionally be of the egg-shell type.

2.17. C False
Airway amyloid is more commonly diffuse, affecting the trachea to the segmental bronchi and resulting in multiple strictures and nodules. These changes can be demonstrated both by CT and conventional tomography or bronchography. Focal amyloid may cause an endobronchial mass indistinguishable from carcinoma.

2.17. D True
This is a feature of the rare parenchymal nodular amyloid. The nodules are typically well-defined and subpleural and can calcify or ossify. They may be single or multiple, and in the latter case bilateral in two-thirds of cases. They are typically of varying sizes, commonly 0.5–5.0 cm, but occasionally up to 15 cm in diameter. Cavitation is a rare feature.

2.17. E True
However, effusions are more likely due to associated cardiac amyloid with cardiac failure. Cardiac amyloid commonly accompanies lung involvement.

(Reference 4)

2.18. A False
Mediastinal lipomatosis typically involves the upper anterior mediastinum and is less commonly seen in a posterior site. The condition is associated with obesity, exogenous steroids and less commonly with Cushing's syndrome. As many as 25% of cases have no predisposing cause.

2.18. B True
The fat in mediastinal lipomatosis typically has CT numbers of −80 to −120 HU. Any areas of high attenuation raise the possibility of haemorrhage, infection, malignant infiltration or liposarcoma. Small foci of residual thymic tissue should not be mistaken for infiltrated fat.

2.18. C False
In mediastinal lipomatosis, tracheal deviation or compression should be absent. The upper mediastinum may have mildly convex lateral borders. Excess fat deposition may also occur in the cardiophrenic angles (pericardial fat pads).

2.18. D False
In omental herniation CT demonstrates a cardiophrenic angle fat density mass which may be differentiated from a pericardial fat pad by the presence of linear soft tissue densities representing omental vessels. Herniation occurs through the foramen of Morgagni and is almost always right-sided.

2.18. E True
Paraoesophageal herniation of lesser omental perigastric fat may present as widening of the paraspinal line or a mass behind the heart on the CXR. CT may demonstrate a mass of fat density containing vessels. Fat may also herniate through the foramen of Bochdelek, in which case the mass is usually left-sided.

(Reference 2)

2.19. A True
Rounded atelectasis is a form of non-segmental pulmonary collapse that may mimic a neoplasm. CT findings include a central air bronchogram, vessels and bronchi converging in a curvilinear fashion into the lower aspect of the mass and adjacent hyperinflated lung.

2.19. B True
Rounded atelectasis usually remains stable on serial CT scans but occasionally very slow growth may occur. The CT features are characteristic in most cases, but if the findings are equivocal CT guided biopsy can be of value.

2.19. C True
Rounded atelectasis is commonly associated with a history of asbestos exposure or other inflammatory processes involving the pleural space. Consequently, other features of asbestos-related disease such as pulmonary fibrosis and pleural plaques may be seen.

2.19. D False
If there is any difference in density of different parts of the mass then increased density may be seen at the periphery since this represents the area of most complete atelectasis.

2.19. E True
The mass of rounded atelectasis is usually round or oval and forms an acute angle with the adjacent thickened pleura. The pleura is typically thickest at its point of contact with the mass.

(Reference 1)

2.20. A False
An obstructing mass in a specific lobar or segmental bronchus may be demonstrated by the additional use of several thin (4–5 mm) sections. Also, on postcontrast scans, a central mass may be distinguished from the distal collapsed lung since the latter usually enhances to a greater degree. However, this is not always the case.

2.20. B True
In LUL collapse there is elevation of the left hilum causing foreshortening of the aortopulmonary window. On CT the collapsed LUL appears as a wedge-shaped soft-tissue-density mass that abuts the chest wall anterolaterally, with the apex of the wedge merging with the left hilum. Crowded air-filled bronchi may be seen within the mass.

2.20. C False
The posterior border of a collapsed RUL is formed by the major fissure, and the minor fissure forms the lateral border. The collapsed lobe appears as a triangular structure between the two. Other CT features include elevation of the hilum, anterior rotation of the right main bronchus and increased expansion of the RML and RLL.

2.20. D False
A collapsed RML will decrease its contact with the lateral chest wall but will maintain contact with the anterior chest wall inferiorly. It is bordered anteriorly by the minor fissure and posteriorly by the major fissure, and abuts the heart medially. If the collapsed lobe is wider centrally than peripherally, a central obstructing mass is probably present.

2.20. **E** True
Both the RLL and LLL collapse posteriorly, medially and caudally, appearing on CT as wedge-shaped paraspinal soft tissue densities. The major fissure forms the lateral border. However, if the inferior pulmonary ligament is incompletely formed, the lower lobe may collapse more completely against the spine and have a rounded appearance.

(Reference 1)

2.21. **A** True
Pleuropulmonary involvement occurs in about 5% of patients with Behçet's disease. CXR features include poorly defined consolidations due to haemorrhage, cavitating lesions due to infarcts, hypovascular areas due to vessel occlusions, pulmonary artery aneurysms and pleural effusions.

2.21. **B** True
Pulmonary involvement by systemic temporal arteritis is rare but the CXR may show diffuse interstitial shadowing and multiple nodules, some of which may cavitate.

2.21. **C** True
In Churg–Strauss syndrome cavitating nodules occur but are unusual and are less frequent than in Wegener's. Other CXR findings are multiple patchy consolidations, lymphadenopathy, diffuse interstitial shadowing and pleural effusions.

2.21. **D** True
The CXR in necrotizing sarcoid angiitis most commonly shows bilateral nodules (in 75% of cases) which are usually located in the lower zones and may cavitate. Consolidations, miliary and interstitial shadows, lymphadenopathy and effusions are less common findings.

2.21. **E** False
In pulmonary alveolar proteinosis, the classic CXR abnormality is bilateral consolidation, which usually has a perihilar or hilar and basal distribution. Less common CXR findings include lobar or multifocal consolidations, a more peripheral than central location and even reticulonodular shadowing.

(Reference 4)

2.22. **A** False
The posterior mediastinum is infrequently involved by lymphoma. The anterior mediastinal and paratracheal nodes are the commonest groups to be involved, particularly by the nodular sclerosing type of Hodgkin's disease, which may arise in the thymus gland. The lymphadenopathy is usually bilateral but asymmetrical. The presence of mediastinal lymphadenopathy without anterior mediastinal involvement would be atypical of lymphoma.

2.22. **B** False
The paracardiac nodes are rarely involved at presentation but may be an important site of recurrence, since they may not be included in the radiation field. Similarly, isolated lung disease is more common at recurrence than at

presentation. In Hodgkin's disease lung involvement at presentation is almost invariably with associated nodal disease. Isolated lung involvement at presentation may occur in up to 50% of cases in NHL.

2.22.   C False
In both Hodgkin's lymphoma and NHL, the enlarged nodes may be discrete or matted together, with either well-defined or ill-defined edges. Low-density areas due to cystic degeneration may be seen and may persist following therapy, when the rest of the nodal disease has resolved. Enhancement is unusual on postcontrast CT.

2.22.   D True
The CXR and CT findings in lymphoma are variable and include focal or patchy consolidations, multiple discrete nodules which may cavitate, linear shadowing, possibly due to spread via bronchopulmonary lymphatics and widespread reticulonodular shadowing, which is an uncommon pattern. Pulmonary shadowing is frequently perihilar or adjacent to the mediastinum in location.

2.22.   E True
Subpleural masses or consolidations are seen with both types of lymphoma. Chest wall invasion and rib destruction are seen occasionally. Pleural effusion is almost always accompanied by mediastinal lymphadenopathy and is usually unilateral. Such effusions are thought to be secondary to venous and lymphatic obstruction and may resolve following mediastinal irradiation.

(Reference 4)

2.23.   A True
Compared to a slice thickness of around 10 mm in conventional chest CT. Narrow collimation reduces volume averaging within the plane of section, resulting in greater spatial resolution. Typically, thin slices at 10–20 mm intervals are used for the evaluation of bronchiectasis.

2.23.   B True
Use of a high-resolution reconstruction technique maximizes spatial resolution by reducing the amount of smoothing of the data. Increased noise within the image is usually only of significance with large subjects and can be compensated for by increasing kV and mA. This technique also increases the visibility of aliasing artefact, which causes fine streaks to radiate from the edges of high-contrast structures such as ribs or vertebral bodies, and is more marked with thin sections.

2.23.   C False
The use of scan times of 2–3 seconds allows the majority of patients to hold their breath during data acquisition. However, these times are still relatively slow compared to cardiac motion, resulting in motion artefacts that are most marked in the lingula and LLL.

2.23.   D False
The interlobular septa are sometimes visualized as very thin straight lines of uniform thickness that are 1–2 cm in length and may contact a pleural surface.

They are normally seen with difficulty and only a few are usually visible. They demarcate the pulmonary lobule, which has a variable shape, typically appearing as a truncated cone in the lung periphery, whilst having a roughly hexagonal shape at the lung base.

2.23. E True
The intralobular arteriole appears as a linear, branching or dot-like density in the centre of the pulmonary lobule or within 1 cm of the pleural surface. In the absence of atelectasis they do not extend to the pleural surface. The intralobular bronchioles have very thin walls and are therefore not normally seen. Pulmonary veins may be seen as branching structures in relation to interlobular septa.

(Reference 8)

2.24. A True
Interstitial oedema also causes smooth thickening of the interlobular septa. However, lymphangitis may be differentiated by the presence of nodular or beaded septal and bronchovascular bundle thickening due to tumour growth. Septal thickening may outline part of or the whole of the pulmonary lobule, and thickened septa peripherally are often associated with thickening of the pleura or adjacent fissure.

2.24. B True
In CFA, septal thickening is accompanied by distortion of the pulmonary lobule. Asbestosis and sarcoidosis may also cause fibrosis, resulting in similar features but differing in their distribution.

2.24. C True
Linear opacities in CFA are also typically in the lower zones, and the pattern of abnormalities is identical in other collagen vascular diseases. Conversely, in sarcoidosis these reticular shadows are typically more central around the bronchovascular bundles and in the mid- and upper zones.

2.24. D False
This finding is consistent with histiocytosis X. In lymphangioleiomyomatosis, thin-walled cystic air-spaces of 0.2–5.0 cm in diameter involve the lung diffusely and are surrounded by relatively normal lung parenchyma. The presence of nodules also favours histiocytosis.

2.24. E True
Cyst development in CFA indicates end-stage disease. These cystic spaces (honeycomb lung) are predominantly located in the subpleural region and lower zones.

(Reference 8)

2.25. A True
The nodular thickening in sarcoidosis is due to non-caseating granulomata along lymphatics in the bronchovascular bundles and less commonly in the interlobular septa and visceral pleura. They are usually irregular and less than 5 mm in diameter.

2.25.   B True
Nodules in silicosis are usually located in the posterior aspects of the upper zones and range from 1 to 10 mm in size. Small nodules are also found in coal-worker's pneumoconiosis, graphite pneumoconiosis and talcosis.

2.25.   C False
Increased parenchymal attenuation is a finding in the active phase of CFA but may also be seen in the subacute phase of extrinsic allergic alveolitis and in alveolar proteinosis. In the latter condition, septal thickening also occurs, producing a characteristic 'crazy paving' appearance.

2.25.   D True
Peripheral subpleural air-space consolidation is also seen in 50% of patients with bronchiolitis obliterans organizing pneumonia (BOOP). CT demonstrates loss of vascular markings and air bronchograms. Other features include nodules located along the bronchovascular bundles and bronchial thickening and dilatation.

2.25.   E False
A combination of consolidation and nodules is also seen with BOOP and bronchoalveolar cell carcinoma. If the consolidation progresses over several months, lymphoma and bronchoalveolar cell carcinoma should be considered in particular.

(Reference 8)

2.26.   A False
Assessment of malignancy versus benignacy by determination of the density of the nodule is difficult since the CT numbers of a lesion depend upon many factors. CT numbers can vary between different scanners and over time within the same scanner. They are also dependent upon differences in patient size, nodule location, kV and the reconstruction algorithm used.

2.26.   B True
Primary tumours of the trachea represent less than 1% of all intrathoracic malignancies. Squamous carcinoma accounts for just over 50% of cases and adenoid cystic neoplasms are the second commonest. Together, the two make up over 90% of all primary tracheal tumours. They characteristically result in extensive local invasion of adjacent mediastinal structures.

2.26.   C True
Carcinoid tumours are part of a spectrum of neuroendocrine lung neoplasms that include both typical and atypical carcinoids as well as small cell carcinoma. Approximately 5% of typical carcinoids and 40–50% of atypical carcinoids will metastasize, being associated with hilar and mediastinal lymphadenopathy. Atypical carcinoids are more likely than typical carcinoids to have irregular margins and show less uniform contrast enhancement.

2.26.   D True
The primary lesion of small cell carcinoma is usually small and difficult to detect on the CXR. The initial abnormality is typically a central mass representing ipsilateral hilar and mediastinal lymphadenopathy. As the disease progresses, CT may demonstrate the nodes as a large mantle of soft tissue infiltrating the mediastinum, which may have a low-density necrotic centre with an enhancing rim on postcontrast scans.

2.26. E False
In one series, CT demonstrated calcification in 10 of 47 hamartomas. Fat density was identified in 28 lesions. The typical CT appearance of a hamartoma is a solitary nodule less than 2.5 cm in diameter with a smooth edge, which may contain fat and calcification. The latter features are best demonstrated by the use of thin sections.

(Reference 6)

2.27. A True
Other CT findings in asbestosis include subpleural short lines (small nodules and thickened single or branching lines less than 2 cm long) and parenchymal bands (2–5 cm linear densities that contact a pleural surface). These changes are predominantly in the posterior aspects of the lower lobes.

2.27. B False
Curvilinear subpleural lines in dependent lung may be seen in other conditions and also in normal subjects, in which case they will disappear if the patient is rescanned in a different position.

2.27. C False
Round atelectasis often demonstrates homogeneous enhancement following intravenous contrast administration. Other CT features are described in question 2.18.

2.27. D False
The paraspinal location is a common site for asbestos-related pleural thickening. CT demonstrates plaques as bilateral soft tissue or calcified pleural thickening which is frequently separated from underlying ribs and extrapleural soft tissues by a thin layer of fat. Circumscribed pleural plaques are the commonest asbestos-related pleural abnormality.

2.27. E True
Sparing of the mediastinal pleura in benign asbestos-related pleural thickening is in contrast to mesothelioma. Diffuse pleural thickening is demonstrated on CT as smooth 3–10 mm pleural thickening, most extensive posteriorly and in the paraspinal regions. Calcification is seen and subpleural interstitial fibrosis is almost invariably associated.

(Reference 8)

2.28. A False
In one study tumour was present in 34.5% of cases of atelectasis with air bronchograms. Absence of an air bronchogram and the presence of mucus-filled bronchi in atelectatic lung strongly favours obstructing tumour.

2.28. B True
Lipoid pneumonia is a cause of persistent consolidation that is typically associated with a history of exogenous mineral oil aspiration. Thin sections through the consolidated area are necessary to ensure that regions of low density are truly due to fat rather than partial volume averaging with adjacent aerated lung.

2.28. C True
This low-attenuation 'halo' is thought to be due to an area of haemorrhagic infarction. Other HRCT features of invasive pulmonary aspergillosis include mass-like consolidation and nodules. Cavitation and an air crescent sign are late findings.

2.28. D True
Bronchial carcinoid typically appears on CT as a smooth nodule located at a bronchial bifurcation and causing splaying of the bronchi. Calcification is demonstrated in 30% of tumours.

2.28. E False
A nodule with a vascular connection is also seen with haematogenous metastases, septic emboli, pulmonary infarction and bronchopulmonary sequestration.

(Reference 8)

2.29. A False
The postpneumonectomy space initially fills with fluid and in the majority of cases remains fluid-filled for many years. In the remainder, the space becomes filled by homogeneous solid material.

2.29. B False
Calcification of the parietal pleura in the postpneumonectomy space may occur, even when the space is fluid-filled and does not indicate infection. CT may demonstrate small locules of gas density that are a feature of empyema formation.

2.29. C False
Tumour recurrence usually manifests as a soft-tissue mass at the pneumonectomy stump. Other features include development of mediastinal or hilar lymphadenopathy and, less commonly, multiple, peripheral soft-tissue density masses projecting into the postpneumonectomy space.

2.29. D True
There is an inverse relationship between the degree of hyperinflation of the contralateral lung and persistence of the postpneumonectomy space. If the contralateral lung expands well then the space tends to become obliterated, while if there is less marked expansion the space may persist and remain fluid-filled.

2.29. E True
After left pneumonectomy, the mediastinum tends to shift to the left, with the aortic arch lying in a more sagittal plane and the right lung herniating both anterior and posterior to the mediastinum. Following a right pneumonectomy, the aortic arch tends to lie in a more coronal plane and the left lung herniates anterior to it into the right hemithorax.

(Reference 1)

2.30. A True
Initially the cysts of PCP are small and thin-walled but may coalesce with time
to form multi-septate, thick-walled cysts which commonly abut the pleural
surface. There may still be residual cysts after clearing of the consolidation but
these usually regress, although underlying parenchymal damage may persist.

2.30. B True
Peripheral enhancement of pulmonary infarcts is possibly due to collateral
blood flow from adjacent bronchial arteries. CT typically shows wedge-shaped
opacities with their bases orientated along a pleural surface. A vessel may be
present at the apex of the infarct. Cavitation is also a feature.

2.30. C True
On CT, metastases typically appear as multiple, smooth, round nodules which
are usually less than 2 cm in diameter and are commonly related to a 'feeding'
pulmonary artery, indicating haematogenous spread. Thin sections may also
demonstrate the metastasis growing within the vessel.

2.30. D False
Cavitary metastases are rare occurring in less than 5% of cases. Most often
they are from primary squamous carcinomas, especially from head, neck,
cervix and bladder. Other causes include gastrointestinal adenocarcinomas and
extrathoracic sarcomas.

2.30. E True
The finding of retrocrural masses in lymphangioleiomyomatosis has been
attributed to either retrocrural lymph node enlargement or dilatation of the
thoracic duct and its tributaries.

(Reference 8)

2.31. A False
Lymphangioleiomyomatosis is confined to women. However, 85% of patients
with pulmonary involvement by tuberous sclerosis are female. Presentation is
usually in the third or fourth decades. There is no sex predominance in
primary pulmonary histiocytosis or in lung involvement by neurofibromatosis.

2.31. B False
Retroperitoneal lymphadenopathy, chylous ascites and occasionally renal
angiomyolipomata are all recognized findings in lymphangioleiomyomatosis.
Chyluria has also been reported, as has chylopericardium and chyloptysis.
Lung involvement may occur many years after presentation with lymph node
disease and chylous ascites.

2.31. C True
Reticulonodular shadowing progresses to a worsening of the linear shadows,
with subsequent cyst and bulla formation. The latter are best demonstrated by
CT (see question 2.25 and answers). The CXR changes are symmetric and may
be either generalized or basally predominant (initially).

2.31. D True
Chylous effusions are found in approximately 75% of patients. They may be unilateral or bilateral and are typically large and recurrent. Pneumothorax occurs in about 40% of cases at some stage. Lymphadenopathy may be seen. In primary pulmonary histiocytosis, lymphadenopathy and pleural effusion are rare, and pneumothorax occurs in about 14%.

2.31. E True
Mean survival after diagnosis is about 4 years, the prognosis being particularly poor in patients with large-volume lungs. Conversely, in pulmonary histiocytosis there is a high rate of spontaneous remission and morbidity is low.

(Reference 4)

2.32. A False
Chylothorax may be either unilateral or bilateral and has no particular differentiating CXR features. It may be preceded by mediastinal widening due to the accumulation of chyle before it spills over into the pleural space.

2.32. B False
Chyle does not irritate the pleura and therefore pleurisy and pleural fibrosis are not features.

2.32. C False
Although chyle has a high fat content it is also protein-rich and therefore CT attenuation values are typically similar to those of other pleural effusions.

2.32. D True
Neoplasms account for approximately 50% of cases, the most common being lymphoma (in 75% of patients). Trauma is the cause in 25%, and 15% are idiopathic. Chylothorax may be a presenting feature of lymphoma. The association of carcinoma with chylothorax is suggestive of mediastinal metastasis. In the neonatal period, the majority of chylous effusions are idiopathic.

2.32. E True
Chylothorax complicates cardiothoracic surgery in 0.2–0.5% of cases, most commonly following operations for Fallot's, PDA and coarctation of the aorta. Miscellaneous causes include heart failure, central venous thrombosis, fibrosing mediastinitis and various developmental anomalies such as thoracic duct atresia and pulmonary lymphangiectasia.

(Reference 4)

2.33. A True
Sarcoid is complicated by pleural effusion in approximately 2% of cases and usually occurs with extensive pulmonary disease or multi-system involvement, typically resolving by 6 months. Approximately one-third of effusions are bilateral and usually small. Resolution is usually complete but may result in pleural thickening in 20% of cases.

2.33. B True
The middle lobe is the commonest lobe to collapse. This may be the result of extrinsic compression by enlarged nodes or due to endobronchial sarcoid. However, although sarcoid may result in massive lymphadenopathy, this nodal enlargement is a rare cause of symptomatic airway narrowing on its own and the latter is usually due to mural granulomas and fibrosis.

2.33. C True
Sarcoid can affect the trachea, bronchi or bronchioles, producing multiple stenoses. The frequency of large airway narrowing is about 5%. Stenoses may be single or multiple and most commonly involve the mainstem, lobar and proximal segmental bronchi. Granulomas may also produce a localized obstructing endobronchial mass.

2.33. D True
The superior vena cava syndrome is a rare complication of sarcoid even in the presence of massive lymphadenopathy, since perinodal fibrosis does not occur. The presence of this finding requires the exclusion of other causes.

2.33. E True
Pulmonary arterial hypertension may be the result of small vessel granulomatous disease causing vascular stenoses or due to cor pulmonale secondary to fibrobullous disease. Large vessel involvement by sarcoid is very rare and usually due to vascular compression by enlarged nodes.

(Reference 4)

2.34. A True
Respiratory tract disease in Wegener's granulomatosis takes the form of either sinusitis, rhinitis or otitis, as well as pulmonary granulomatous disease. Renal involvement occurs in 85% of patients. A limited form of disease has been described that involves the lung, with or without upper respiratory tract disease, but without renal or other systemic involvement.

2.34. B False
Pulmonary radiographic changes are present in approximately three-quarters of patients at time of diagnosis. The most characteristic features are discrete focal opacities that range from nodular masses to ill-defined areas of consolidation. Pulmonary nodules commonly resolve with or without treatment over a period of weeks to months and may heal with or without scar formation.

2.34. C False
A single nodule occurs in 33% of cases, and in the remainder multiple nodules are found. The nodules are typically 2–4 cm in diameter but may reach up to 10 cm. Pulmonary consolidations are seen in approximately 30% of patients and may be single or multiple. They range from small inhomogeneous patches to homogeneous segmental or lobar consolidations.

2.34. D True
Tracheal stenosis is associated with nasal disease and is commoner in women. Airway stenosis is typically subglottic, smooth or irregular and 3–4 cm in length. Distal airway stenoses are less common. CT shows abnormal soft tissue

in the tracheal rings which themselves may be abnormally thickened and calcified.

2.34.   E False
Cavitation in nodules and consolidations occur in about 40% of cases at presentation. The cavities may be unilocular or multilocular. Cavitation within nodules occurs spontaneously, typically resulting in a thick-walled cavity. Long-standing cavities may be thin-walled. Secondary infection can result in the development of air–fluid levels.

(Reference 4)

2.35.   A True
Intrathoracic extension of the thyroid is usually anterior or lateral to the trachea but may occur posteriorly between the trachea and the oesophagus in about 25% of cases. The continuity between the mediastinal mass and the thyroid gland is an important diagnostic feature, both on the CXR and on chest CT. Development of primary intrathoracic goitre from ectopic thyroid tissue is extremely rare.

2.35.   B False
Both benign and malignant thyroid masses may calcify. Dense, amorphous, well-defined calcification is seen with benign goitres or medullary carcinoma, whereas calcifications within papillary or follicular carcinomas tend to be very fine. The longer a goitre has been present, the more likely it is that calcification will be demonstrated.

2.35.   C True
Areas of higher-than-muscle attenuation represent iodine-containing normal thyroid tissue which shows enhancement on postcontrast scans. CT also demonstrates areas of reduced attenuation that do not enhance, and calcification.

2.35.   D True
Typically, CT demonstrates goitres as well-defined mixed-attenuation para-tracheal masses that are limited laterally by the brachiocephalic vessels and are either partly or wholly located behind the arteries to the head and neck. In some cases, the mediastinal mass is connected to the cervical thyroid by a thin vascular or fibrous pedicle that is not identified by CT.

2.35.   E True
High signal on T2-WSE MRI is seen with cystic changes in the gland. MRI will not demonstrate calcifications.

(Reference 4)

2.36.   A True
Other causes of thymic masses include thymolipoma, malignant lymphoma, thymic carcinoid, germ cell tumours and thymic carcinoma. Thymomas are very uncommon in patients under 20 years of age. They are typically located anterior to the ascending aorta and above the pulmonary outflow tract, and may project into the middle mediastinum. Less commonly, they are situated adjacent to the left or right heart borders or in the cardiophrenic angle.

2.36. B True
Other associations of thymoma include non-thymic carcinomas, haematological cytopenias and myasthenia gravis. The latter occurs in 30–40% of patients with thymoma, whereas 10–17% of patients with myasthenia have thymoma.

2.36. C False
Calcification is seen in 6–28% of thymomas on the CXR. The CT features include a well-defined, mixed-attenuation, upper mediastinal mass which is usually spherical or has lobulated contours. The calcification is either punctate or curvilinear and may be seen in both benign and invasive thymomas.

2.36. D False
Thymoma typically spreads by direct extension through its capsule (then referred to as invasive thymoma), through the mediastinal fat to the pleura and pericardium. Extension may also occur through the diaphragm. Pleural 'drop' metastases may also occur. Until mediastinal invasion has occurred, it is not possible to differentiate benign and invasive thymoma.

2.36. E False
On T2-WSE MRI, the signal intensity of the tumour increases (compared to T1-weighted values) to approach that of the mediastinal fat. Therefore, early spread is better assessed on T1-WSE images where fat has high signal and thymoma has signal intensity similar to muscle.

(Reference 4)

2.37. A False
Thymic cysts are usually simple cysts within a normal or hyperplastic gland. They may occur in thymoma, thymic germ cell tumours, as well as treated Hodgkin's disease. On plain radiographs the cysts are indistinguishable from any other cause of thymic mass but CT typically demonstrates their water density.

2.37. B False
A focal mass is a rare CT appearance for thymic hyperplasia. Usually the gland is of normal size or CT may demonstrate enlargement of both lobes of a normally shaped gland. Thymic hyperplasia is associated with myasthenia and thyrotoxicosis.

2.37. C False
The CT demonstration of an enlarged gland that has retained its normal shape is consistent with thymic hyperplasia rather than recurrent lymphoma. Thymic rebound hyperplasia is most commonly seen in children and young adults and may occur in a variety of situations, such as recovery from burns, surgery and tuberculosis or during and after antineoplastic chemotherapy.

2.37. D False
Fat density is seen in only 25–50% of benign cystic teratomas on CT. Water density areas are common and peripheral calcification, teeth and bone may all be demonstrated. The masses are typically well-defined, rounded or lobulated and grow slowly. However, rapid increase in size may occur with haemorrhage. Rarely the lesion may be found in the posterior mediastinum.

2.37. E False
Calcification is rare and fat density absent in germ cell tumours. CT typically demonstrates a homogeneous or mixed-density, lobulated, anterior mediastinal mass. Mediastinal lymphadenopathy, pulmonary and pleural metastases may all be demonstrated, as may SVC obstruction. Other causes of poorly defined anterior mediastinal masses include thymoma, lymphoma, metastatic adeno-carcinoma and metastatic teratoma.

(Reference 4)

2.38. A True
High attenuation in bronchogenic cysts is due to the presence of milk of calcium. CT may also demonstrate homogeneous soft-tissue density due to high protein content while other cysts show water density. Mixed-attenuation masses may result from haemorrhage, which may cause rapid enlargement of the cyst. Calcification in the wall of the cyst is rare.

2.38. B True
Bronchogenic cysts may also arise in the anterior mediastinum (rarely). However, they are typically related to the lower trachea or proximal mainstem bronchi and frequently cause displacement of the oesophagus. They typically have smooth, round or oval contours but approximately 10% have a lobulated outline.

2.38. C False
Pericardial cysts occur more commonly in the right cardiophrenic angle than the left, but they may be found higher in the mediastinum and occasionally extend into the superior mediastinum. They are usually unilocular but may contain trabeculae and occasionally be truly multiloculated. CT demonstrates a well-defined water or soft-tissue density mass in contact with the heart. Calcification is exceptional.

2.38. D False
CT typically demonstrates a neurenteric cyst as a smooth, round, water-density mass either in the posterior mediastinum or in a paraspinal location. Associated vertebral anomalies are common and include hemivertebrae, sagittal clefts in the vertebral body (causing butterfly vertebra and spina bifida). The vertebral anomaly may lie above the level of the cyst. Cyst communication with the subarachnoid space can occur.

2.38. E False
In adults, most thoracic pseudocysts are secondary to chronic pancreatitis. CT typically shows the cysts as posterior mediastinal masses, commonly displacing the oesophagus and associated with left-sided or bilateral pleural effusions.

(Reference 4)

2.39. A True
Castleman's disease is a variety of benign lymph node hyperplasia presenting usually as asymptomatic, occasionally massive lymphadenopathy in young adults. The disease may be multifocal and can involve extrathoracic lymph nodes, including cervical and retroperitoneal node groups. In the chest,

mediastinal and proximal nodes are typically enlarged. Calcification has been reported.

2.39. B True
Aggressive fibromatosis is a locally invasive tumour of fibrous origin that does not metastasize and is rarely seen in the mediastinum. CT demonstrates a muscle attenuation mass on precontrast scans which becomes hyperdense postcontrast. Aggressive fibromatosis typically affects the soft tissues of the extremities, neck and trunk.

2.39. C False
Lymphangioma appears on CT as a water to soft-tissue-density mass typically located in the superior mediastinum that shows no enhancement. It commonly extends into the thorax from a primary location in the neck. Occasionally, lymphangiomas may invade the chest wall, resulting in rib erosion.

2.39. D True
Nerve sheath tumours may have a high myelin content which results in their appearing as low-attenuation masses on precontrast CT. However, being vascular, they can enhance following intravenous contrast. Calcification is an unusual feature. The majority of intrathoracic nerve sheath tumours are schwannomas, the remainder being neurofibromas.

2.39. E True
Mediastinal paraganglionomas are either chemodectomas (mostly aortic body tumours) or phaeochromocytomas. Aortic body tumours may be located lateral to the brachiocephalic artery, anterolateral to the aortic arch, at the angle of the ductus arteriosus, or superolateral to the right pulmonary artery. Phaeochromocytomas may occur in the posterior mediastinum or adjacent to the heart, particularly in the wall of the left atrium.

(Reference 4)

2.40. A False
The average coronal measurement of the adult trachea on CT is 18.2 mm for men and 15.2 mm for women. The mean cross-sectional area is 272 mm for men and 194 mm for women. The normal trachea may appear circular, triangular, elliptical or C-, D- or U-shaped on axial CT scans. In children, the trachea is usually round.

2.40. B True
In tracheobronchomegaly (the Mounier–Kuhn syndrome), there is diffuse dilatation of the trachea and main bronchi, with associated bronchiectasis. The trachea develops a corrugated appearance due to redundant mucosa prolapsing between the cartilaginous tracheal rings. This feature is better seen on plain radiographs or conventional tomography than on CT.

2.40. C False
Sabre-sheath trachea is a narrowing of the trachea in the coronal plane below the level of the thoracic inlet. The sagittal diameter of the trachea is either normal or slightly increased. A coronal diameter less than two-thirds the

sagittal diameter allows the diagnosis to be made. CT also demonstrates calcification of the tracheal cartilages. There is an association with COAD.

2.40.   D True
Relapsing polychondritis is an autoimmune condition affecting the cartilage of the ear, nose, airway and joints. Laryngeal and tracheal inflammation occurs in 50% of cases. CT shows narrowing of the trachea, wall thickening and dense cartilage calcification. Wall thickening may extend into the main and segmental bronchi and bronchiectasis may occur. Uncommonly, diffuse tracheal dilatation due to tracheomalacia is seen.

2.40.   E True
Laryngeal papillomatosis in adults more commonly affects the lower respiratory tract. Radiographs and CT typically show nodular thickening involving part or all of the tracheal wall. Distal disease is manifest by solid or cavitating lung nodules with thin or thick walls, dilated and irregular bronchi and peripheral atelectasis or consolidation.

(Reference 8)

3 Liver, biliary tract, pancreas, spleen

3.1. Concerning ^{99m}Tc-sulphur colloid scintigraphy of the liver
A the radionuclide is taken up by normally functioning hepatocytes.
B metastases cannot be distinguished from simple cysts.
C on the anterior view, the spleen should not appear more active than the liver.
D absent liver activity is a feature of acute alcoholic hepatitis.
E increased bone marrow activity is diagnostic of impaired liver function.

3.2. Concerning ultrasound in the assessment of parenchymal liver disease
A diffuse fatty infiltration results in increased prominence of the portal vein walls.
B increased parenchymal echogenicity is seen in glycogen storage disease.
C multiple echogenic areas are a recognized finding in focal fatty infiltration.
D multiple hypoechoic lesions are a recognized finding in cirrhosis.
E surface nodularity is a feature of cirrhosis.

3.3. Concerning CT of parenchymal liver diseases
A diffuse fatty infiltration may cause the liver to have lower attenuation than the spleen on precontrast CT.
B focal fatty infiltration typically results in displacement of vessels.
C there is an association between cirrhosis and increased liver density on precontrast CT.
D cirrhosis is associated with reduction in size of the right lobe.
E cirrhosis results in uniformly increased parenchymal enhancement on postcontrast CT.

3.4. Concerning ultrasound of focal liver lesions
A a mass with a central, stellate group of echoes is diagnostic of focal nodular hyperplasia (FNH).
B anechoic areas within a mass are a feature of adenoma.
C the presence of solid areas in a multiseptate, cystic mass indicates a diagnosis of biliary cystadenocarcinoma rather than cystadenoma.
D lipoma is a cause of a hyperechoic mass.
E hypoechoic masses without distal acoustic enhancement are a feature of lymphoma.

3.5. Concerning imaging of hepatic haemangiomas
A distal acoustic enhancement is a recognized finding on ultrasound.
B a hypoechoic 'halo' is a characteristic ultrasound feature.
C they typically show increased activity during the dynamic phase of ^{99m}Tc-labelled RBC scintigraphy.
D they are usually hypodense on precontrast CT.
E on postcontrast CT, failure of complete enhancement excludes the diagnosis.

3.6. Concerning CT in the assessment of focal liver lesions
A focal nodular hyperplasia is a recognized cause of a hyperdense mass on immediate postcontrast CT.
B hyperdense regions are a feature of hepatic adenoma on precontrast CT.
C hepatic lymphoma typically results in multiple hypodense masses on precontrast CT.
D hepatic infarcts typically appear as central, hypodense lesions on precontrast CT.
E enhancement is a typical feature of primary angiosarcoma on postcontrast CT.

3.7. Which of the following are true of imaging of hepatocellular carcinoma?
A Less than 50% of lesions show gallium-67 citrate uptake.
B A solitary hyperechoic mass is a recognized finding on ultrasound.
C CT following intra-arterial Lipiodol typically results in increased attenuation of hepatocellular carcinoma compared to liver parenchyma.
D A central hypodense scar is a recognized finding on CT.
E Demonstration of portal vein invasion differentiates it from adenoma.

3.8. Concerning imaging of liver abscesses
A an echogenic rim is a typical ultrasound feature of amoebic abscess.
B on ultrasound, acoustic shadowing may occur distal to a pyogenic abscess.
C the majority of pyogenic abscesses are hypodense on precontrast CT.
D uniform liver activity on gallium-67 citrate scintigraphy excludes the diagnosis of pyogenic liver abscess.
E multiple, small, hypoechoic lesions on ultrasound are a recognized feature of hepatic candidiasis.

3.9. Concerning CT in the assessment of liver metastases
A peripheral enhancement on postcontrast scans is a recognized finding.
B delayed enhancement of the centre of a lesion excludes the diagnosis.
C metastases appear hyperdense on delayed high-dose iodine CT.
D metastases typically appear hypodense on CT arterial portography.
E metastases appear uniformly hyperdense on CT hepatic arteriography.

3.10. Recognized imaging features of Budd–Chiari syndrome include which of the following?
A Generalized reduction of liver activity on sulphur colloid scintigraphy.
B Increased bone marrow activity on sulphur colloid scintigraphy.
C Enhancement of the main hepatic veins on postcontrast CT.
D Greater enhancement of the caudate lobe relative to the remainder of the liver on postcontrast CT.
E 'Comma'-shaped intrahepatic vessels demonstrated on MRI.

3.11. Concerning MRI in the assessment of the liver
A focal fatty infiltration is a cause of a hypointense lesion on T1-WSE scans.
B haemochromatosis results in diffusely increased signal intensity on T2-WSE scans.
C regenerating nodules of cirrhosis are a cause of hypointense masses on T2-WSE scans.
D on T1-WSE scans, areas of parenchymal hypointensity are a recognized feature of portal vein occlusion.
E simple cysts are uniformly hypointense on T1-WSE scans.

3.12. Concerning MRI in the assessment of focal liver lesions

A   focal nodular hyperplasia is a recognized cause of a hyperintense mass on STIR sequences.
B   haemangiomas are typically hyperintense on T2-WSE scans.
C   adenomas are typically isointense on all pulse sequences.
D   areas of high signal intensity on T1-WSE scans are a recognized feature of hepatocellular carcinoma.
E   a hypointense rim is a feature of hepatocellular carcinoma on T2-WSE scans.

3.13. Concerning ultrasound and magnetic resonance imaging of hepatic metastases

A   leiomyosarcoma is a recognized cause of anechoic metastases on ultrasound.
B   a hypoechoic rim is a recognized ultrasound feature.
C   portal vein invasion is not a recognized feature.
D   a lesion with a relatively hyperintense rim on T2-WSE scans is a recognized appearance.
E   metastases have shorter T1 relaxation times than normal liver.

3.14. Concerning CT in the assessment of blunt hepatic trauma

A   hepatic contusions typically appear as poorly defined, hypodense areas.
B   a linear hypodense lesion paralleling the portal vein is a recognized appearance of a hepatic laceration.
C   a well-defined water-density mass is a recognized finding following deep hepatic laceration.
D   a hypodense rim around the IVC is a recognized feature.
E   subcapsular haematomas typically cause indentation of the underlying liver parenchyma.

3.15. Concerning imaging of liver transplantation complications

A   loss of hepatic artery diastolic flow is diagnostic of acute rejection.
B   on CT, a low-attenuation rim around the central portal veins is a specific feature of acute rejection.
C   non-anastomotic biliary strictures are a recognized complication of hepatic artery thrombosis.
D   biliary strictures are typically smooth and focal.
E   on CT, a fluid density intrahepatic mass is a recognized finding.

3.16. Concerning ^{99m}Tc-HIDA scintigraphy in the assessment of the biliary tract

A   the presence of renal activity indicates biliary obstruction.
B   activity in the gallbladder excludes acute calculous cholecystitis.
C   activity in the gallbladder excludes chronic cholecystitis.
D   intrahepatic cholestasis can be diagnosed.
E   space-occupying lesions in the liver can be demonstrated.

3.17. Concerning cholangiography of the intrahepatic biliary tree
A   elongation of the ducts is a recognized finding in cirrhosis.
B   a normal cholangiogram excludes primary biliary cirrhosis.
C   biliary cystadenomas typically communicate with the biliary tree.
D   diminished branching is a recognized abnormality in metastatic disease to the liver.
E   multiple strictures are a feature of cholangiocarcinoma.

3.18. Concerning ultrasound imaging of the biliary tract
A   a left main hepatic duct diameter of 2 mm is diagnostic of obstruction.
B   increased diameter of the common bile duct (CBD) after a fatty meal is a normal finding.
C   acoustic shadowing by gallstones is related to stone size.
D   gallbladder wall thickening is a recognized finding in AIDS.
E   echogenic bile occurs in patients receiving total parenteral nutrition.

3.19. Concerning imaging of sclerosing cholangitis
A   echogenic tissue around the portal tracts is a recognized ultrasound finding.
B ^{99m}Tc-HIDA scintigraphy may demonstrate focal areas of increased activity.
C   the cystic duct is invariably normal at cholangiography.
D   cholangiography usually demonstrates strictures of both intrahepatic and extrahepatic ducts.
E   pseudodiverticula of the extrahepatic ducts are a recognized abnormality on cholangiography.

3.20. Concerning imaging in the diagnosis of Mirizzi's syndrome
A   dilatation of the intrahepatic ducts is typical.
B   the CBD is usually over 10 mm in diameter.
C   an impacted stone in the gallbladder neck is a recognized finding.
D   a soft tissue mass in the gallbladder fossa excludes the diagnosis.
E   cholangiography typically demonstrates a medial, extrinsic compression on the common hepatic duct.

3.21. Concerning the imaging features of cholangiocarcinoma
A   delayed enhancement relative to liver parenchyma is a feature of intrahepatic cholangiocarcinoma.
B   focal dilatation of the intrahepatic bile ducts is a feature.
C   an infiltrating hypodense hilar mass is a recognized CT appearance.
D   ultrasound invariably demonstrates a mass at the site of biliary obstruction.
E   lobar atrophy is a recognized feature.

3.22. Concerning CT in the assessment of the biliary tract
A   sclerosing cholangitis typically results in non-uniform dilatation of the intrahepatic ducts.
B   enhancement of the wall of the CBD is a feature of sclerosing cholangitis on postcontrast CT.
C   dilated bile ducts are indistinguishable from portal vein radicles on precontrast CT.
D   non-calcified CBD calculi cannot be demonstrated.
E   CT is less sensitive than ultrasound in demonstrating gallstones.

3.23. Concerning MRI in the assessment of the biliary tract
A the signal intensity of bile in the gallbladder varies with the fasting state of the patient.
B bile is typically hypointense on T1-WSE scans in patients with acute cholecystitis.
C gallbladder carcinoma is typically hyperintense to liver on T2-WSE scans.
D gallstones invariably have low signal intensity on all pulse sequences.
E the scirrhous subtype of cholangiocarcinoma is typically hypointense to liver on T2-WSE scans.

3.24. Concerning the ultrasound features of acute cholecystitis
A thickening of the gallbladder wall is a specific finding.
B sonolucency within the gallbladder wall is a recognized finding.
C intraluminal membranes are a recognized finding.
D hyperechoic shadowing foci in the gallbladder wall are a feature.
E pericholecystic anechoic areas indicate perforation of the gallbladder.

3.25. Concerning the ultrasound features of cholesterolosis and adenomyomatosis
A cholesterol polyps typically cast an acoustic shadow.
B gallbladder wall thickening is a feature of adenomyomatosis.
C anechoic foci in the gallbladder wall are a finding in adenomyomatosis.
D focal adenomyomatosis typically appears as a mass in the gallbladder neck.
E a stricture of the gallbladder is a feature of adenomyomatosis.

3.26. Concerning the imaging features of gallbladder tumours
A inflammatory polyps are typically hyperechoic.
B direct extension into the liver is an atypical feature of carcinoma.
C ultrasound demonstrates calculi in association with carcinoma in less than 50% of cases.
D intrahepatic bile duct dilatation is a recognized feature of carcinoma.
E metastases to the gallbladder typically produce hyperechoic masses on ultrasound.

3.27. Concerning the ultrasound features of acute pancreatitis
A the pancreas appears diffusely enlarged in over 80% of cases.
B a focal hypoechoic mass in the pancreatic body is a typical feature.
C dilatation of the main pancreatic duct is a recognized finding.
D an intrapancreatic phlegmon usually appears as a hyperechoic mass.
E hypoechoic areas surrounding the splenic vein are a feature.

3.28. Regarding CT in the assessment of acute pancreatitis
A the gland invariably appears abnormal.
B thickening of Gerota's fascia may be seen.
C a pancreatic phlegmon typically has attenuation values of fluid.
D the degree of pancreatic necrosis can be assessed on postcontrast CT.
E gas within a peripancreatic collection is diagnostic of abscess.

3.29. Recognized imaging features of chronic pancreatitis include which of the following?
A   At ERCP, the main pancreatic duct is invariably abnormal.
B   At ERCP, loss of normal tapering of the main duct is a feature.
C   Stenoses involving the main duct are typically longer than 2 cm.
D   A focal, non-calcified mass is a recognized CT finding.
E   Pancreatic parenchymal atrophy is a recognized CT feature.

3.30. Concerning imaging of chronic pancreatic pseudocysts
A   enhancement of the wall is a feature on postcontrast CT.
B   the presence of internal echoes on ultrasound indicates superadded infection.
C   at ERCP, contrast extravasation into the pseudocyst occurs in over 90% of cases.
D   CT attenuation values of the cyst fluid are typically over 30 HU.
E   a cystic mass in the left interlobar fissure of the liver is a finding on CT.

3.31. Concerning the ultrasound features of pancreatic neoplasms
A   a solid hyperechoic mass is a recognized appearance of microcystic adenoma.
B   central echogenic foci with distal acoustic shadowing are a feature of microcystic adenoma.
C   macrocystic (mucinous) adenoma usually appears as a unilocular cyst.
D   calcification is not a feature of macrocystic adenoma.
E   metastasis from melanoma is a recognized cause of a cystic mass.

3.32. Concerning the CT features of pancreatic adenocarcinoma
A   calcification in a pancreatic mass excludes the diagnosis.
B   they are typically hypodense on precontrast scans.
C   the majority are located in the pancreatic head.
D   enlarged peripancreatic lymph nodes at diagnosis are an atypical finding.
E   associated pancreatic atrophy is a recognized feature.

3.33. Concerning PTC and ERCP in the assessment of pancreatic adenocarcinoma
A   obstruction to the CBD usually occurs in its proximal half.
B   complete obstruction of the CBD is atypical.
C   obstruction of the CBD and pancreatic duct is diagnostic.
D   the main pancreatic duct between the ampulla and the carcinoma is typically normal.
E   complete obstruction of the pancreatic duct is the commonest finding.

3.34. Concerning the imaging features of pancreatic islet cell tumours
A   the majority are hyperechoic to pancreas at intraoperative ultrasound.
B   an isoechoic lesion with a surrounding echogenic 'halo' is a recognized appearance at intraoperative ultrasound.
C   central low attenuation on precontrast CT is a typical feature of a non-functioning tumour.
D   the majority of functioning tumours are less than 2 cm in size at presentation.
E   calcification is commoner in malignant neoplasms.

3.35. Concerning MRI of the pancreas
A the normal pancreas is approximately isointense to the liver on T1-WSE scans.
B reduced signal intensity on T1-WSE scans is a typical feature of ageing.
C an islet cell tumour is a recognized cause of a hyperintense mass on T2-WSE scans.
D uncomplicated pseudocysts are typically hypointense to pancreas on T1-WSE scans.
E non-haemorrhagic acute pancreatitis and haemorrhagic acute pancreatitis cannot be differentiated.

3.36. Concerning CT and MRI of the normal spleen
A the spleen is completely surrounded by fluid in patients with ascites.
B inhomogeneous enhancement on immediate postcontrast CT scans is always abnormal.
C on T1-WSE MRI, the spleen is isointense to the liver.
D a splenic index (volume) of 700 cm^3 is abnormal.
E accessory spleens (splenunculi) typically show no enhancement on postcontrast CT.

3.37. Concerning imaging of infectious conditions of the spleen
A focal uptake of gallium-67 citrate indicates splenic abscess.
B multiple, small hypoechoic lesions are a recognized feature of splenic candidiasis.
C candidal microabscesses typically enhance on postcontrast CT.
D gas is seen in the majority of splenic abscesses on CT.
E hydatid disease is a cause of a solitary splenic cyst.

3.38. Concerning imaging of focal benign lesions of the spleen
A wall calcification is not a feature of epidermoid cysts.
B post-traumatic (false) cysts characteristically have a trabeculated wall.
C a solitary hyperechoic mass is a feature of haemangioma on ultrasound.
D peripheral enhancement is a feature of haemangioma on postcontrast CT.
E lymphangiomatosis is a cause of multiple hypodense masses on precontrast CT.

3.39. Concerning imaging of malignant lesions of the spleen
A splenomegaly in patients with non-Hodgkin's lymphoma indicates splenic involvement.
B lymphoma is a recognized cause of a solitary hypoechoic mass on ultrasound.
C uptake of gallium-67 citrate is a feature of splenic lymphoma.
D lymphoma is a cause of multiple low-density lesions on postcontrast CT.
E splenic metastases are invariably associated with splenomegaly.

3.40. Concerning imaging of the spleen

A   splenic infarcts are typically initially hypoechoic on ultrasound.

B   splenic infarct is a recognized cause of a hyperdense lesion on precontrast CT.

C   sickle-cell disease is a cause of diffusely increased splenic density on precontrast CT.

D   subcapsular haematomas characteristically cause distortion of the splenic parenchyma.

E   increased blood clearance half-time of heat-damaged ^{51}Cr-labelled RBCs is a feature of hyposplenism.

Liver, biliary tract, pancreas, spleen: Answers

3.1. A False
Technetium-labelled sulphur colloid is taken up by the cells of the reticuloendothelial system in the liver (Kupffer cells), spleen and bone marrow. Therefore, on a normal scan, the liver and spleen are well visualized and a small amount of activity is seen in the bone marrow. However, gallium-67 citrate is taken up by normal hepatocytes, as are the ^{99m}Tc-labelled iminodiacetic acid derivatives (^{99m}Tc-HIDA).

3.1. B True
The majority of space-occupying lesions in the liver will appear as focal 'cold' areas on the colloid scan and metastases, abscesses, cysts and hepatoma cannot be differentiated. However, focal nodular hyperplasia will show uniform uptake of isotope in about 60% of cases. About 25% of adenomas will also demonstrate uniform colloid uptake.

3.1. C True
The spleen is a posterior organ and therefore should not appear more active than the liver on anterior views. If the spleen does appear 'hotter' on anterior views, diffuse reduction in hepatic uptake may be present, signifying parenchymal liver disease.

3.1. D True
Absent uptake of colloid by the liver in acute alcoholic hepatitis is probably due to a direct toxic effect on the Kupffer cells. There will be increased uptake by the spleen and bone marrow. These scan findings are virtually pathognomonic of this condition.

3.1. E False
Increased bone marrow activity may be seen in isolation when there is marrow hyperfunction, as occurs with polycythaemia rubra vera or chronic anaemia. The features of parenchymal liver disease include inhomogeneous reduction of hepatic uptake with increased splenic and bone marrow activity. Focal 'cold' areas may be present, and causes include regenerating nodules, metastases and multifocal hepatoma.

(References A, C, D, E, 5; B, 19)

3.2. A False
Diffuse fatty infiltration results in increased echogenicity of the liver parenchyma and consequently decreased prominence of the normally bright portal vein walls. Also, smaller vessels may be compressed due to expansion of the parenchyma against a relatively non-distensible liver capsule.

3.2.   B True
Glycogen storage disease is associated with fatty infiltration which is the result of increased lipid deposition in the hepatocytes. The commonest cause is obesity. Other causes include alcohol abuse, diabetes, steroids, chemotherapy, malnutrition and parenteral nutrition.

3.2.   C True
Focal fatty infiltration may appear as a single or multiple discrete areas of echogenicity, and a common appearance is of a relatively angular or band-like area with geographic borders that typically does not distort the liver outline. Focal fatty sparing appears as a relatively hypoechoic area of normal echotexture within an otherwise echogenic liver due to fatty infiltration. Common sites are anterior to the gallbladder fossa and portal vein.

3.2.   D True
In cirrhosis, there is altered parenchymal echotexture ranging from mild coarsening to multiple masses. However, the liver may also appear entirely normal. Multiple hypoechoic masses may be seen due to regenerating nodules, but the differentiation from neoplasm may be impossible, necessitating biopsy.

3.2.   E True
Other sonographic features of cirrhosis include increased parenchymal echogenicity and selective enlargement of the caudate lobe, with a caudate/right lobe ratio of 0.65, highly suggestive of cirrhosis.

(Reference 11)

3.3.   A True
On precontrast CT, the attenuation value of normal liver parenchyma is 5–10 HU greater than that of the spleen. Diffuse fatty infiltration results in reduced attenuation value of the liver, either equal to or less than that of the spleen. On postcontrast CT, the spleen enhances by at least 10 HU more than the liver throughout the bolus and non-equilibrium phase.

3.3.   B False
Focal fatty infiltration appears as areas of low attenuation on both pre- and postcontrast CT and may be difficult to distinguish from tumour. However, on dynamic postcontrast scans, a normal pattern of hepatic vasculature in the region of low attenuation strongly supports a diagnosis of fatty infiltration. Fat-suppressed MRI can also be used to distinguish between the two conditions.

3.3.   C True
Causes of increased parenchymal attenuation on precontrast CT include haemochromatosis, haemosiderosis, iodine deposition secondary to drugs (amiodarone) and glycogen storage disease. Both haemochromatosis and haemosiderosis can progress to cirrhosis. The attenuation values of liver in these conditions can be increased to 70–100 HU, in contrast to values of 55–60 HU for normal liver on precontrast scans.

3.3.   D True
Typical CT features in cirrhosis are a nodular hepatic outline (in macronodular cirrhosis), and relative atrophy of the right lobe with hypertrophy of the left and caudate lobes. Other findings include ascites, splenomegaly and portosys-

temic varices (in the presence of portal hypertension), regenerating nodules, fatty infiltration and complicating hepatocellular carcinoma.

3.3. E False
On postcontrast CT, cirrhosis is associated with inhomogeneous and reduced enhancement of the liver parenchyma. Hepatic and portal vein radicles are compressed and often difficult to visualize.

(References A, B, C, D, 19; E, 2)

3.4. A False
Focal nodular hyperplasia (FNH) is a benign liver tumour composed of hepatocytes and Kupffer cells sometimes arranged around a central stellate scar and without a capsule. Lesions may be pedunculated and are multiple in 20% of cases. On ultrasound, FNH has variable echogenicity and a central echogenic scar is occasionally seen. Central scars may also be identified by ultrasound in fibrolamellar hepatoma, haemangioma, hepatic necrosis and adenoma with associated haemorrhage.

3.4. B True
Liver cell adenomas are typically solitary lesions that are most commonly seen in young women using oral contraceptives. They are also associated with type 1 glycogen storage disease (von Gierke's disease). Ultrasound demonstrates a mass of variable echotexture with areas of increased and decreased echogenicity due to haemorrhage and necrosis.

3.4. C False
Biliary cystadenoma is a rare neoplasm that occurs usually in middle-aged women. It may be locally recurrent and dedifferentiate into cystadenocarcinoma. Ultrasound demonstrates a predominantly cystic mass with multiple septations. Mural nodules and papillary projections may also be seen and do not necessarily indicate malignant change.

3.4. D True
Lipomas are rare liver tumours that may be associated with tuberous sclerosis. Ultrasound demonstrates a homogeneous, well-defined, densely hyperechoic mass that may mimic a haemangioma. However, a unique feature is discontinuity of the visualized diaphragm deep to the mass because of the slower speed of sound in fat compared to normal liver.

3.4. E True
Both Hodgkin's and non-Hodgkin's lymphoma (NHL) most commonly appear as multiple, poorly defined hypoechoic masses, but diffuse heterogeneity of liver texture may also be seen. 'Target' or echogenic lesions are commoner with NHL than Hodgkin's disease, but ultrasound usually cannot differentiate between the two. Leukaemia usually involves the liver via microscopic infiltration, and ultrasound is usually normal.

(Reference 19)

3.5. A True
Haemangiomas are the commonest benign liver tumours, with an incidence of 4–7%. They are most commonly located in the posterior segment of the right lobe, are usually solitary and less than 3 cm in diameter. They are multiple in approximately 10% of cases. Ultrasound typically shows a well-defined, uniformly hyperechoic mass (in 70–80% of cases), but hypoechoic central areas may be seen due to fibrosis or necrocis.

3.5. B False
A peripheral hypoechoic 'halo' is not a feature of haemangioma. Other ultrasound appearances include a hypoechoic lesion (in 15–20% of cases). They may also appear hypoechoic or isoechoic in livers affected by diffuse fatty infiltration. Occasionally they reach 8–15 cm in size (then referred to as a giant haemangioma) and may rarely show rapid growth.

3.5. C False
The characteristic features of haemangioma on ^{99m}Tc-labelled RBC scans are decreased perfusion during the dynamic phase and increased activity in the blood pool phase. These scan appearances are mimicked only by angiosarcoma. Increased blood pool activity alone may also be seen with small hepatocellular carcinomas and hypervascular metastases and is therefore not diagnostic of haemangioma.

3.5. D True
Haemangiomas may appear hyperdense precontrast against a background of diffuse fatty infiltration. Occasionally, calcified phleboliths may be seen. On dynamic postcontrast CT, characteristic findings are initial peripheral enhancement, which may be nodular, progressing to uniform centripetal enhancement on delayed scans. Eventually, the lesion may become isodense with liver. Uniform enhancement usually occurs within 15 minutes of injection.

3.5. E False
The characteristic enhancement pattern of haemangioma occurs in about 55% of cases. In the rest the enhancement pattern is not diagnostic. Patchy peripheral or central enhancement may be seen in the early postcontrast phase, mimicking metastasis. Giant haemangiomas may have central, stellate, low-attenuation areas due to cystic degeneration, liquefaction or fibrosis. These areas will not enhance.

(Reference 19)

3.6. A True
The lesions of FNH may range from 1–20 cm in diameter. On precontrast CT, the mass may be isodense to liver, becoming hyperdense on early postcontrast scans. On delayed scans, the lesion typically is isodense, with a low attenuation central scar being seen in less than 50% of cases. Differential diagnosis includes metastasis, adenoma and fibrolamellar hepatocellular carcinoma.

3.6. B True
The major complication of adenoma is internal haemorrhage, which may be life-threatening. Within the first few days of such a complication, precontrast CT may show areas that are relatively hyperdense, which progress to

hypodense cystic areas. Uncomplicated lesions show similar enhancement patterns to FNH, except that a central scar may not be seen. Also, a fibrous capsule will sometimes be identified.

3.6. C False
In most patients with lymphoma, there is diffuse periportal tumour infiltration and CT may demonstrate either a normal-sized liver or hepatomegaly, but no focal masses. However, patients with AIDS may develop a multifocal malignant lymphoma with hepatic, splenic and renal masses demonstrated on CT. The masses may show rim enhancement. The differential diagnosis includes metastatic Kaposi's sarcoma.

3.6. D False
Hepatic infarcts may occur secondary to arterial or venous thrombi, or traumatic arterial disruption. CT demonstrates wedge-shaped areas of low attenuation in a peripheral, vascular distribution. Central hypodense areas within hepatic infarcts may occur due to cystic change or bile accumulation.

3.6. E True
Primary hepatic angiosarcoma is a rare, highly malignant tumour that occurs in patients who have been exposed to thorium oxide (Thorotrast) several decades earlier. Postcontrast CT demonstrates a poorly defined, enhancing mass. Patchy areas of high attenuation will be seen in the liver and spleen due to thorium deposition in the reticuloendothelial system.

(Reference 19)

3.7. A False
Hepatocellular carcinoma shows gallium-67 citrate uptake in about 95% of cases, and approximately 50% will show uptake that is greater than that of normal liver. Uptake similar to normal liver also occurs with adenoma and FNH, but both may also show greater uptake. Hepatocellular carcinoma may also show uptake of ^{99m}Tc-HIDA on late scans, whereas it will result initially in a focal defect.

3.7. B True
Hepatocellular carcinoma may be solitary, multiple or diffusely infiltrating. The echogenicity of tumours varies with histological composition and size. Solid tumours without necrosis are typically hypoechoic. Tumours with partial necrosis show mixed echogenicity, whereas hyperechoic appearances are attributable to fatty change or sinusoidal dilatation. Also, small tumours tend to be hypoechoic and echogenicity increases with increasing size. A hypoechoic 'halo' may be seen.

3.7. C True
Lipiodol is taken up by hepatocellular carcinoma after intra-arterial injection into either the hepatic artery or coeliac axis. Scanning 1–4 weeks after the injection demonstrates the tumour as hyperdense mass(es) relative to the remainder of the liver. This technique may be particularly useful in patients

with cirrhosis, since small carcinomas will retain the contrast medium whereas regenerating nodules will not.

3.7. D True
A central scar is a feature of fibrolamellar hepatocellular carcinoma, which is a less aggressive tumour typically occurring in younger patients without pre-existing liver disease. It is usually a solitary, lobulated tumour with imaging features similar to FNH.

3.7. E True
Unless there is evidence of portal or hepatic vein tumour thrombi, solitary uncomplicated hepatocellular carcinoma cannot be distinguished from adenoma by CT. Typical CT features include an inhomogeneously enhancing mass following contrast. A fibrous capsule is commonly seen in Asian patients but is infrequently seen in Western countries. Tumour thrombus produces an enhancing, expanding filling defect in the portal or hepatic veins. Internal haemorrhage, rupture into the peritoneal cavity and biliary obstruction are all features.

(Reference 19)

3.8. A False
Amoebic colitis is complicated by liver abscess in approximately 8% of cases. These lesions typically show little inflammatory response and therefore appear on ultrasound as well-defined, round or ovoid, hypoechoic masses which lack rim echoes to suggest an abscess wall. They are often located peripherally in the right lobe of the liver. Recognized complications include intraperitoneal rupture, subdiaphragmatic abscess with diaphragmatic rupture, and spread to the brain.

3.8. B True
Acute pyogenic abscesses may appear as solid hypoechoic or isoechoic masses, whereas older abscesses are usually hypoechoic with an irregular hyperechoic wall. Ultrasound may demonstrate heterogeneous internal echoes and, occasionally, a fluid–debris level may be seen. Acoustic enhancement is usually seen but if gas is present within the abscess acoustic shadowing may occur. Pyogenic liver abscesses are commoner in the right lobe of the liver.

3.8. C True
Pyogenic hepatic abscesses are typically hypodense, rim-enhancing lesions on postcontrast CT. The attenuation of their contents depends upon the viscosity of the pus, debris and associated haemorrhage. Air–fluid levels may be seen with gas-forming organisms.

3.8. D False
Approximately 80% of pyogenic liver abscesses will take up gallium-67 citrate to an equal or greater degree than liver parenchyma. For this reason, a sulphur colloid scan should also be performed to avoid false negative gallium scans. Apparently uniform liver uptake of gallium-67 citrate with a photopenic area

on sulphur colloid scintigraphy is suggestive of a focal gallium-avid lesion, such as abscess or tumour.

3.8. E True
Hepatosplenic candidiasis may produce a characteristic ultrasound picture of multiple small 'target' lesions, caused by a hypoechoic rim due to fibrous tissue surrounding a central echogenic core of inflammatory cells. Other appearances include the presence of a small hypoechoic focus within the echogenic centre, or lesions that are completely hypoechoic or hyperechoic. These ultrasound features may also be observed in AIDS patients with hepatic infection from *Pneumocystis carinii.*

(References A, C, E, 19; B, D, 20)

3.9. A True
On postcontrast CT, metastases may appear uniformly hyperdense, uniformly hypodense, or hypodense with an enhancing rim. The latter two appearances are characteristic of metastases from colon, lung, breast and pancreas. The hypodense region represents hypovascular or necrotic areas of the tumour, whereas rim enhancement occurs in the peripheral, vascularized, viable tumour.

3.9. B False
Delayed enhancement of the centre of rim-enhancing tumours may occur as a result of contrast diffusion into both viable and necrotic areas of the tumour. This explains the reduced sensitivity of infusion-enhanced liver CT as opposed to bolus-enhanced dynamic CT in the detection of metastases. Hypervascular metastases may become isodense to liver parenchyma on dynamic enhanced CT. Therefore, patients with known hypervascular primary tumours (e.g. carcinoid, islet cell, renal, thyroid) may benefit from pre- and postcontrast CT of the liver.

3.9. C False
In the normal situation, approximately 1–2% of injected contrast medium will be excreted by the liver. On scans obtained several hours after the injection, functioning hepatocytes will appear hyperdense relative to tumour and also compared with spleen, muscle and blood vessels. Metastases will appear hypodense. However, FNH, which contains functioning hepatocytes, may enhance using this technique.

3.9. D True
CT arterial portography (CTAP) is performed by positioning a catheter into either the superior mesenteric artery or splenic artery and scanning the liver dynamically approximately 7–10 seconds after injection of contrast medium. The liver will enhance brightly due to its dominant blood supply by the portal vein, and lesions will appear hypodense since they gain blood supply via the hepatic artery.

3.9. E False
CT hepatic arteriography (CTHA) is performed by positioning a catheter in the proper hepatic artery and scanning the liver dynamically after injection of contrast medium. With this technique, liver tumours will demonstrate rim

enhancement due to their dominant blood supply from the hepatic artery. Some lesions may have a hypodense centre due to internal necrosis.

(Reference 19)

3.10.   A True
Budd–Chiari syndrome is chronic hepatic venous congestion caused by either hepatic venous outflow obstruction or microvenous intrahepatic occlusion. Generalized reduction of colloid uptake by the liver will occur if all the hepatic veins are occluded. More characteristically, sparing of the caudate lobe results in increased activity within it, seen as a central, posteriorly located 'hot' area within the liver.

3.10.   B True
Other scintigraphic features include increased splenic activity, wedge-shaped peripheral defects due to infarcts, and hepatomegaly. The posterior 'hot' spot seen in Budd–Chiari syndrome distinguishes it from an anterior 'hot' spot that can occur with SVC obstruction.

3.10.   C True
The hepatic veins will enhance normally if there is only microvenous occlusion, as may occur in patients receiving chemotherapy or radiotherapy, and also in bone marrow transplant recipients. Main hepatic vein outflow obstruction may be due to tumour thrombus from hepatocellular carcinoma, renal cell carcinoma and cholangiocarcinoma, or due to polycythaemia, congenital webs or thrombophlebitis.

3.10.   D True
A characteristic finding on dynamic postcontrast CT is hepatomegaly with enhancement of the deep central portion of the liver, including the hypertrophied caudate lobe. There is abnormally low attenuation of the periphery of the liver, especially the right lobe. In macrovenous occlusion, central hepatic veins will not be identified during the dynamic phase of the study.

3.10.   E True
MRI features of Budd–Chiari syndrome include reduced calibre or absence of the hepatic veins, comma-shaped intrahepatic collateral vessels, reduced calibre of the IVC, and ascites. Chronic Budd–Chiari syndrome is indistinguishable from other causes of cirrhosis.

(References A, B, 20; C, D, E, 19)

3.11.   A False
Conventional spin-echo (SE) techniques are poor at detecting diffuse fatty infiltration. However, focal fatty infiltration has been demonstrated as areas of increased signal intensity on both T1- and T2-WSE sequences. MRI many be of value in distinguishing focal fatty infiltration from tumour, which can be a problem with CT. Compared with metastases, focal fat never has as low signal intensity on T1-WSE scans or as high signal on T2-WSE scans.

3.11. B False
The MR signal intensity of the liver with iron overload (due to haemosiderosis or haemochromatosis) depends upon the concentration of iron present. Ferric iron exerts a paramagnetic effect on the liver parenchyma with shortening of T1 and T2 relaxation times. With small amounts of iron, the liver may appear bright on T1-WSE scans due to exclusive T1 shortening effects. With increasing iron content, the liver becomes dramatically hypointense on all pulse sequences because of predominant T2 shortening effects.

3.11. C True
Regenerating nodules contain haemosiderin which has superparamagnetic effects and results in the nodules appearing hypointense. These nodules will be better demonstrated on GE images due to the increased sensitivity of these sequences to magnetic susceptibility effects. Large nodules (sometimes termed adenomatous hyperplastic nodules) may be distinguished from hepatocellular carcinoma by their signal characteristics on T2-WSE scans. They appear hypointense whereas hepatocellular carcinoma is typically hyperintense.

3.11. D True
Chronic hypoperfusion due to portal vein occlusion may result in the affected areas appearing hypointense on T1-WSE scans and hyperintense on T2-WSE scans relative to normal liver. Similar segmental intensity differences can be seen secondary to radiation hepatitis, probably reflecting both parenchymal damage and radiation-induced vascular effects.

3.11. E True
Simple cysts appear as well-defined, markedly hypointense lesions on T1-WSE scans and hyperintense on T2-WSE scans. The presence of high protein content or haemorrhage will alter the signal characteristics. Hydatid cysts have produced characteristic findings on MRI. They appear as well-defined, multiloculated lesions with a surrounding rim that is hypointense on all pulse sequences, representing the fibrous pericyst. The cyst contents are usually hypointense on T1-WSE images and hyperintense on T2-WSE images.

(Reference 19)

3.12. A True
FNH has a characteristic appearance on MRI. It tends to be isointense or slightly hyperintense to liver on all SE pulse sequences and mildly hyperintense on STIR (short T1 inversion recovery) sequences. A central scar may be seen, which is characteristically hyperintense on T2-WSE scans. The lesion is poorly defined, consistent with the lack of capsule histologically.

3.12. B True
Cavernous haemangiomas are characteristically uniformly hypointense on T1-WSE scans (although less so than simple cysts) and markedly hyperintense on T2-WSE scans (similar to simple cysts). Heterogeneity of signal intensity may be due to central fibrosis, thrombosis, calcification and haemorrhage, especially in giant haemangiomas. The enhancement pattern following Gd-DTPA is similar to that seen on CT.

3.12. C False
MRI of liver cell adenomas is non-specific. They are usually mildly hyperintense on T2-WSE scans but may be hypointense to liver. On T1-WSE scans, they are usually isointense or slightly hyperintense. They may have a surrounding hypointense capsule. Internal haemorrhage will alter the signal characteristics depending upon its age.

3.12. D True
Hyperintensity on T1-WSE scans has been reported in 31–47% of hepatocellular carcinomas. This is related to fat within the tumour in some cases. Other liver masses that occasionally appear hyperintense on T1-WSE scans include fat-containing adenomas, adenomatous nodules, and rare fatty tumours such as angiomyolipoma, myelolipoma, lipoma and fat-containing metastases. Hepatocellular carcinoma is almost always hyperintense on T2-WSE scans.

3.12. E True
The pseudocapsule associated with slower growing, less aggressive hepatocellular carcinomas is formed by compressed liver parenchyma. On T2-WSE scans, it may have a double-layered appearance, consisting of an outer hyperintense rim composed of compressed vascular channels and bile ducts, and an inner hypointense layer of fibrous tissue. Other magnetic resonance features include daughter nodules and tumour thrombi (seen with more aggressive tumours), and intratumoral septa.

(Reference 19)

3.13. A True
Hepatic metastases may be hyperechoic, hypoechoic, anechoic or mixed on ultrasound. Metastases from colon carcinoma are frequently hyperechoic, whereas necrotic tumours, especially leiomyosarcoma metastases, are commonly anechoic. Septations, mural nodules and areas of wall thickening may also be seen.

3.13. B True
Some metastases will have a hypoechoic 'halo', caused in most cases by a rim of compressed normal liver parenchyma around a rapidly growing lesion. Ultrasound may also demonstrate calcification in metastases, most commonly seen with mucin-producing lesions from colloid carcinomas of the colon. However, calcified metastases may also occur from breast, stomach, ovarian, pancreatic islet-cell and melanoma primaries.

3.13. C False
Although portal vein invasion is common with primary hepatocellular carcinoma, portions of the portal system may be invaded in up to 8% of cases of metastatic disease.

3.13. D True
On T2-WSE scans, hepatic metastases have been described as having amorphous, target and halo signs. In one study, one of these signs was seen in 92% of patients with metastases. The amorphous appearance refers to a featureless, inhomogeneous mass with indistinct, irregular borders. The target sign refers to a lesion with a relatively brighter centre, probably related to

central necrosis. The halo sign refers to a lesion with a relatively hyperintense periphery, corresponding to peritumoral oedema.

3.13. E False
Metastases have prolonged T1 and T2 relaxation times relative to normal liver parenchyma and therefore appear hypointense on T1-WSE scans and hyperintense on T2-WSE scans. The signal intensity on T2-WSE scans can be very bright, particularly in hypervascular metastases, and may make distinction from haemangioma difficult.

(Reference 19)

3.14. A True
True hepatic contusions are relatively uncommon lesions. Precontrast CT usually demonstrates a poorly defined, hypodense area that typically does not disrupt portal or hepatic venous structures. Pathologically, a contusion represents areas of microscopic haemorrhage, necrosis and oedema.

3.14. B True
On postcontrast CT, hepatic lacerations typically appear as linear or branching, hypodense lesions paralleling the portal or hepatic veins. Higher attenuation areas of clot and haematoma may be seen centrally within the areas of laceration. Hepatic lacerations may be classified as simple or complex, depending upon their degree of branching and their configuration. They may also be subclassified as superficial, perihilar or deep.

3.14. C True
Deep hepatic lacerations extend to the first or second-order branches of the major portal veins and have a higher incidence of post-traumatic complications such as biloma or haemobilia. Central stellate lacerations have a higher incidence of bile duct injuries. If treated conservatively, they often result in low-attenuation collections of bile (bilomas) centrally within the liver.

3.14. D True
Extension of laceration to the bare area of the liver may result in extravasation of blood around the IVC, producing a pericaval 'halo' sign, as well as extraperitoneal haematoma that may surround the right adrenal gland. Right hepatic lobe injuries may also be associated with right adrenal parenchymal haemorrhage, which is usually of no clinical significance. Periportal low-density fluid may also be seen, and is due to either blood or lymphoedema.

3.14. E True
Subcapsular haematomas of the liver have a characteristic lenticular configuration and their mass effect causes a scalloped impression along the peripheral aspect of the involved hepatic segment. The density of acute subcapsular haematomas gradually reduces with time, but their size may initially increase over the first few weeks due to osmotic effects.

(Reference 19)

3.15. A False
Acute rejection is the commonest significant complication following hepatic transplantation, occurring in up to one-third of all cases. There is no correlation between loss of hepatic artery diastolic flow and acute allograft rejection. Doppler ultrasound is of value in identifying hepatic artery occlusion, which is the commonest vascular complication and also in detecting pseudoaneurysm formation.

3.15. B False
There is no correlation between hypodense rims around central portal veins and acute rejection. Low attenuation around peripheral portal veins is a CT finding in rejection but its clinical value is limited since the sign has both low sensitivity and specificity. Histological examination of a biopsy from the transplant is necessary to diagnose rejection.

3.15. C True
The donor biliary tree receives its blood supply solely from the hepatic artery, and arterial occlusion can result in ischaemia to the bile ducts with consequent formation of non-anastomotic strictures and bile leaks. Hepatic artery occlusion can also result in sepsis, liver infarction and allograft failure. Ischaemia can manifest on CT as lobar or peripheral areas of low attenuation.

3.15. D True
Biliary complications are the second commonest cause of allograft dysfunction. These complications include obstruction, most commonly due to strictures, and bile leaks. Strictures may be anastomotic or non-anastomotic and appear on cholangiography as focal narrowings with smooth walls. Other causes of obstruction include malposition or occlusion of the T-tube, and redundancy of the CBD which may result in kinking, mucocele of the cystic duct remnant and stones in the CBD.

3.15. E True
Intrahepatic fluid collections include biloma, due to bile leakage, haematomas and abscesses. CT cannot differentiate between these. Bile leaks can be identified by demonstration of contrast extravasation at cholangiography or by the use of ^{99m}Tc-HIDA scintigraphy. Extrahepatic fluid collections can be the result of bile leak, haematoma, abscess, loculated ascites or seroma.

(Reference 19)

3.16. A False
Normally a small amount (less than 15%) of ^{99m}TC-HIDA is excreted by the kidneys, the amount increasing with increasing degrees of hepatic dysfunction. On a normal scan, both renal pelves may be visualized and the right renal pelvis can mimic the gallbladder. However, it can be distinguished by its earlier filling and emptying and its posterior location on the lateral view.

3.16. B True
In a normal HIDA scan, there is homogeneous activity in the liver in the first 15–20 minutes following injection. After this the common duct is visualized, followed by the gallbladder (at about 20–25 minutes) and duodenum. In acute

calculous cholecystitis, the cystic duct is obstructed and no activity will be identified in the gallbladder.

3.16. C False
If activity is not seen in the gallbladder initially, imaging should be continued for up to 4 hours. The identification of gallbladder activity on later images is consistent with a diagnosis of chronic cholecystitis. In rare cases of acute acalculous cholecystitis, the gallbladder may fill with isotope and result in a false negative scan.

3.16. D True
Both intra- and extrahepatic cholestasis can be demonstrated on the HIDA scan. In the former case, there is decreased but progressive uptake of isotope but no excretion into the bile duct will be seen. With extrahepatic obstruction, the HIDA scan will demonstrate delayed or absent gut activity. On the parenchymal phase, filling defects due to dilated ducts may be seen.

3.16. E True
The HIDA scan is as sensitive as sulphur colloid scanning for the detection of metastases, which will appear as focal 'cold' areas. Any lesion that contains functioning hepatocytes may show uptake of HIDA. Consequently, adenoma, FNH, hepatocellular carcinoma and regenerating nodules can all appear active. Washout of activity will also occur with FNH and regenerating nodules since these contain bile ducts.

(References A, B, C, D, 5; E, 19)

3.17. A True
In the early stage of cirrhosis, when fatty infiltration predominates, the ducts may be elongated and stretched, due to hepatomegaly. This cannot be differentiated from other causes of diffuse infiltrative diseases of the liver (e.g. lymphoma). In milder cases of fatty change, the cholangiogram will appear normal. In late disease, as the liver shrinks, the ducts become crowded together and tortuous. Duct displacement may occur due to regenerating nodules.

3.17. B False
In primary biliary cirrhosis, cholangiographic abnormalities indicate advanced disease. The extrahepatic ducts are normal but may contain filling defects due to inspissated bile. The intrahepatic ducts may be normal, narrowed, or both narrowed and shortened with decreased branching. The ducts may be tortuous and of mildly varying calibre. These changes correlate well with the development of cirrhosis.

3.17. C False
Biliary cystadenomas rarely communicate with larger intrahepatic bile ducts. This can be demonstrated by cholangiography, which may also demonstrate intraductal prolapse of solid components of the tumour. In cases of extrahepatic biliary cystadenoma, or carcinoma, proximal ductal dilatation and a partially obstructing extraluminal mass will be demonstrated.

3.17. D True
In cases of metastatic disease, cholangiography may also demonstrate stretching and attenuation of the intrahepatic ducts. These features are non-specific, occurring with other causes of diffuse malignant infiltration of the liver, including lymphoma, leukaemia, diffuse hepatocellular carcinoma and multifocal cholangiocarcinoma.

3.17. E True
In diffuse cholangiocarcinoma (10%), focal or widespread, beaded strictures or long segment constrictions are recognized findings. The features can mimic those of sclerosing cholangitis. With extrahepatic cholangiocarcinoma, cholangiography may demonstrate a complete obstruction (in approximately 70%), an irregular, long or short stricture, or a polypoid filling defect.

(Reference 20)

3.18. A False
Dilated intrahepatic bile ducts are manifest on ultrasound as the 'parallel channel' sign. However, the main left and right hepatic ducts can be seen and measure up to 2 mm in diameter. Normal ducts proximal to the common hepatic duct average 20% of the width of the accompanying portal vein but should not measure more than 40% of the diameter of the accompanying vein. Also, dilatation does not equal obstruction.

3.18. B False
Following a fatty meal, or after injection of cholecystokinin (CCK), the diameter of the common duct should either remain unchanged or diminish. This is associated with increased bile flow, gallbladder contraction and relaxation of the sphincter of Oddi. In patients with partial or complete biliary obstruction, the diameter of the duct should increase by at least 1 mm.

3.18. C True
The ultrasound diagnosis of gallstones requires the demonstration of three criteria: hyperechoic foci, acoustic shadowing and mobility. Acoustic shadowing is best demonstrated with the highest-frequency transducer possible, with the focal zone set to the depth of the stone. The stone size must be large enough in relation to the ultrasound beam width to produce acoustic shadowing.

3.18. D True
Acalculous cholecystitis is a recognized complication of AIDS. This may be due to infection by *Cryptosporidium* species or cytomegalovirus, both of which have been found in the duodenal contents and bile of AIDS patients. Dilatation of the common bile duct is also a feature, either due to infection, or associated with papillary stenosis.

3.18. E True
Echogenic bile is associated with prolonged fasting and is commonly seen in patients in the intensive care unit and those receiving TPN. The echogenicity is thought to be related to calcium bilirubinate within the bile. Very echogenic

bile with heterogeneous echotexture may be seen in patients receiving somatostatin therapy for acromegaly. They may also develop gallstones which resolve after withdrawal of the medication.

(References A, C, 19; B, D, E, 11)

3.19. A True
In primary sclerosing cholangitis, periportal echogenicity on ultrasound is a manifestation of extensive wall thickening of the intrahepatic bile ducts. The wall of the CBD may also be thickened and echogenic. Other ultrasound features include non-shadowing, intraluminal echogenic material due to pus, sludge or desquamated bile duct epithelium. The differential diagnosis of these appearances includes cholangiocarcinoma.

3.19. B True
Biliary scintigraphy may demonstrate focal 'hot' spots that may appear at about 15 minutes and can persist for up to 1 hour. These represent isotope within dilated intrahepatic bile ducts. Delayed clearance of isotope from various segments of the liver may also be seen.

3.19. C False
Cystic duct abnormalities have been reported in up to 20% of patients with sclerosing cholangitis. Uncommonly the gallbladder wall may be irregularly thickened. Irregularities of the pancreatic duct have also been reported, even in the absence of pancreatitis.

3.19. D True
In sclerosing cholangitis, especially when associated with ulcerative colitis, the extrahepatic ducts may be normal. Isolated involvement of the extrahepatic ducts is very unusual.

3.19. E True
Pseudodiverticula are less commonly seen in the intrahepatic ducts and are almost pathognomonic of sclerosing cholangitis. Cholangiographic features include smooth dilatation of ducts proximal to high-grade strictures, ductal mural irregularities, long smooth strictures and decreased duct branching. Strictures are typically short and annular with a predilection for bifurcations.

(Reference 20)

3.20. A True
Mirizzi's syndrome refers to a benign inflammatory condition in which a mass, secondary to a gallstone impacted in the cystic duct, neck of the gallbladder or cystic duct remnant causes extrinsic compression of the common hepatic duct, resulting in dilatation of the intrahepatic ducts. It is thought to occur more commonly in patients with an abnormally low insertion of the cystic duct into the common hepatic duct, with the two sharing a common sheath.

3.20. B False
The site of obstruction by the inflammatory mass is the entry of the cystic duct into the common hepatic duct. The CBD will therefore be of normal calibre, or may appear smaller than usual. Recognized complications of Mirizzi's

syndrome include stricturing of the common hepatic duct and cholecystobiliary fistula.

3.20.   C True
This may be demonstrated by ultrasound or CT. The demonstration of dilated intrahepatic and common hepatic bile ducts associated with a stone in the gallbladder fossa suggests the diagnosis. A calcified stone at the site of the porta hepatis may also be seen on plain films.

3.20.   D False
An inflammatory mass can dominate the ultrasound or CT picture in Mirizzi's syndrome to such an extent that carcinoma of the gallbladder or pancreas and cholangiocarcinoma cannot be excluded. Ultrasound may also demonstrate an irregular cavity adjacent to the gallbladder. This may or may not contain an extruded stone.

3.20.   E False
Cholangiography in patients with Mirizzi's syndrome typically shows medial deviation of the common hepatic duct with a smooth, extrinsic lateral compression, and dilatation of the intrahepatic ducts. In cases of complete occlusion, a curved stenosis of the common hepatic duct at the expected site of insertion of the cystic duct may be seen.

(Reference 20)

3.21.   A True
Cholangiocarcinoma is the commonest primary malignancy of the bile ducts. The CT appearance depends upon the location and morphology of the tumour. Intrahepatic lesions have a non-specific appearance. They are usually hypodense on pre- and postcontrast CT, but may become hyperdense to liver parenchyma on delayed postcontrast scans.

3.21.   B True
Cholangiocarcinoma is located in the proximal extrahepatic ducts in approximately 50% of cases, the rest arising in the intrahepatic ducts or distal extrahepatic ducts. Bile duct dilatation is a typical feature, and when the level of obstruction is between the confluence of the main left and right ducts and the pancreatic head, there should be a strong suspicion of cholangiocarcinoma.

3.21.   C True
Proximal cholangiocarcinomas may infiltrate throughout the porta hepatis and extend into the liver parenchyma, making it difficult to distinguish from primary or metastatic liver tumours. Hilar cholangiocarcinomas may be either hypodense on pre- and postcontrast CT, or may appear as enhancing masses. More distal lesions may result in thickening of the bile duct wall, which can be demonstrated by CT.

3.21.   D False
Sonographic features that suggest the diagnosis of cholangiocarcinoma include focal bile duct dilatation, frequently without evidence of an obstructing mass, and without extrahepatic duct dilatation. About 10–25% of tumours are

located at the junction of the right and left main hepatic ducts and are termed Klatskin tumours. These tumours have a particularly poor prognosis.

3.21. E True
Lobar atrophy occurs in response to chronic biliary obstruction and is manifest by crowding of bile ducts in the atrophic lobe. Other ultrasound features include intraduct echoes, which may be non-shadowing and due to excessive mucin production by the tumour, or shadowing due to stones formed in dilated obstructed ducts. Diffuse cholangiocarcinoma may produce increased echogenicity of the bile duct walls.

(References A, B, C, D, 19; E, 20)

3.22. A True
CT in cases of sclerosing cholangitis usually demonstrates non-uniform biliary dilatation, with segmentation and beading of ducts. Focal, discontinuous areas of minimal duct dilatation are also seen in this condition, but may also occur in diffuse cholangiocarcinoma.

3.22. B True
Abnormalities of the extrahepatic bile ducts in sclerosing cholangitis demonstrated by CT include duct wall thickening, often with marked contrast enhancement, mural nodules and duct stenoses. Significant intrahepatic bile duct dilatation is not typical of sclerosing cholangitis and should raise the possibility of a cholangiocarcinoma.

3.22. C False
Dilated intrahepatic bile ducts appear as branching linear or circular structures of near water density on precontrast CT, which enlarge as they approach the junction of the right and left hepatic ducts. Rarely, infiltrating periductal neoplasms may mimic dilated ducts. Dilated ducts can be distinguished from portal vein radicles on precontrast CT since the latter have blood density. Inspissated bile may have high density and cause confusion.

3.22. D False
The detection of CBD stones by CT is dependent upon the composition of the stone. Calcified stones are easily detected if narrow collimation is used. However, non-calcified CBD stones can also be demonstrated as soft-tissue-density filling defects surrounded by low-attenuation bile, producing a 'target' sign, or as soft tissue densities in the dependent part or the duct, with a crescent of low-density bile above.

3.22. E True
Gallstones may be demonstrated on CT as densely or faintly calcified filling defects in the low-density bile, or as a rim of calcification around an otherwise low-density stone. Cholesterol stones may also be seen if they are of lower attenuation than the surrounding bile, but those that are isodense to bile will be missed by CT.

(References A, C, D, E, 2; B, 19)

3.23.   A True
In fasted patients, concentrated bile within the gallbladder appears hyperintense to the liver on T1-WSE scans, but non-concentrated bile has the signal characteristics of water. When the gallbladder contains both fresh and concentrate bile, a fluid–fluid level may be seen due to the differing specific gravities of the two, with the hypointense fresh bile layering above the hyperintense concentrated bile.

3.23.   B True
In patients with symptomatic cholecystitis, bile tends to be hypointense on T1-WSE scans, possibly reflecting the poor concentrating ability of the diseased gallbladder. However, there is no apparent difference in T1 and T2 relaxation times of bile from patients with acute and chronic cholecystitis.

3.23.   C True
MRI findings in cases of gallbladder carcinoma include an irregular mass surrounding a low-signal-intensity gallstone. The mass becomes relatively hyperintense to liver on T2-WSE scans. Loss of a clear plane between the mass and the liver may indicate invasion. If the tumour has a large desmoplastic component, areas of the tumour mass may remain hypointense on all pulse sequences.

3.23.   D False
In general, gallstones produce no measurable signal on MRI, regardless of chemical composition, reflecting their solid nature and the associated relative immobility of hydrogen protons. Occasional high signal on T2-WSE scans may be seen within gallstones due to bile within clefts. Stones may also exhibit a laminated appearance. Higher signal intensity may also be seen with cholesterol stones.

3.23.   E False
Cholangiocarcinoma has no specific MRI characteristics. Two different appearances have been described on T2-WSE scans. Well-differentiated tumours demonstrate hyperintensity relative to the liver, while the scirrhous subtype shows minimal hyperintensity. Dilatation of the intrahepatic ducts may be seen. These typically have low signal intensity on T1-WSE scans and appear hyperintense on T2-WSE scans.

(Reference 19)

3.24.   A False
In a normally distended state, gallbladder wall thickening is present if the wall measures more than 2 mm in thickness. Measurements are most reliable in the transverse axis and from the superficial wall. Gallbladder wall thickening is a non-specific finding and may also be seen with hepatitis, cirrhosis, hypoproteinaemia, congestive heart failure, nephrotic syndrome and myeloma.

3.24.   B True
The normal gallbladder wall appears as a well-defined, thin echogenic line. In acute cholecystitis the gallbladder wall may contain irregular anechoic regions (30% of cases), but this is not specific and similar appearance may occur with varices in the gallbladder wall. These can be differentiated by the use of colour

Doppler ultrasound. Other ultrasound findings in acute cholecystitis include a positive sonographic Murphy's sign, dilatation of the gallbladder and demonstration of an impacted calculus in the gallbladder neck.

3.24. C True
Gangrenous cholecystitis results in wall necrosis and microabscess formation and may progress to perforation and empyema. Sloughed mucosa appears on ultrasound as a thin membrane within the lumen parallel to the gallbladder wall. Intraluminal membranes and coarse non-shadowing intraluminal echoes may also be seen in haemorrhagic cholecystitis, which is another complication of acute cholecystitis.

3.24. D True
Emphysematous cholecystitis is an uncommon but severe form of acute cholecystitis, commonly occurring in diabetics and associated with infection by gas-forming organisms. Gas typically accumulates in the gallbladder wall and lumen, and sometimes in the bile ducts and gallbladder fossa. Ultrasound demonstrates highly hyperechoic collections with 'dirty' shadowing and reverberation emanating from the wall and lumen of the gallbladder.

3.24. E False
Anechoic areas may be seen around the gallbladder in uncomplicated acute cholecystitis. Pericholecystic abscess due to perforation may result in a pericholecystic collection of mixed echogenicity. A defect may rarely be demonstrated in the gallbladder wall.

(Reference 19)

3.25. A False
Cholesterol polyps typically appear on ultrasound as small, echogenic masses that are adherent to the gallbladder wall. They are not mobile and do not cast an acoustic shadow. They are the commonest cause of a fixed gallbladder wall mass, other causes including adenoma, papilloma and metastasis.

3.25. B True
Gallbladder wall thickening is a non-specific feature of adenomyomatosis, with or without Rokitansky–Aschoff sinuses. Mild thickening of the gallbladder wall may also be seen in the diffuse form of cholesterolosis, and may be the only ultrasound abnormality.

3.25. C True
Anechoic spaces within the gallbladder wall in adenomyomatosis are due to fluid bile-filled Rokitansky–Aschoff sinuses. However, they usually appear echogenic if filled by sludge, debris or calculi and may then be associated with acoustic shadowing. They may be better demonstrated on post-fatty meal sonography.

3.25. D False
The focal form of adenomyomatosis is the most common type and usually appears as a soft-tissue mass in the region of the gallbladder fundus, which may be mostly extraluminal, or focal thickening of the wall of the gallbladder.

3.25. E True
Segmental adenomyomatosis is caused by a band-like constriction of the gallbladder, separating it into compartments. The differential diagnosis includes a congenital septum, but this usually appears thinner than the constriction caused by adenomyomatosis.

(Reference 20)

3.26. A False
Benign tumours of the gallbladder include adenomas and papillomas. Inflammatory polyps are rare lesions that may be solitary or multiple and are nearly always found in association with gallstones. On ultrasound, they are relatively hypoechoic.

3.26. B False
Gallbladder carcinoma is a rare tumour occurring more commonly in women and with highest incidence in the sixth and seventh decades of life. Histologically, the tumour may be scirrhous, papillary or mucinous. Local extension at the time of diagnosis is common. The most common ultrasound findings are a mass filling the gallbladder, focal or generalized wall thickening, or a fungating tumour.

3.26. C False
Gallbladder carcinoma is associated with gallstones and chronic inflammation in approximately 75% of cases. Other associations are with inflammatory bowel disease, polyposis coli and choledochal cyst. Xanthogranulomatous cholecystitis may produce ultrasound findings that are indistinguishable from gallbladder carcinoma, including calculi (in 96% of cases) and wall thickening (in 70%), which may be irregular.

3.26. D True
Local invasion of gallbladder carcinoma into the porta hepatis may cause proximal biliary dilatation. Other features include fistula formation to adjacent organs, retroperitoneal lymphadenopathy and liver metastases. Metastatic spread to nodes around the pancreatic head may be mistaken for a primary pancreatic carcinoma.

3.26. E True
Metastases to the gallbladder are most commonly from malignant melanoma and may appear on ultrasound as multiple flat or polypoid lesions, or as hypoechoic nodules within the gallbladder wall. The gallbladder serosa may be involved in peritoneal carcinomatosis.

(References A, B, C, E, 19; D, 2)

3.27. A False
The pancreas may appear normal in about 29% of cases of acute pancreatitis. However, due to associated interstitial oedema, the gland may appear diffusely enlarged and hypoechoic in approximately half of the remaining patients. The appearances of the gland usually revert to normal in days to weeks.

3.27. B False
Focal masses are demonstrated in about half the cases of acute pancreatitis in which the pancreas appears abnormal. The abnormal area is usually, but not always, hypoechoic. Focal masses arise in the head in 60% of cases and in the tail in the remainder. Focal enlargement of the body alone is not a feature of inflammatory diseases of the pancreas.

3.27. C True
Pancreatic ductal dilatation may be associated with focal enlargement of the pancreatic head. However, this is a variable finding and the duct may also be compressed due to diffuse intraparenchymal swelling. In these cases, mild dilatation may be seen during the convalescent phase.

3.27. D False
Intrapancreatic masses in acute pancreatitis may be due to acute fluid collections, phlegmon or focal haemorrhage. A phlegmon is an inflammatory mass that appears on ultrasound as an irregular, hypoechoic mass. The sonographic appearance of phlegmon closely resembles that of carcinoma. Pancreatic haemorrhage may appear as a focal hyperechoic mass.

3.27. E True
The most common extrapancreatic sites for fluid collections include the lesser sac, anterior pararenal spaces, transverse mesocolon and the perirenal space. However, acute inflammatory exudate may also dissect along perivascular planes and encase the splenoportal venous system. This feature appears as a hypoechoic area surrounding the splenic vein, most commonly in its retropancreatic portion, and may be associated with subsequent splenic vein thrombosis.

(References A, B, 20; D, E, 21; C, 20 and 21)

3.28. A False
CT will demonstrate a normal-appearing pancreatic gland in one-third or more of patients with acute pancreatitis. CT abnormalities include diffuse or focal glandular enlargement, contour irregularity, focal areas of decreased attenuation, presumably due to oedema or necrosis, and changes in the peripancreatic fat and parietal peritoneal fasciae.

3.28. B True
Gerota's fascia represents the anterior perinephric fascia and will appear thickened when inflammatory exudate extends into the anterior pararenal space. As the amount of exudate increases, this space fills with fluid and extends superiorly and inferiorly within the space. Fluid may also extend into the posterior pararenal space and, less commonly, the perirenal space will be involved.

3.28. C False
A pancreatic phlegmon refers to 'a solid mass of indurated pancreas and adjacent retroperitoneal tissues due to oedema, infiltration by inflammatory cells and tissue necrosis'. A phlegmon is a common CT finding in complicated pancreatitis and typically appears as a mass of low soft-tissue attenuation with heterogeneous areas due to necrosis, altered retroperitoneal fat and haemorrhage. Its distribution is similar to that seen with exudative fluid collections.

3.28.   D True
Dynamic postcontrast CT of the pancreas allows the degree of ischaemic and necrotic tissue to be assessed, and this is related to ultimate prognosis. In one study of 58 patients, all 36 with uncomplicated pancreatitis showed either increased or normal pancreatic enhancement, whereas 22 with haemorrhagic, necrotizing pancreatitis showed a decrease in pancreatic enhancement.

3.28.   E False
Pancreatic or peripancreatic abscesses may develop from either a phlegmon or from secondary infection of pancreatic or peripancreatic fluid collections. Gas within such a collection is suggestive of abscess formation and may be seen in 33–66% of cases. The attenuation value of an abscess is generally greater than that of a sterile fluid collection. However, gas may also be seen in a collection in the rare situation when there is a fistula to the gastrointestinal tract.

(Reference 2)

3.29.   A False
In mild chronic pancreatitis, the main pancreatic duct may be normal. The earliest pancreatographic changes are slight ectasia or clubbing of the side branches. There may be narrowing of the junctions of the side branches and the main duct, and side branch filling may be impaired.

3.29.   B True
In moderate pancreatitis, there is dilatation, tortuosity, rigidity and stenoses of the main pancreatic duct. Branch ducts show irrregular distribution, cystic dilatation, partial stenosis or obstruction, and wall rigidity. Main duct dilatation may be associated with filling defects due to calculi.

3.29.   C False
Stenoses of the main pancreatic duct are typically short segment and, if over 1 cm long, carcinoma should be suspected. Multifocal stenoses with associated dilatation may result in a beaded appearance to the duct. Complete occlusion of the duct is uncommon in pancreatitis but may occur due to fibrosis, calculi, abscess or pseudocyst.

3.29.   D True
A focal non-calcified mass is an occasional finding in chronic pancreatitis, being indistinguishable from pancreatic carcinoma by CT criteria. Other CT findings include pancreatic duct dilatation (66%), pancreatic calcifications (50%), focal pancreatic enlargement (32%), biliary dilatation (29%), and alterations of peripancreatic fat or fasciae (16%). The gland may appear normal in about 7% of cases.

3.29.   E True
Parenchymal atrophy may occur in up to 54% of cases of chronic pancreatitis and may or may not be associated with fatty replacement of the gland. However, fatty replacement is not a specific feature of chronic pancreatitis and may occur as a normal ageing process and in cystic fibrosis.

(References A, B, C, 20; D, E, 21)

3.30.   A True
Pseudocysts are a complication of both acute and chronic pancreatitis and represent fluid collections surrounded by a fibrous capsule. They may occur either within the pancreas, in the peripancreatic region or in the mediastinum. The wall of a pseudocyst is typically well defined and smooth, with a thickness of 3–4 mm. Irregularity and thickening of the wall may be seen if haemorrhage or infection has occurred.

3.30.   B False
On ultrasound, the contents of the pseudocyst may be entirely echo-free, but varying amounts of internal echoes or even a fluid–debris level may be seen. These features do not necessarily indicate infection. Internal septations may also occur, mimicking a cystic neoplasm. Acoustic shadowing may also be seen due to gas within the cyst or calcification of the wall.

3.30.   C False
Approximately one-half of pseudocysts will not fill at pancreatography, making ultrasound and CT more sensitive in their detection. Pseudocysts may cause smooth displacement of the pancreatic duct or complete obstruction. The point of obstruction is usually blunt rather than irregular, as is seen with carcinoma. Pseudocysts may also cause obstruction of the common bile duct, either by compression, or due to inflammation and fibrosis. Extravasation of contrast medium into a pseudocyst may result in infection and is an indication to abandon the procedure.

3.30.   D False
The CT numbers of the cyst contents is variable, but in uncomplicated cases is usually close to 0 HU. However, it may be as high as 25–30 HU. Increased attenuation is seen with infection and haemorrhage.

3.30.   E True
Pseudocysts usually remain peripancreatic in location but may also be found in the transverse mesocolon, in the mesentery of the small bowel or sigmoid colon, in the retroperitoneum and the pelvis. They can dissect along the psoas muscle and present as a groin mass. They may also extend into the spleen or liver, especially under the surface of the left lobe. Extension into the thorax via the aortic or oesophageal hiatus is recognized, and presentation as a neck mass has been reported.

(Reference 20)

3.31.   A True
Microcystic adenoma is a benign pancreatic neoplasm that is most commonly seen in middle-aged and elderly females. Pathologically, it consists of multiple cysts that range from 1 mm to 2 cm in diameter. When the cysts are small, ultrasound shows a homogeneous, solid, well-defined mass that is commonly hyperechoic relative to the pancreas. CT may demonstrate a solid, hypodense mass that shows marked, homogeneous enhancement on postcontrast scans.

3.31.   B True
Microcystic adenoma characteristically has a central, stellate fibrous scar, which may occasionally calcify. When the cysts are larger, ultrasound demonstrates a multicystic, lobulated mass with a hyperechoic central zone. In

this form, enhancement will be seen in the solid portions of the mass on postcontrast CT. The tumour can arise in any part of the pancreas and has a mean diameter of 10 cm.

3.31.   C False
Mucinous (macrocystic) adenoma is a potentially malignant tumour of the pancreas that occurs most commonly in middle-aged women. The tumour is seen in the tail or body of the gland in 70–90% of cases and in the head in the remainder. Ultrasound more commonly demonstrates a multilocular, cystic mass with good through transmission of sound. Internal septa and nodular or papillary projections are also seen.

3.31.   D False
Plain radiographs may demonstrate tumour calcification in 16% of cases, with the calcification occurring peripherally in the cyst walls. CT of mucinous cyst-adenomas typically demonstrates a well-defined round or oval cystic mass with density values close to water. Multiple small cysts may be seen along the internal surface of the main cyst. The cyst walls enhance on postcontrast CT. In the absence of invasion or metastases, benign and malignant variants cannot be differentiated.

3.31.   E True
Metastases from lung, ovary and melanoma can all produce cystic lesions. Other causes of cystic pancreatic neoplasms include the mucinous duct ectatic type of mucinous cystadenoma (typically located in the uncinate process), papillary cystic tumour (typically occurring in young women and commonly located in the tail), and rarely, cystic islet cell tumours, and lymphangioma.

(Reference 21)

3.32.   A False
Pancreatic adenocarcinoma accounts for approximately 95% of primary malignant pancreatic tumours. Calcification rarely occurs in these tumours but a carcinoma may arise in a calcified gland.

3.32.   B False
Adenocarcinomas are usually isodense to the remainder of the gland on precontrast CT, and therefore are recognized when they cause a focal or generalized deformity of the gland contour. Several CT features should be regarded as suspicious for the presence of a tumour when a mass is not seen. With ageing, fatty replacement results in reduced density of the gland and an area of homogeneous soft-tissue density is suggestive. Dilatation of the main pancreatic duct in the tail and body but not in the neck, and rounding of the uncinate process are also suspicious findings.

3.32.   C True
Approximately 60% of pancreatic adenocarcinomas arise in the head of the gland, with 15% and 5% occurring in the body and tail respectively. The gland may be diffusely enlarged in 20% of cases. Those arising in the head tend to

present earlier due to involvement of the CBD. On postcontrast CT, most tumours enhance to a lesser degree than the rest of the gland.

3.32. D False
Because of the pancreas's rich lymphatic supply and lack of capsule, metastases to regional lymph nodes (peripancreatic, periaortic, pericaval and periportal) occur early in the disease. Encasement of the superior mesenteric artery (SMA) manifests on CT as loss of the normal fat plane between the artery and the body of the gland, and although strongly suggestive of carcinoma, it is not diagnostic.

3.32. E True
Tumours located in the pancreatic head may produce changes in the body and tail that are demonstrable by CT. These include pancreatic duct dilatation, cyst due to obstruction, associated pancreatitis and parenchymal atrophy. CT may also demonstrate hepatic metastases, dilatation of the CBD, and splenic vein occlusion.

(Reference 2)

3.33. A False
Pancreatic adenocarcinoma most commonly causes CBD obstruction in its distal third (the intrapancreatic portion). The obstruction is usually abrupt and rounded (cigar-shaped) but the terminal portion of the duct may be smoothly tapered or conical. A nipple or beak at the point of obstruction is especially characteristic. Biliary tract obstruction can also occur at higher levels, anywhere from the common hepatic duct to the porta hepatis. This may be due to direct tumour extension or nodal metastases. Chronic pancreatitis typically results in long incomplete strictures.

3.33. B False
Although complete obstruction to the flow of contrast is frequent at ERCP, the CBD is usually patent though encased by tumour, and guide wires for drainage procedures can almost always be passed. Occasionally metastatic disease to the head of the pancreas can result in bile duct obstruction or encasement that is indistinguishable from primary adenocarcinoma.

3.33. C False
The 'double duct sign' (obstruction of both common bile and main pancreatic ducts) may also be seen with benign disease, particularly chronic pancreatitis. However, if the individual appearances of the obstructed ducts is typical of malignancy, and the obstructed or encased parts of the ducts are separated by less than 10 mm, the likelihood of carcinoma is very high, with metastases and lymphoma being the only reasonable differential diagnoses.

3.33. D True
In carcinoma, the duct distal to the lesion (between the ampulla and the tumour) may be a little dilated but otherwise appears normal, whereas it will show changes of pancreatitis when the obstruction is due to the latter

condition. The terminal portion of the duct is usually irregular in carcinoma but smooth in pancreatitis. Both pancreatic metastases and lymphoma can cause features identical to carcinoma.

3.33.   E True
The second most common ERCP finding is localized encasement of the main pancreatic duct. The encasement is generally 1–2 cm long and irregular, with a sharp transition from the normal or slightly dilated distal duct. The main duct and side branches proximal to the stricture are dilated and may show the changes of pancreatitis. Disorganization of side branches and pooling of contrast medium are other features.

(Reference 20)

3.34.   A False
On intraoperative ultrasound (IOUS), the normal pancreatic parenchyma is of high echogenicity and most islet cell tumours appear relatively hypoechoic. However, approximately 10% are either isoechoic or hyperechoic. This may be related to the normal increase in parenchymal echogenicity with ageing, and it has been noted that iso- and hyperechoic lesions are found in patients under 30 years of age.

3.34.   B True
Isoechoic islet cell tumours may only be identified at IOUS due to the presence of an echogenic halo, focal calcification, or distortion of the contour of the gland. Hyperechoic tumours may be associated with distal acoustic shadowing.

3.34.   C False
Approximately 15% of islet cell tumours are non-functioning, causing symptoms due to their size or to metastases. Consequently, they are large at the time of diagnosis, with 30% being over 10 cm in diameter. Unlike adenocarcinoma, islet cell tumours do not show central necrosis due to their vascularity. Other differentiating features include the presence of calcification (occurring in 20% of cases, compared to less than 2% in adenocarcinoma), hyperdensity on postcontrast scans and hyperdense liver metastases.

3.34.   D True
Functioning islet cell tumours included insulinoma, gastrinoma, VIPoma, somatostatinoma and glucagonoma. These tumours are usually small at diagnosis, with 90% being less than 2 cm in diameter. Also, 60% of gastrinomas and 10% of insulinomas are multiple. On precontrast CT, they may cause no distortion of the gland outline, but characteristic hyperdensity on dynamic postcontrast scans allows them to be seen.

3.34.   E True
The incidence of malignancy varies according to the type of tumour, with approximately 90% of glucagonomas, 60% of gastrinomas, VIPomas and somatostatinomas, and 10% of insulinomas being malignant. Also malignant tumours tend to be larger than benign tumours.

(References A, B, 21; C, D, 2; E, 1)

3.35. A True
The normal pancreas is of medium signal intensity on T1-WSE scans, being isointense or slightly hypointense to the liver. On T2-WSE scans the gland is slightly hyperintense to the liver. The normal pancreatic duct is difficult to visualize but may be seen in thin patients using high-resolution surface coils. The common bile duct can be seen in the head of the pancreas as a hyperintense 'dot' on T2-WSE scans.

3.35. B False
If there is atrophy with fatty replacement of the pancreas, which is a common degenerative process seen in the elderly, the intensity of the gland will increase, becoming similar to that of the surrounding fat. Tissue contrast between the pancreas and retroperitoneal fat is better on T1-WSE scans compared to T2-WSE scans.

3.35. C True
Islet cell tumours may appear hyperintense on T2-WSE scans due to their hypervascularity. Other causes of hyperintense areas include tumour necrosis and haemorrhage. Pancreatic adenocarcinomas typically have prolonged T1 and T2 relaxation times compared to the normal gland, but in general magnetic resonance signal intensities are not reliable in the differentiation of normal and diseased pancreatic tissue and in distinguishing between inflammatory and neoplastic lesions.

3.35. D True
Uncomplicated pancreatic pseudocysts have similar signal intensities to water, appearing uniformly hypointense on T1-WSE scans, and hyperintense on T2-WSE scans. Their T1 signal intensity may increase if they have a high protein content, or if they are complicated by haemorrhage or infection. In these situations, their signal intensity may overlap with that of a solid lesion.

3.35. E False
Fluid collections associated with non-haemorrhagic acute pancreatitis typically have low-to-medium signal intensity on T1-WSE scans, whereas effusions associated with haemorrhagic pancreatitis may show areas of high signal on T1-WSE scans due to the presence of subacute haematoma.

(References A, D, 3; B, E, 2; C, 2 and 3)

3.36. A False
The spleen, like the liver, has a bare area that is not covered by peritoneum. This corresponds to approximately 2–3 cm of the surface that lies between the leaves of the splenorenal ligament. Ascites and other intraperitoneal, left upper quadrant fluid collections tend to cover all surfaces of the spleen except this small area.

3.36. B False
Dynamic scanning following a rapid intravenous bolus of contrast medium commonly results initially in a heterogeneous pattern of splenic enhancement, due to variable blood flow in different areas of the spleen. If scanning is delayed for a minute or more, or if the contrast medium is administered by a slow infusion, enhancement of the spleen is more uniform.

3.36.   C False
On T1-WSE MRI, the signal intensity of the spleen is less than that of the liver and slightly greater than that of muscle. The spleen is well contrasted with the surrounding hyperintense fat. On T2-WSE scans, the spleen is hyperintense relative to the liver.

3.36.   D True
The splenic index is calculated by multiplying the length, width and depth of the spleen, each expressed in centimetres. The length is calculated by summating the number of CT slices on which the spleen can be visualized. The width is taken as the maximum splenic diameter that can be measured on any transverse image and the depth is taken at the splenic hilum. Using these criteria, a normal splenic size corresponds to an index of $120–480 \, cm^3$.

3.36.   E False
Accessory spleens occur in up to 40% of the population and are usually located near the splenic hilum, but may also be seen in the suspensory ligament or the pancreatic tail. Rarer sites include the gastric or intestinal wall, the greater omentum or mesentery, or the pelvis and scrotum. The pre- and postcontrast CT characteristics of splenunculi are similar to those of the spleen. Splenunculi may hypertrophy following splenectomy, reaching 5 cm or more.

(Reference 2)

3.37.   A False
Gallium-67 citrate may demonstrate a splenic abscess as a focal area of increased activity, but is non-specific since gallium can localize in a variety of tumours also. Absence of gallium uptake in a lesion does not exclude abscess, since an infected infarct may appear 'cold'. Abscesses can cause a focal defect in the spleen on sulphur colloid scans, or may result in splenomegaly, both non-specific findings.

3.37.   B True
Splenic candidiasis typically occurs in association with hepatic infection (see question 3.8E). Patients are usually immunocompromised, and such infections are particularly common in leukaemic patients on treatment. As with liver lesions, the hypoechoic lesions may contain a hyperechoic centre representing fungus growing within the microabscess. The spleen may also appear diffusely hypoechoic.

3.37.   C False
CT in patients with splenic candidiasis demonstrates multiple, small hypodense lesions that do not enhance on postcontrast scans. On ^{99m}Tc-sulphur colloid scans, hepatosplenomegaly is seen together with multiple small defects in the liver and spleen. The microabscesses may show either increased or decreased activity on gallium scans.

3.37.   D False
On CT, a splenic abscess is seen as an irregular area of low density that may be focal, or may replace most of the splenic parenchyma. Gas is seen in the minority of abscesses and may occasionally be seen in the splenic or portal

veins. On postcontrast CT, the abscess will not enhance and becomes better defined.

3.37. E True
Echinococcal infection of the spleen is usually associated with lung and/or liver involvement. The resulting cysts may be solitary or multiple, and can demonstrate peripheral, ring-like calcification. The presence of multiple cysts, daughter cysts within the main lesion, and cysts elsewhere in the body indicate the diagnosis.

(Reference 22)

3.38. A False
Epidermoid cysts represent 'true' congenital cysts of the spleen and are characterized pathologically by an epithelial lining, allowing differentiation from 'false' or post-traumatic cysts. Both types of cyst have similar imaging features. However, epidermoid cysts usually present in children or young adults, are well-defined and may have a trabeculated wall. Calcification is an uncommon feature.

3.38. B False
Both epidermoid and post-traumatic cysts are large lesions with a mean diameter of 13 cm. Post-traumatic cysts tend to present in older individuals and typically have smooth inner walls. Curvilinear or punctate calcifications may be seen in 50% of these cysts. Ultrasound may show some internal echoes due to debris. No enhancement is seen on CT.

3.38. C True
Haemangioma is the commonest primary neoplasm of the spleen and is usually small and single. Multiple lesions are referred to as splenic haemangiomatosis and may be part of a generalized angiomatosis also involving the liver and bones. Ultrasound demonstrates single or multiple hyperechoic masses in the case of purely solid lesions, or complex solid-cystic lesions, due to a combination of necrosis, haemorrhage and serous fluid. Calcification may be seen.

3.38. D True
CT may demonstrate two patterns in cases of splenic haemangioma: solid and cystic. Both the former and the solid parts of the latter type are hypodense or isodense on precontrast CT, but show peripheral to central enhancement on postcontrast CT, as with liver lesions. MRI also demonstrates similar features to liver haemangiomas, with the lesion being hypointense on T1- and hyperintense on T2-WSE scans.

3.38. E True
Splenic lymphangiomatosis is a very rare condition that classically involves almost the entire spleen. Occasionally, a solitary lesion (lymphangioma) will be seen. Cyst wall calcification is a feature. Ultrasound demonstrates multiple hypoechoic masses that may contain septations and debris. CT demonstrates a single or multiple hypodense, non-enhancing masses.

(Reference 22)

3.39.   A False
As a rule, splenomegaly is not a reliable sign of splenic involvement by lymphoma. Up to one-third of patients with non-Hodgkin's lymphoma and splenomegaly have no involvement histologically. Also, in the absence of splenomegaly, approximately one-third of patients with any type of lymphoma will have microscopic splenic disease. If the liver is involved, the likelihood of splenic involvement increases, even if no splenic involvement can be recognized radiographically.

3.39.   B True
Four appearances of splenic lymphoma have been described on gross pathology; (1) homogeneous enlargement without masses; (2) miliary masses; (3) 2–10 cm masses; and (4) a large solitary mass. Ultrasound usually shows a single or multiple hypoechoic masses of variable size that are often poorly defined. Anechoic and hyperechoic lesions have been described.

3.39.   C True
Serial gallium imaging can be used to monitor the progress of therapy and to detect disease recurrence. However, there are several major pitfalls to gallium imaging in lymphoma. Not all sites accumulate gallium, and lesions smaller than 1–2 cm may not be demonstrated. Also, gallium uptake is not specific for lymphoma, and activity in overlying colon and liver may result in problems of interpretation.

3.39.   D True
CT in patients with splenic involvement by lymphoma may show a normal sized spleen, multiple focal areas of reduced attenuation, or just splenomegaly (the commonest appearance). Small splenic deposits may be demonstrated with increased sensitivity using contrast agents such as ethiodol oil emulsion-13 (EOE 13). Splenic lymphoma often appears as foci of increased intensity on T2-WSE MRI.

3.39.   E False
Splenic metastases may or may not cause splenomegaly. CT demonstrates single or multiple low-density areas, which may have central regions of lower density due to necrosis. The commonest primary tumours to metastasize to the spleen are melanoma, lung and breast carcinoma.

(References A, B, C, 22; D, E, 2)

3.40.   A True
Causes of splenic infarction include sickle cell disease, embolization in patients with cardiac valve lesions, atherosclerosis, splenic artery aneurysm, splenic torsion and mass lesions such as pancreatic carcinoma. The ultrasound features vary with the age of the infarction, being hypoechoic acutely and becoming more echogenic with time.

3.40.   B True
In the hyperacute stage of splenic infarction (day 1), CT usually shows an area of reduced attenuation on precontrast scans that may enhance in a patchy fashion. However, a large hyperdense lesion may also be seen precontrast. With time, the infarct becomes better defined and hypodense. A scar may

eventually form, causing a defect in the splenic contour. Infarcts may be wedge-shaped and peripheral, irregular in outline and multiple.

3.40.   C True
Sickle cell disease may result in many splenic abnormalities. In homozygous patients, multiple splenic infarctions usually result in autosplenectomy and complete loss of splenic function by the age of 5 years. Hyperdensity of the spleen on precontrast CT may be due to a combination of diffuse calcification and increased iron deposition due to haemolysis and multiple transfusions.

3.40.   D True
Subcapsular haematomas on postcontrast CT are seen as peripheral areas of low attenuation that may cause flattening or bulging of the splenic contour. Perisplenic haematomas appear as low-density collections that surround the spleen. CT in cases of splenic trauma may also show lacerations, which appear as low-density bands with associated splenic parenchymal defects and free intraperitoneal blood.

3.40.   E True
Causes of hyposplenism associated with splenic atrophy include coeliac disease, dermatitis herpetiformis, inflammatory bowel disease and thyrotoxicosis. Hyposplenism associated with a normal or enlarged spleen may be seen in sarcoidosis and amyloidosis. Hyposplenism is best demonstrated by scintigraphy using ^{51}Cr-labelled RBCs. The normal blood clearance half-time is 10–16 minutes, but may be increased to several hours in functionally asplenic patients.

(Reference 22)

4 Gastrointestinal tract and abdominal cavity

4.1. Concerning imaging in the assessment of chronic sialadenitis
A on sialography, a beaded appearance to the main duct is a feature.
B on sialography, areas of non-filling of the gland parenchyma are most commonly due to associated neoplasm.
C parotid duct calculi, when present, are typically located at the opening of the main duct.
D calculi are the commonest cause of obstruction.
E on postcontrast CT, an enhancing mass is a recognized finding.

4.2. Which of the following are true of Sjögren's syndrome?
A Bilateral salivary gland involvement is usual.
B Punctate sialectasis is a recognized finding at sialography.
C The main parotid duct is normal at sialography in the majority of cases.
D CT typically demonstrates bilateral parotid gland enlargement.
E Lymphoma is a recognized cause of a mass lesion seen on CT.

4.3. Concerning the imaging features of parotid gland neoplasms
A malignant tumours usually have ill-defined borders.
B pleomorphic adenomas are usually hyperdense to the normal glands on precontrast CT.
C a tumour demonstrated in the superficial lobe of the parotid gland is most likely benign.
D bilateral parotid gland masses are a recognized finding in adenolymphoma.
E on CT, anteromedial displacement of the parapharyngeal fat plane is a recognized feature of parotid tumours.

4.4. Concerning imaging of the salivary glands
A the parotid and submandibular glands are isointense on T1-WSE MRI.
B sialosis is associated with increased attenuation of the parotid glands on precontrast CT.
C infiltration of the masseter muscle is a recognized feature of parotid haemangioma.
D multiple cysts in the parotid glands are a recognized finding in AIDS.
E the CT attenuation of the parotid glands typically increases with age.

4.5. Concerning scintigraphy in the assessment of the salivary glands
A following intravenous ^{99m}Tc-pertechnetate, uptake by the sublingual glands is identified in the majority of cases.
B salivary gland obstruction can be demonstrated following intravenous ^{99m}Tc-pertechnetate.
C recurrent parotitis is a cause of absent uptake of ^{99m}Tc-pertechnetate.
D uptake of ^{99m}Tc-pertechnetate is a feature of adenolymphoma.
E bilateral parotid uptake of gallium-67 citrate is a recognized finding in sarcoidosis.

4.6. Concerning scintigraphy in the assessment of the gastrointestinal tract

A the oesophageal transit test (OTT) should be performed with the patient standing.

B activity in the oesophagus following intravenous ^{99m}Tc-pertechnetate is a feature of Barrett's oesophagus.

C duodenogastric reflux of bile can be demonstrated during an HIDA scan.

D following intravenous ^{99m}Tc-pertechnetate, activity within a Meckel's diverticulum typically precedes that in the stomach.

E following ^{99m}Tc-pertechnetate, activity in the pelvis (apart from that in the bladder) indicates a Meckel's diverticulum.

4.7. Concerning CT in the assessment of the oesophagus

A identification of gas in the lumen is usually abnormal.

B with optimal distension, a wall thickness of 8 mm is considered within normal limits.

C absence of a visible fat plane between the oesophagus and trachea is abnormal.

D concentric thickening of the oesophageal wall is a typical appearance of carcinosarcoma.

E leiomyomas are typically well-defined masses.

4.8. Concerning CT in the assessment of squamous carcinoma of the mediastinal oesophagus

A the primary tumour is usually not identified by CT.

B displacement of the trachea away from the spine by a tumour mass is a recognized sign of airway invasion.

C invasion of the aorta is identified in most cases at diagnosis.

D demonstration of a fat plane between the mass and the pericardium on all sections excludes pericardial invasion.

E a soft-tissue mass in the gastrohepatic ligament is a recognized finding.

4.9. Concerning CT in the assessment of the stomach

A assessment of the gastric antrum is best in the left lateral decubitus position.

B with optimal gastric distension, the stomach wall should measure less than 5 mm in thickness.

C the wall thickness of the gastric cardia typically appears greater than that of the body.

D the gastrohepatic ligament is located adjacent to the lesser curvature of the stomach.

E any soft-tissue density structure greater than 8 mm in diameter in the gastrohepatic ligament is diagnostic of lymphadenopathy.

4.10. Concerning CT in the assessment of benign gastric lesions

A ulceration is not a feature of leiomyoma.

B leiomyomas usually enhance on postcontrast scans.

C gastric duplications typically fill with orally administered contrast medium.

D gastric wall thickening in Ménétrier's disease typically involves the antrum.

E emphysematous gastritis is characterized by linear streaks of intramural air.

4.11. Concerning CT in the assessment of adenocarcinoma of the stomach
A   the commonest finding is an area of wall thickening.
B   enhancement on postcontrast CT is a feature of linitis plastica.
C   poor definition of the serosal surface of the stomach is a feature of transmural extension.
D   lymphadenopathy in the hepatoduodenal ligament is a recognized finding.
E   a mass in the transverse mesocolon is a recognized finding.

4.12. Which of the following are true concerning the CT findings in gastric malignancies?
A   Diffuse thickening of the entire stomach wall is more suggestive of lymphoma than adenocarcinoma.
B   Gastric lymphoma is usually associated with regional lymphadenopathy.
C   Calcification is a recognized feature of leiomyosarcoma.
D   Metastasis to the periaortic lymph nodes is characteristic of leiomyosarcoma.
E   Diffuse thickening of the gastric wall is a recognized feature of metastases to the stomach.

4.13. Concerning the imaging findings in the oesophagus and stomach in a patient with AIDS
A   mucosal fold thickening is a recognized finding in candidal oesophagitis
B   diffuse mucosal ulceration is typical of cytomegalovirus (CMV) oesophagitis.
C   mucosal destruction is characteristic of oesophageal Kaposi's sarcoma.
D   wall thickening at the gastro-oesophageal junction is a recognized CT finding.
E   diffuse thickening of the stomach wall is a recognized CT finding.

4.14. Concerning CT in the assessment of the duodenum
A   with maximal distension, the upper limit of wall thickness is 4 mm.
B   wall thickening adjacent to an intraluminal mass is not a typical feature of benign neoplasms.
C   primary adenocarcinoma is typically associated with local lymph node enlargement at diagnosis.
D   encasement of the duodenum is a recognized finding with peripancreatic lymphadenopathy.
E   intramural haematoma is a recognized cause of inhomogeneous thickening of the duodenal wall.

4.15. Concerning CT in the assessment of the small bowel
A   intussusception characteristically appears as a homogeneous soft-tissue density mass.
B   primary adenocarcinoma is the commonest cause of a mass involving the terminal ileum.
C   haematogenous metastases typically appear as masses on the mesenteric surface of the bowel wall.
D   lymphoma is a recognized cause of focal bowel wall thickening.
E   lobulated mesenteric masses are a recognized finding in non-Hodgkin's lymphoma.

4.16. Concerning CT in the assessment of the small bowel mesentery and the transverse mesocolon

A the mesentery normally has a CT density similar to that of subcutaneous fat (-100 to $-160\,HU$).

B the transverse mesocolon lies in the same axial plane as the uncinate process of the pancreas.

C increased attenuation of the mesenteric fat occurs in hypoalbuminaemia.

D encasement of mesenteric vessels by soft-tissue density is a feature of systemic amyloid.

E an increased volume of mesenteric fat is found in pseudotumoral lipomatosis.

4.17. Concerning CT in the assessment of the small bowel mesentery

A a spiculated mass is a feature of small bowel carcinoid.

B fat necrosis is a recognized cause of diffuse increased density of the mesentery.

C fat necrosis is a recognized cause of a calcified soft-tissue density mass.

D the small bowel mesentery is the commonest site of mesenteric cysts.

E a desmoid tumour is a recognized cause of a poorly defined soft-tissue density mesenteric mass.

4.18. Concerning the imaging findings in the duodenum, small bowel and mesentery in a patient with AIDS

A thickening of the distal ileum is a feature of CMV enteritis.

B cryptosporidiosis typically produces fold thickening in the distal ileum.

C small bowel dilatation is a recognized feature of *Mycobacterium avium intracellulare* (MAI) enteritis.

D small bowel Kaposi's sarcoma is characterized by submucosal nodules with normal intervening mucosal folds.

E retroperitoneal lymph nodes greater than 2 cm in size are indicative of AIDS-related lymphoma.

4.19. Concerning CT in the assessment of Crohn's disease

A disease limited to the small bowel mucosa will be demonstrated in the majority of patients.

B bowel wall thickening indicates active disease.

C an inner ring of low attenuation within the bowel wall is a specific feature.

D soft-tissue density in the perirectal fat is a recognized finding.

E increased volume of mesenteric fat is a recognized finding.

4.20. Concerning CT in the assessment of ulcerative colitis

A the thickened colonic wall typically has homogeneous attenuation.

B a 'target' appearance to the rectum is a recognized finding.

C rectal wall thickening is typically greater than 2 cm.

D enlarged perirectal lymph nodes are a recognized feature.

E gas within the colonic wall is a recognized finding.

4.21. Concerning scintigraphy in the assessment of inflammatory bowel disease (IBD) and intra-abdominal sepsis
A   uptake of gallium-67 citrate by the colon indicates active IBD.
B   imaging of IBD with ^{111}In-labelled WBCs should be commenced 24 hours after injection.
C   ^{111}In-labelled WBCs can differentiate abscess from a focus of active IBD.
D   steroid treatment is a recognized cause of a false negative ^{111}In-labelled WBC study.
E   renal activity within 48 hours of gallium-67 citrate injection is a normal finding.

4.22. Concerning CT in the assessment of the colon
A   with optimal distension, the upper limit of colonic wall thickness is 4 mm.
B   colonic dilatation is a finding in pseudomembranous colitis.
C   pseudomembranous colitis is associated with enhancement of the colonic wall on postcontrast scans.
D   diminished enhancement of the colonic wall is a feature of ischaemic colitis on postcontrast scans.
E   wall thickening of the descending colon is a typical finding in typhlitis (neutropenic colitis).

4.23. Concerning CT in the assessment of diverticular disease
A   the colonic wall is typically thicker than 4 mm.
B   increased density of the pericolic fat indicates complicating carcinoma.
C   gas density anterior to the liver is a recognized finding.
D   caecal diverticulitis typically results in focal thickening of the caecal wall.
E   CT guided transgluteal pelvic abscess drainage is ideally performed below the level of the sacrospinous ligament.

4.24. Concerning CT in the assessment of the appendix
A   the normal appendix cannot be identified.
B   in acute appendicitis, increased density of the pericaecal fat is the commonest CT abnormality.
C   gaseous distension of the appendix is a feature of gangrenous appendicitis.
D   a pericaecal phlegmon typically has water density.
E   a mucocele of the appendix typically has soft-tissue density on precontrast scans.

4.25. Concerning CT and MRI in the assessment of rectosigmoid carcinoma
A   presence of gas within a large bowel mass indicates diverticular abscess rather than carcinoma.
B   loss of fat planes between a carcinoma and adjacent muscle indicates invasion.
C   following abdominoperineal resection (APR), postoperative fibrosis is characterized by a streaky soft-tissue density in the presacral space.
D   recurrent rectal carcinoma following APR usually occurs within 2 years.
E   following APR, a presacral mass that enhances with Gd-DTPA on T1-WSE MRI is more likely to be tumour than fibrosis.

4.26. Concerning CT in the assessment of malignant tumours of the anal canal

A the primary tumour usually appears as a hyperdense mass on postcontrast scans.
B gas in the perineum is a recognized finding.
C absence of a fat plane between the tumour and the vagina indicates invasion.
D enlarged superficial inguinal nodes are a feature.
E common iliac lymphadenopathy is a recognized feature.

4.27. Concerning the imaging findings in the large bowel in patients with AIDS

A toxic megacolon is a recognized feature of CMV colitis.
B diffuse mucosal granularity is a feature of CMV colitis.
C reduced density of the colonic wall on CT is a recognized finding in CMV colitis.
D widespread colonic ulceration is a typical finding in colonic Kaposi's sarcoma.
E Perirectal masses are a recognized CT finding.

4.28. Concerning ultrasound of the gastrointestinal tract

A the bowel submucosa normally appears echogenic.
B the ileum and jejunum cannot be distinguished.
C the thickness of the bowel wall is normally less than 5 mm.
D bowel wall thickening invariably results in loss of the normal bowel wall layers.
E a long segment of bowel wall thickening excludes neoplasm as the cause.

4.29. Concerning ultrasound in the assessment of inflammatory conditions of the gastrointestinal tract

A acute appendicitis cannot be diagnosed in the absence of an appendicolith.
B an appendix diameter of 4 mm indicates appendicitis.
C in acute diverticulitis, oedema of the pericolic fat appears as a hypoechoic mass.
D inflamed and non-inflamed diverticula cannot be differentiated.
E an echogenic mesenteric mass is a recognized finding in Crohn's disease.

4.30. Concerning ultrasound in the assessment of the gastrointestinal tract

A ulceration in a gut wall mass cannot be demonstrated
B mucosal ulceration in ulcerative colitis can be demonstrated.
C a complex solid/cystic gut wall mass is a recognized appearance of leiomyoma.
D lymphoma of the bowel typically results in a hypoechoic mass.
E a 'pseudokidney' sign is diagnostic of carcinoma.

4.31. Which of the following are true concerning transrectal endosonography?

A   Primary rectal carcinoma typically appears as a superficial hypoechoic mass.

B   Normal sized perirectal lymph nodes can be identified in the majority of patients.

C   Microscopic metastases can be demonstrated in involved perirectal lymph nodes in 80% of cases.

D   Local recurrence of rectal carcinoma typically appears as an area of mucosal destruction.

E   Metastases to the rectum typically involve the deep layers of the rectal wall.

4.32. Concerning the imaging of intra-abdominal lymph nodes

A   CT demonstrates metastases in normal sized nodes in 50% of cases.

B   on T1-WSE MRI, normal nodes are isointense to muscle.

C   calcification within enlarged nodes indicates a benign aetiology.

D   mesenteric lymph node involvement by tumour cannot be identified by lymphangiography.

E   low-density nodes are invariably malignant.

4.33. Concerning CT in the assessment of intra-abdominal lymphadenopathy

A   retrocrural lymphadenopathy is most commonly due to ovarian carcinoma.

B   retroperitoneal lymph nodes should be considered enlarged if measuring more than 10 mm in short-axis diameter.

C   gastrohepatic lymphadenopathy is a recognized finding in carcinoma of the pancreas.

D   portal lymph nodes are located entirely anterior to the portal vein.

E   portal lymphadenopathy is a recognized finding in primary biliary cirrhosis.

4.34. Concerning CT of intra-abdominal lymph nodes

A   superior mesenteric artery nodes are considered enlarged if measuring over 10 mm in short-axis diameter.

B   pancreaticoduodenal nodes lie between the duodenum and pancreatic head.

C   pancreaticoduodenal nodes are considered enlarged if measuring 5 mm in short-axis diameter.

D   perisplenic lymphadenopathy is a recognized feature of metastatic carcinoma of the stomach.

E   lymphadenopathy in the small bowel mesentery is a typical feature of carcinoma of the sigmoid colon.

4.35. Concerning CT in the assessment of intra-abdominal tumour spread

A   metastases to the omentum result in soft-tissue density between the bowel and anterior abdominal wall.

B   pancreatic carcinoma is a cause of a mass encasing the splenic flexure of the colon.

C   intraperitoneal seeding of tumour most commonly involves the left paracolic gutter.

D   intraperitoneal seeding of tumour is invariably associated with ascites.

E   pseudomyxoma peritonei typically results in multiple, enhancing masses on postcontrast scans.

4.36. Concerning the imaging of intra-abdominal fluid collections
A   CT can differentiate benign and malignant ascites by differences in attenuation values.
B   benign transudative ascites typically results in a relatively large lesser sac collection.
C   in a supine patient, fluid initially collects in the paracolic gutters.
D   biloma is a recognized cause of a left upper quadrant fluid collection.
E   transudative ascites has low signal intensity on T1-WSE MRI scans.

4.37. Which of the following are true of imaging of the psoas muscle?
A   The cephalad portion of the psoas major normally appears round on axial imaging.
B   The psoas minor muscle appears as a mass lateral to the psoas major.
C   Lymphoma involving the muscle is typically associated with reduced density on precontrast CT.
D   Neoplastic involvement is associated with increased signal intensity on T1- and T2-WSE MRI scans.
E   On CT, reduced density within an enlarged muscle is a feature of abscess.

4.38. Concerning the CT features of primary tumours of the retroperitoneum
A   lymphangioma is a recognized cause of a homogeneous fat-density mass.
B   absence of fat density excludes a diagnosis of liposarcoma.
C   a homogeneous mass of near water density is a recognized appearance of liposarcoma.
D   leiomyosarcoma is characterized by areas of central low attenuation.
E   paraganglionomas are typically located posterolateral to the kidney.

4.39. Which of the following are true of retroperitoneal fibrosis?
A   The mass is confined to the pelvis.
B   Displacement of the aorta and IVC is a characteristic feature.
C   Multiple soft-tissue masses are a recognized finding on CT.
D   On precontrast CT, a mass that is hyperdense to muscle is a recognized finding.
E   High signal intensity on T2-WSE MRI excludes the diagnosis.

4.40. Concerning imaging of the small and large bowel following pelvic radiotherapy
A   mucosal ulceration of the small bowel is the commonest finding on small bowel barium studies.
B   separation of small bowel loops is not a feature.
C   large bowel strictures are most common at the rectosigmoid junction.
D   streaky soft-tissue density in the perirectal fat is a finding on CT.
E   increased signal intensity of the rectal submucosa on T2-WSE scans is the earliest MRI finding.

Gastrointestinal tract and abdominal cavity: Answers

4.1. A True
Sialography in cases of chronic sialadenitis typically demonstrates dilatation of the main parotid or submandibular ducts due to a combination of obstruction and true ectasia, secondary to infection. Alternating areas of dilatation and stenosis may result in a beaded appearance to the main ducts.

4.1. B False
With more severe disease, there is pruning of the peripheral ducts and acini, and in the most extreme cases only the main primary and secondary ducts are dilated and filled with contrast medium. Varying degrees of sialectasis may be demonstrated. Areas of parenchymal non-filling in patients with chronic sialadenitis are more commonly due to inflammatory masses or abscesses.

4.1. C False
Parotid duct calculi tend to be situated at the gland hilum or where the gland crosses the anterior border of the masseter muscle. About 40% of parotid duct calculi are radiolucent and may be demonstrated by CT or ultrasound.

4.1. D True
Calculi are most commonly identified in the submandibular gland and 80% of these are radio-opaque. Less common causes of duct obstruction include inflammation and stenosis of the duct orifice secondary to oral ulceration, irritation from a sharp tooth or a poorly fitting denture, or a stricture secondary to spontaneous or surgical passage of a stone.

4.1. E True
Inflammatory masses in cases of chronic sialadenitis may appear on precontrast CT as poorly defined lesions that can be difficult to differentiate from malignant tumours. However, on postcontrast scans these masses tend to enhance to a greater degree than tumour and they never extend beyond the capsule of the gland.

(References A, B, C, D, 39; E, 35)

4.2. A True
Sjögren's syndrome is a combination of benign lymphoepithelial sialadenopathy, keratoconjunctivitis sicca and a rheumatoid-type arthritis. The disease typically occurs in middle-aged females and is usually bilateral.

4.2. B True
Sialographic findings in patients with Sjögren's syndrome include varying degrees of contrast extravasation. The earliest and commonest pattern consists of punctate sialectasis, appearing as small (less than 1 mm) collections of contrast medium distributed throughout the gland.

4.2. C False

With disease progression, the collections of extraductal contrast increase in size (globular sialectasis) and become more irregular (cavitary sialectasis), with non-uniform distribution throughout the gland. The main duct may then appear deformed and dilated. The final stage (destructive sialectasis) shows contrast medium dissecting through the gland parenchyma with no recognizable peripheral ductal system.

4.2. D False

In advanced cases of Sjögren's syndrome, precontrast CT shows the parotid (and submandibular) glands to be markedly atrophic, hyperdense, and with heterogeneous attenuation and multiple calculi. Low-density cystic areas, representing sialectasis, may also be seen.

4.2. E True

Mass lesions demonstrated on ultrasound, CT or MRI may be due to lymphoma or the benign lymphoepithelial lesion. The latter consists of isolated clusters of epithelial and myoepithelial cells surrounded by a dense lymphocytic infiltrate. In Sjögren's syndrome, there is an increased risk of predominantly non-Hodgkin's lymphoma, with 1–6% of patients being affected per year.

(References A, 35; B, C, D, E, 39)

4.3. A True

Malignant parotid gland tumours may appear as relatively hyperdense well-defined masses when small but usually show an infiltrative growth pattern as they enlarge, resulting in poorly defined borders. They typically have heterogeneous echotexture and density on ultrasound and CT respectively and intermediate signal intensity on T1- and T2-WSE MRI.

4.3. B True

The parotid glands have relatively low attenuation on precontrast CT due to their high fat content and therefore mass lesions tend to appear hyperdense to the normal gland. Pleomorphic adenomas appear as well-defined masses that may have a lobulated border. They are usually hypoechoic on ultrasound and hypointense to normal gland on T1-WSE MRI, becoming hyperintense with increased T2 weighting. They are the most common benign tumours.

4.3. C True

The majority of salivary gland tumours are located in the superficial lobe of the parotid gland and approximately 80% of these are benign. Malignant tumours are more commonly located in the deep lobe of the parotid gland, the submandibular gland or the minor salivary glands. The various types of malignancy involving the salivary glands include mixed malignant tumour, mucoepidermoid carcinoma, adenoid cystic carcinoma, adenocarcinoma, squamous cell carcinoma, lymphoma and metastases.

4.3. D True

Adenolymphoma (Warthin's tumour) accounts for about 8% of all parotid tumours. On ultrasound, it is characteristically markedly hypoechoic, with occasional internal septa. On CT and MRI, it has a similar appearance to a pleomorphic adenoma. About 20% are multiple and they may be bilateral. They are most commonly found in elderly men.

4.3. E True
The deep lobe of the parotid gland extends posteriorly around the ramus of the mandible, and enlargement of a tumour causes anteromedial displacement of the parapharyngeal fat plane, allowing differentiation from other masses arising in the parapharyngeal space. As the tumour enlarges, it extends anterior to the styloid process and carotid sheath.

(Reference 39)

4.4. A False
Since the normal parotid glands have a higher fat content than the submandibular glands, they will appear relatively hyperintense on T1-WSE magnetic resonance scans. MRI has been reported to show the facial nerve within the parotid gland, allowing identification of the superificial and deep lobes of the gland. The submandibular glands have approximately muscle attenuation on precontrast CT and intermediate signal intensity on MRI.

4.4. B False
Sialosis is a benign condition associated with cirrhosis, chronic alcoholism and malnutrition. CT typically demonstrates enlarged, fat-density parotid glands containing thickened septa of soft-tissue attenuation.

4.4. C True
Haemangioma is the commonest parotid gland tumour in young children. It may be localized within the gland or infiltrate local musculature. Postcontrast CT demonstrates an enhancing lesion consisting of serpiginous vascular channels or spaces.

4.4. D True
Patients with AIDS may develop benign lymphoepithelial cysts that have similar imaging features to adenolymphoma. Kaposi's sarcoma and non-Hodgkin's lymphoma may rarely involve the salivary glands in these patients. Another cause of multiple soft-tissue-density masses in the parotid glands is sarcoidosis, which can also produce a single mass.

4.4. E False
With increasing age there is fatty replacement of the gland parenchyma resulting in reduced CT attenuation. However, with age-related fatty replacement, the glands are either of normal size or only slightly enlarged, allowing differentiation from sialosis.

(References A, 3; B, C, D, E, 39)

4.5. A False
Scintigraphy of the major salivary glands can be performed after the intravenous administration of ^{99m}Tc-pertechnetate, which is taken up and secreted by functioning salivary gland epithelium. Dynamic imaging over the upper neck will demonstrate gradually increasing uptake of isotope in the parotid and submandibular glands to a similar degree to that seen in the

thyroid. Activity in the mouth represents saliva. A small amount of activity is normally seen in the paranasal sinuses but the sublingual glands are not identified. By drawing regions of interest around the glands, time/activity curves can be obtained for each gland.

4.5. B True
Approximately 10 minutes after injection of isotope, a sialogogue is given to assess emptying of the gland. In the normal case with patent ducts, there is rapid secretion of activity and a steep downward slope in the time/activity curve. When the main duct is obstructed, the time/activity curve shows progressive increase in activity.

4.5. C True
Destruction of salivary gland epithelium results in reduced or absent uptake of ^{99m}Tc-pertechnetate. This may occur in cases of recurrent parotitis, if the gland is replaced by tumour, and in patients with Sjögren's syndrome.

4.5. D True
Tumours that contain functioning salivary gland epithelium will show uptake of ^{99m}Tc-pertechnetate. These tumours include adenolymphoma (Warthin's tumour) and benign oxyphilic adenomas. Functioning tumours are always benign. A non-functioning tumour may manifest as a photopenic area that distorts the normal tissue, or as absent uptake.

4.5. E True
A small amount of gallium-67 citrate activity in the salivary glands can be a normal finding. Increased activity in the parotid glands with associated activity in the lacrimal glands is strongly suggestive of sarcoidosis. Radiation sialitis (commonly following upper mantle radiotherapy for lymphoma) is also a cause of increased gallium-67 citrate activity, especially in the submandibular glands.

(Reference 5)

4.6. A False
The oesophageal transit test (OTT) is performed with the patient lying supine beneath a gamma camera to exclude the effects of gravity. Images are obtained dynamically as the patient performs multiple swallows. Typically, 90% of the swallowed bolus is cleared from the oesophagus within 15 seconds. The pattern of dysmotility can also be assessed. An 'adynamic' pattern is characteristic of achalasia or scleroderma, whereas diffuse oesophageal spasm typically results in an 'incoordinate' pattern. However, there is a wide variation in the pattern observed.

4.6. B True
Barrett's oesophagus represents the presence of ectopic gastric columnar epithelium in the lower third of the oesophagus, which, like other forms of ectopic gastric mucosa, will accumulate intravenously injected ^{99m}Tc-pertechnetate. However, interpretation of the results may be made difficult due to the presence of activity in swallowed saliva and refluxed gastric contents.

4.6. C True
Intravenously administered ^{99m}Tc-HIDA is taken up by the liver and passed into the biliary tract, from where it normally enters the duodenum and small bowel. Demonstration of activity in the left upper quadrant may indicate duodenogastric reflux. The test may be improved by the administration of a labelled fatty meal to outline the position of the stomach and stimulate gallbladder contraction.

4.6. D False
A Meckel's diverticulum represents a failure of obliteration of the proximal part of the vitelline duct. Its endodermal lining has the potential to differentiate into gastric, duodenal, pancreatic or colonic mucosa. Bleeding is a potential complication of those that are lined by gastric mucosa. These can be investigated using ^{99m}Tc-pertechnetate. A Meckel's diverticulum is manifest as an area of increased activity, usually in the right lower quadrant. An important diagnostic feature is that it becomes visible synchronously with the stomach.

4.6. E False
Possible causes of a false-positive Meckel's scan include activity in the uterus, renal pelvis, or activity secreted by the stomach. In the first case, uterine activity is visualized earlier than the stomach. The ureter or renal pelvis can be differentiated on lateral views due to their posterior position (a Meckel's diverticulum will appear in the mid- or anterior half of the abdomen). If secreted activity is present the scan can be repeated using a combination of an H2-antagonist, pentagastrin and glucagon.

(References A, B, D, E, 14; C, 5)

4.7. A False
The cervical oesophagus usually does not contain air when the patient is scanned supine, but air can be identified in the thoracic oesophagus in about 65% of normal cases. However, the demonstration of a fluid-filled oesophagus or an air–fluid level is unusual and may indicate the presence of a distal obstruction or motility disorder.

4.7. B False
With optimal distension, an oesophageal wall thickness greater than 3 mm is usually abnormal. A luminal diameter greater than 10 mm is also suspicious of abnormality. Oesophageal wall thickening is a non-specific finding and can be seen with neoplasm, following sclerotherapy for varices, due to reflux or infective oesophagitis, and also with oesophageal intramural pseudodiverticulosis.

4.7. C False
In the normal case, the oesophagus lies directly posterior to the trachea and left main stem bronchus. A thin fat plane may or may not be present between the oesophagus and the airway, and therefore absence of fat cannot be used as an indication of abnormality.

4.7. D False
Carcinosarcoma is a rare form of oesophageal malignancy, accounting for less than 1% of primary oesophageal carcinomas. This tumour, as well as the rare pseudosarcoma, characteristically produces large polypoid intraluminal masses. Over 95% of primary oesophageal tumours are squamous cell carcinomas, with adenocarcinomas accounting for the majority of the rest.

4.7. E True
Leiomyomas are the commonest benign tumours of the oesophagus and appear on CT as well-defined, oval or round masses that may cause eccentric thickening of the oesophageal wall. They tend to be of homogeneous soft-

tissue density and do not infiltrate the surrounding fat. They cannot be reliably distinguished from leiomyocarcomas.

(References A, B, C, E, 2; D, 38)

4.8. A False
The majority of primary oesophageal carcinomas are advanced at the time of diagnosis and CT is abnormal in virtually all cases. Findings include an intraluminal mass, wall thickening (often sufficient to cause a soft-tissue mass) and dilatation of the lumen proximal to the tumour.

4.8. B True
Tumours of the upper and middle third of the oesophagus may invade the trachea or left main stem bronchus. CT features of invasion include anterior displacement of the airway away from the spine by a tumour mass, and a posterior indentation on the airway by a tumour mass with the scan obtained in held inspiration. In expiration, the posterior border of the airway may normally be concave inward, and this sign cannot be used. Displacement of the airway as a solitary finding is not diagnostic of invasion.

4.8. C False
Aortic invasion by oesophageal carcinoma is uncommon, being identified at postmortem in only 2% of 2440 patients dying from the disease. The descending aorta normally abuts the oesophagus and, therefore, absence of a fat plane between the two structures is not a sign of invasion. An invasive carcinoma will increase the amount of contact between the tumour mass and the aorta such that if there is more than 90° of contact between the tumour and the aortic circumference, invasion is likely. Less than 45° of contact is normal.

4.8. D True
Fat planes may normally be absent between the oesophagus and the posterior pericardium. The best CT sign of pericardial invasion is if there is obliteration of fat at the level of the tumour but fat can be identified above and below the level of the tumour. If no fat is seen at all, CT is indeterminate.

4.8. E True
Tumours of the lower third of the oesophagus may spread to lymph nodes in the upper abdomen, predominantly at the level of the coeliac axis and in the gastrohepatic ligament. Liver metastases will also be identified on scans through the upper abdomen.

(References A, 2; B, C, D, E, 38)

4.9. A True
CT examination specifically aimed at the stomach can be performed after gastric distension with air, and if the site of abnormality is known the patient can be positioned to demonstrate this region optimally. Assessment of the gastric antrum and lesser curvature is best achieved in the left lateral decubitus and supine positions, respectively, while lesions in the stomach fundus, posterior gastric wall and greater curvature are best seen in the prone position.

4.9. B True
Optimal gastric distension is achieved when there is effacement of the normal mucosal folds. In this situation, the stomach wall usually measures about 3 mm in thickness, and measurements above 5 mm should be considered abnormal. The mucosal surface should appear smooth and the perigastric fat has uniform low density.

4.9. C True
The gastric cardia appears thicker than the remainder of the stomach since the cardia is orientated obliquely to the axial CT slice. The normal oesophagogastric junction can produce a smooth bulge along the medial border of the cardia, and posterior/posterolateral thickening of the wall of the cardia has been described in patients with a sliding hiatus hernia.

4.9. D True
The gastrohepatic ligament is identified on CT as a fat density structure that extends from the lesser curvature of the stomach to the medial surface of the liver. It is continuous with the fissure for the ligamentum venosum. Other perigastric ligaments identified by CT include the gastrocolic ligament (greater omentum) and gastrosplenic ligament. These represent pathways for direct extension of gastric carcinoma.

4.9. E False
The gastrohepatic ligament contains the left gastric artery, coronary vein and left gastric lymph nodes. It should not contain any soft tissue structure greater than 8 mm in diameter. Such structures may represent either enlarged lymph nodes or varices, which can be distinguished on postcontrast scans.

(Reference 38)

4.10. A False
Leiomyomas are the commonest benign tumours of the stomach, arising in the submucosa. They may grow either toward the gastric lumen or have a predominantly extragastric component. Tumours greater than 5 cm in diameter may ulcerate. Both ulceration and calcification can be seen on CT, the former as a small collection of air related to the surface of the tumour.

4.10. B True
Regardless of size, leiomyomas tend to be of homogeneous muscle density on precontrast CT scans and show uniform enhancement to about 1.5 times their precontrast density on postcontrast scans. The tumour may be pedunculated or broad-based. Leiomyoblastoma is a rare variant of leiomyoma and has similar CT features, although multiseptate cystic masses have also been reported. These tumours may rarely metastasize to the liver or lymph nodes.

4.10. C False
Gastric duplications are usually located on the greater curvature of the stomach and represent approximately 4% of all gastrointestinal duplication cysts. They rarely communicate with the stomach lumen. CT demonstrates a

near-water-density spherical mass inseparable from the stomach wall. Gastric diverticula arise usually from the posterior surface of the fundus, and if oral contrast is not administered they may be mistaken for an adrenal or other retroperitoneal mass.

4.10.   D False
Menetrier's disease is associated with thickened gastric folds up to 2.5 cm thick, located predominantly in the fundus and body of the stomach. Other benign causes of gastric wall thickening include Crohn's disease, chronic granulomatous disease of childhood, and opportunistic infections, all predominantly affecting the antrum, and also graft-versus-host disease, amyloidosis, eosinophilic gastritis and the Zollinger–Ellison syndrome.

4.10.   E False
CT in patients with emphysematous gastritis demonstrates irregular thickening of the gastric wall associated with multiple intramural gas bubbles. Linear streaks of intramural gas are typically seen in cases of non-infectious gastric emphysema, such as occurs with gastric outlet obstruction.

(References A, B, D, E, 38; C, 2)

4.11.   A True
Adenocarcinoma is by far the commonest primary malignancy of the stomach. CT appearances vary according to the gross morphology of the tumour. The commonest finding is an area of focal wall thickening. The attenuation value of the thickened wall is usually similar to that of muscle but may occasionally be of near-water density. Rarely, stippled calcifications may be identified. Other CT appearances include a plaque-like or exophytic intraluminal mass which may be ulcerated and have a broad or narrow mural attachment.

4.11.   B True
Linitis plastica represents a form of diffuse infiltrating gastric carcinoma which results in more diffuse wall thickening. Enhancement on postcontrast CT is a characteristic feature.

4.11.   C True
With gross invasion of the gastric wall and serosa, CT demonstrates a blurred serosal contour or strand-like soft tissue densities extending into the perigastric fat. The probability of transmural extension appears to correlate closely with the thickness of the gastric wall, and it has been suggested that transmural extension is almost assured when wall thickness is greater than 2 cm.

4.11.   D True
Distal gastric carcinomas can invade the pancreas and metastasize to the infrapyloric nodes and the hepatoduodenal ligament, where the CBD may be obstructed. Lesser curvature carcinomas can extend to the porta hepatis via the gastrohepatic ligament, causing high biliary tract obstruction. Other sites of

lymphatic metastases include the splenic hilum (from greater curvature lesions) and the cardiophrenic angles.

4.11. E True
Direct spread of gastric carcinoma can result via the gastrocolic ligament and less commonly the transverse mesocolon, to the transverse colon, and to the spleen via the gastrosplenic ligament. Intraperitoneal spread is particularly common from scirrhous carcinoma of the stomach and may result in diffuse enhancement of the peritoneal surfaces of the small bowel or colon on postcontrast scans. Liver and lung are the sites of haematogenous spread.

(References A, C, 38; B, 2; D, E, 2 and 38)

4.12. A True
The stomach is the commonest site of gastrointestinal tract lymphoma. Lymphoma accounts for approximately 3–5% of all primary gastric malignancies and the CT appearances can be indistinguishable from adenocarcinoma. Findings include focal or diffuse wall thickening, a mural mass and ulceration. In contrast to adenocarcinoma, lymphoma more commonly produces thickening of the entire gastric wall.

4.12. B True
The majority of patients with gastric lymphoma have clearly defined lymphadenopathy in the greater omentum, gastrohepatic ligament or gastrosplenic ligament. Widespread retroperitoneal lymphadenopathy may also be present. Lymphoma can also extend through the stomach wall to involve the perigastric fat and adjacent organs.

4.12. C True
Leiomyosarcoma accounts for approximately 1–3% of primary malignant tumours of the stomach. It has a better prognosis than adenocarcinoma. CT features include a relatively large spherical or ovoid mass (up to 15 cm in diameter) with a significant extragastric component. Areas of low attenuation representing necrosis are frequent. Deep ulceration and communication with the gastric lumen can also be identified, as can gas or calcification within the mass.

4.12. D False
Leiomyosarcomas typically metastasize via a haematogenous route to the lungs, liver and bone. Direct extension to adjacent organs and intraperitoneal seeding are also seen, but metastasis to perigastric and para-aortic lymph nodes is unusual.

4.12. E True
Metastases to the stomach may be blood-borne, most commonly from malignant melanoma, lung and breast carcinoma. CT may demonstrate ulcerated or excavated mural nodules. Breast carcinoma metastatic to the stomach may produce diffuse thickening of the stomach wall that is indistinguishable from primary adenocarcinoma. The stomach may also be

involved by direct spread of tumour from the pancreas, oesophagus or transverse colon.

(References A, C, D, E, 38; B, 2)

4.13. A True
Candidal infection is the commonest cause of oesophagitis in AIDS patients. Classically, on double-contrast barium studies, there is diffuse ulceration resulting in a 'shaggy' outline to the oesophagus. Other recognized findings include cobblestoning, plaques and thickened folds. Plaques may result in longitudinally orientated, linear filling defects.

4.13. B False
CMV oesophagitis may display several unique appearances that can allow differentiation from candidal oesophagitis. A frequent finding is a well-defined, diamond-shaped ulcer surrounded by a lucent region due to oedema, against a background of normal oesophageal mucosa. Other characteristic features include ulceration at the gastro-oesophageal junction and the formation of giant oesophageal ulcers, due to a combination of infection and ischaemic necrosis of the mucosa. Discrete ulceration may also be seen with herpes simplex oesophagitis.

4.13. C False
Lesions of Kaposi's sarcoma (KS) are submucosal in origin and appear on double-contrast barium studies as discrete submucosal nodules along the length of the oesophagus. Lesions may also be identified in the hypopharynx.

4.13. D True
CMV is the commonest cause of gastritis in AIDS patients and has a predilection for the gastro-oesophageal junction and the juxtapyloric antrum. CT demonstrates mural thickening at these sites. Thickening at the gastro-oesophageal junction may be associated with increased density of the lesser omental fat. Progressive disease can result in stricture formation at both sites.

4.13. E True
Gastric KS is usually associated with skin lesions. Barium studies demonstrate well-defined submucosal nodules usually measuring from 0.5 to 2.0 cm. The overlying mucosa may be breeched but the intervening mucosa is normal. Progressive infiltration of the submucosa results in gross thickening of the stomach wall, which is a non-specific finding that may also occur with lymphoma.

(Reference 7)

4.14. A False
With optimal distension of the duodenal lumen, the wall should not measure more than 1 mm in thickness. Gas contrast CT examination of the duodenum can be performed by scanning the patient in the left lateral decubitus position following administration of an effervescent agent and intravenous Buscopan or glucagon.

4.14. B True
Benign tumours of the duodenum include leiomyoma, Brunner's gland adenoma and adenomatous polyp. CT typically demonstrates a soft-tissue density intramural or intraluminal mass. The absence of adjacent wall thickening is a useful sign in differentiating benign from malignant lesions.

4.14. C False
Primary adenocarcinoma of the duodenum appears on CT as an intrinsic mass with associated wall thickening but usually without local lymph node enlargement. Liver metastases may be present. Lymphoma of the duodenum causes annular or eccentric wall thickening accompanied by bulky regional and more widespread lymphadenopathy.

4.14. D True
Metastatic disease involving the duodenum is most commonly due to direct involvement by adjacent carcinomas, particularly of the head of the pancreas. Right colon carcinomas may extend to the duodenum via the transverse mesocolon, and gastric and biliary primaries can also invade the duodenum. Duodenal encasement by peripancreatic lympadenopathy is most commonly due to lymphoma or metastatic lung or breast carcinoma.

4.14. E True
CT findings following blunt trauma to the duodenum include fluid or gas in the right anterior pararenal space. Intramural haematomas appear as a region of inhomogeneous wall thickening with narrowing of the duodenal lumen. A curvilinear area of high density may be apparent just inside the margin of the mass, related to the site of acute or subacute haemorrhage.

(Reference 38)

4.15. A False
Intussusception may have a variable CT appearance depending upon the severity and duration of the abnormality. Initially CT demonstrates a 'target' sign. This can progress to a sausage-shaped mass with alternating layers of low and high attenuation due to mesenteric fat and bowel wall respectively. A kidney-shaped mass may result, an appearance that has been correlated with severe oedema and ischaemia.

4.15. B False
Primary adenocarcinomas of the small bowel most commonly involve the duodenum (50% of cases) or proximal jejunum. CT usually demonstrates an annular constricting mass that may be associated with proximal bowel dilatation. Less commonly, a polypoid mass or extensive bowel wall thickening with ulceration is seen, the latter appearance mimicking lymphoma.

4.15. C False
Metastatic involvement of the small bowel may be via direct spread from tumours of adjacent organs, by intraperitoneal seeding or by haematogenous spread, typically from malignant melanoma, or lung and breast carcinomas.

The latter appear on CT as round masses on the antimesenteric surface of the small bowel wall.

4.15. D True
Purely infiltrative small bowel lymphoma results in marked thickening of the bowel wall which may be either focal or diffuse. Spread to regional lymph nodes and the mesentery are frequent, and mesenteric involvement is seen in 50% of patients with non-Hodgkin's lymphoma but in only 4% of patients with Hodgkin's lymphoma.

4.15. E True
Mesenteric disease in patients with non-Hodgkin's lymphoma may appear on CT as lobulated masses that displace bowel, which tend to have irregular borders and may contain areas of central low density due to necrosis. Multiple small masses may also be seen, representing mesenteric lymph nodes. Other findings include encasement of the SMA and SMV by a mantle of tumour tissue, and fistulae between adjacent bowel loops.

(References A, B, D, E, 38; C, 2)

4.16. A True
On CT, the small bowel mesentery appears as a fat-density area central to the small bowel loops. The SMA, SMV and jejunal and ileal vessels can be seen within it as linear or round structures of soft-tissue density. Normal mesenteric lymph nodes less than 10 mm in diameter can occasionally be identified.

4.16. B True
The transverse mesocolon appears on CT as a structure of fat density extending anteriorly from the pancreas to the margin of the colonic wall. It extends from the duodenocolic ligament on the right to the phrenicocolic and splenorenal ligaments on the left. The branches of middle colic artery and vein can be identified within it.

4.16. C True
Hypoalbuminaemia (usually due to cirrhosis or the nephrotic syndrome) is the commonest cause of diffuse mesenteric oedema. Less common causes include ischaemia and venous or lymphatic obstruction. The CT findings in mesenteric oedema include increased density of the mesenteric fat, poor definition of segmental mesenteric vessels, relative sparing of the retroperitoneal fat, and associated oedema of the subcutaneous tissues.

4.16. D True
Mesenteric involvement by amyloid results in encasement of mesenteric vessels and increased density of the mesenteric fat. However, these findings are non-specific and can be seen with other conditions that cause diffuse mesenteric infiltration, including peritonitis, metastatic carcinoma and peritoneal meso-thelioma.

4.16. E True
Pseudotumoral mesenteric lipomatosis is a benign condition resulting in excessive proliferation of normal mesenteric fat. It may be idiopathic or can occur with obesity, Cushing's syndrome or steroid therapy. Barium studies

may show displacement of small bowel loops but CT will confirm that the cause is either focal or diffuse accumulation of normal fat rather than a tumour or other mass.

(Reference 2)

4.17.   A True
The characteristic CT finding in a patient with small bowel carcinoid is a soft-tissue density mesenteric mass in the right lower quadrant that causes displacement of small bowel loops. It has a spiculated appearance due to radiating strands of soft tissue secondary to the desmoplastic reaction produced by the mass. There may be associated lymphadenopathy and liver metastases.

4.17.   B True
Fat necrosis involving the mesentery (also termed panniculitis or lipodystrophy) may be idiopathic or may be associated with trauma, surgery, infection or foreign body. The CT appearance depends upon whether the process is diffuse or focal. Diffuse panniculitis results in a generalized increase in the density of the involved fat, which usually also contains strands of linear soft-tissue density.

4.17.   C True
Nodular or mass-like fat necrosis appears on CT as single or multiple relatively well-defined masses that contain areas of near-fat density interspersed with areas of soft-tissue or water density. The areas of higher attenuation are due to inflammatory infiltrate, oedema and fibrosis. Calcification in the central necrotic areas is also a recognized finding. The differential diagnosis includes liposarcoma or fat-containing teratoma.

4.17.   D True
Mesenteric cysts appear on CT as well-defined, near-water-density abdominal masses that may contain thin soft-tissue-density septa. A thin wall may be identified. The density of the cyst contents may be increased in the presence of infection or haemorrhage, or if the cyst contains mucinous fluid. The transverse mesocolon and omentum are less common sites of occurrence. Multiple cysts are occasionally found.

4.17.   E True
The mesenteric desmoid tumour is a non-encapsulated, locally invasive form of fibromatosis which occurs most commonly in patients with Gardener's syndrome who have undergone abdominal surgery. CT shows a soft-tissue mass displacing local structures. Its margin is usually well defined but may be irregular, reflecting its infiltrative nature. Desmoid tumours have low signal intensity on T1-WSE MRI and also relatively low signal intensity on T2-WSE scans.

(Reference 2)

4.18.   A True
CMV infection of the small bowel results in a diffuse enteritis or ileocolitis which may be complicated by perforation. Barium studies may reveal narrowing of the terminal ileum with multiple submucosal nodules and

ulceration. Caecal disease may be associated. CT reveals uniform thickening of the distal ileum and absence of mesenteric lymphadenopathy, which helps to distinguish it from lymphoma.

4.18 B False
Cryptosporidiosis produces an enteritis which predominantly involves the duodenum and proximal small intestine. Barium studies may show thickened mucosal folds in the duodenum and proximal small bowel with evidence of hypersecretion. The findings are non-specific and may also be seen with infections such as giardiasis, strongyloidosis and isosporiasis.

4.18. C True
In the small bowel, MAI infection may produce radiographic findings similar to Whipple's disease, consisting of mild dilatation with thick undulating folds and fine nodularity. CT reveals mesenteric and retroperitoneal lymphadenopathy, splenomegaly and ascites. The small bowel folds may appear thickened, without associated wall thickening. Central low density within the enlarged nodes is a characteristic finding helping to distinguish MAI from KS and AIDS-related lymphoma.

4.18. D True
The nodules of KS are usually 1–2 cm in size and the preservation of intervening mucosal folds helps to distinguish KS from lymphoma, which is typified by diffuse submucosal infiltration. However, distortion of the folds may also be due to intramural haemorrhage, which is a rare complication of KS. Lymphoma may produce focal masses with deep ulcers and fistula formation.

4.18. E False
Causes of retroperitoneal and mesenteric lymphadenopathy in patients with AIDS include persistent generalized lymphadenopathy (PGL) as part of the AIDS-related complex (ARC), MAI infection, KS and AIDS-related lymphoma. PGL typically produces mild lymphadenopathy (0.5–1.0 cm in size). KS also produces mild enlargement, and enhancement following contrast is a rare but distinctive finding in these nodes. Bulky lymphadenopathy usually indicates lymphoma or MAI infection, and the two are indistinguishable by imaging, unless central low density is identified (suggesting MAI).

(Reference 7)

4.19. A False
The earliest radiographic manifestations of small bowel Crohn's disease include diffuse mucosal granularity, lymphoid hyperplasia and aphthous ulcers. These lesions are beyond the spatial resolution of CT, although CT can demonstrate pseudopolyposis.

4.19. B False
Bowel wall thickening occurs due to a combination of oedema, fibrosis inflammation and lymphangiectasis. CT demonstrates mural thickening ranging from 1 to 3 cm. The abnormality may be discontinuous (skip lesions) and is most commonly observed in the terminal ileum. However, the bowel wall may be thickened by fibrosis in patients with inactive disease.

4.19. C False
The thickened bowel wall more commonly has a homogeneous appearance. However, a 'double-halo' appearance has been described in the small bowel and colon in Crohn's disease. This consists of a low, near-water or fat-density inner ring corresponding to severe mucosal oedema or submucosal fat deposition, and a higher-density outer ring due to the muscularis propria and serosa. These findings have also been decribed in ulcerative colitis, radiation enteritis, pseudomembranous colitis, ischaemic colitis, mesenteric venous thrombosis and acute pancreatitis. Significant enhancement of the inflamed mucosa and bowel wall may be seen on postcontrast scans, and the intensity of enhancement has been reported to correlate with clinical activity.

4.19. D True
Perianal abnormality is not uncommon in patients with Crohn's disease. CT findings include increased density of the perianal fat and in the fat of the ischiorectal fossa, rectal wall thickening, fistulae and sinus tracts and abscesses.

4.19. E True
Mesenteric abnormalities in patients with Crohn's disease include abscess, diffuse inflammation and fibro-fatty proliferation (the most common cause of bowel loop separation). Fibro-fatty mesenteric proliferation appears on CT as increased density of the mesenteric fat between bowel loops (to -70 to $-90\,HU$) and absence of soft-tissue mass or fluid collections. Diffuse mesenteric inflammation also results in increase in fat density but has no clearly defined borders.

(References A, B, C, D, 38; E, 2)

4.20. A False
In acute ulcerative colitis, changes are limited to the colonic mucosa and CT is usually normal. In chronic disease, involvement of the muscularis mucosa and submucosa results in mural thickening and lumen narrowing. On CT, the thickened bowel wall often has an inhomogeneous attenuation with regions of fat density and near water density due to oedema.

4.20. B True
The 'target' appearance in the rectum (and also in other areas of the colon) consists of an inner and outer ring of soft-tissue density and a middle ring of either near fat or water density. The 'target' sign can also be seen in long-standing Crohn's disease without active inflammation.

4.20. C False
In one study, the mean rectal wall thickness in patients with ulcerative colitis was approximately 1 cm, whereas that in patients with long-standing Crohn's disease was 2.7 cm.

4.20. D True
Perirectal abnormalities that may be identified on CT in patients with ulcerative colitis include enlarged lymph nodes (also seen in Crohn's disease) and changes in the perirectal fat. Fatty proliferation will result in an increase in

the presacral space. On CT, this fat is characterized by the presence of streaky soft-tissue densities and an abnormal attenuation value that is 20–40 HU greater than normal fat.

4.20. E True
Severe acute ulcerative colitis may be complicated by toxic megacolon and CT may show thinning of the colonic wall. Air within the colonic wall (pneumatosis coli) associated with toxic dilatation is often better demonstrated by CT than by conventional radiography.

(Reference 38)

4.21. A False
Scintigraphic localization of infection and inflammation utilizes either gallium-67 citrate or ^{111}In-labelled WBCs. The former is unsuitable for investigating inflammatory bowel disease (IBD) since the colon is one of the normal sites of gallium accumulation. Other disadvantages of gallium scanning are that it usually requires 48 hours to make a diagnosis and bowel preparation is necessary. Gallium will detect both infection and inflammation.

4.21. B False
[111]In-labelled WBCs show normal uptake in the liver, spleen (particularly) and the bone marrow. Lack of normal bowel uptake and low background activity in the abdomen make them suitable for investigating IBD. In active IBD, labelled WBCs migrate through the wall into the lumen of the bowel during the course of the study. Typically, imaging is commenced at 2 hours and repeated at 24 hours.

4.21. C True
In the typical case of active IBD, the scan is positive at 2 hours, and the 24-hour image shows either normal appearances or significantly less activity. The site of disease must be assessed from the initial images since passage of labelled WBCs into the lumen and distal propagation may result in false localization of disease site if only delayed images are obtained. In cases of abscess, there is progressive accumulation of labelled WBCs at the abscess site over 24 hours.

4.21. D True
A false-negative labelled WBC study may also occur if the patient is immunosuppressed by chemotherapy, if the patient is neutropenic, if effective antibiotic therapy is being given, and if images are only obtained at 24 hours, since the scan appearances may have reverted to normal.

4.21. E True
Imaging with gallium-67 citrate earlier than 48 hours will always show some renal activity and this should not be misinterpreted as renal infection/inflammation. However, renal activity at 72 hours is always abnormal.

(Reference 5)

4.22. A True
Colonic wall thickening is a non-specific finding and can be seen in inflammatory bowel disease, infection, ischaemia, haemorrhage, radiation injury, AIDS and graft-versus-host disease. CT may not be able to distinguish any of these in the absence of clinical history.

4.22. B True
Causes of pseudomembranous colitis include antibiotic treatment, due to *Clostridium difficile* enterotoxin (commonest), and also uraemia, ischaemia and heavy metal poisoning. CT findings include colonic dilatation, mural thickening and oedema of haustral folds.

4.22. C True
Other CT findings in cases of pseudomembranous colitis include marked enhancement of the colonic wall, thickening of fascial planes and small bowel dilatation. Intramural gas and ascites are less frequently seen. The disease may occasionally be focal.

4.22. D True
Irregular thickening of the colonic wall is the commonest CT finding in ischaemic colitis, and is attributed to oedema and haemorrhage of the mucosa and submucosa. Haemorrhage into the mesentery may also be seen. Rarer, more specific findings include absent or diminished wall enhancement on postcontrast CT, pneumatosis intestinalis and gas in the mesenteric or portal venous system.

4.22. E False
Typhlitis (neutropenic colitis) is a necrotizing inflammation of the caecum and ascending colon that develops in the presence of profound neutropenia, particularly in patients receiving chemotherapy for acute leukaemia. Perforation is a recognized complication. CT findings include diffuse thickening of the involved colonic wall, which may contain low-density areas, pneumatosis, thickened fascial planes and pericolic fluid.

(Reference 38)

4.23. A True
Other CT features of diverticular disease include the demonstration of air-filled diverticula adjacent to the thickened bowel wall. The presence of rectally administered contrast material within a thickened bowel wall indicates an intramural sinus tract which is also a recognized finding.

4.23. B False
Diverticulitis results in pericolonic inflammation/abscess formation which appears on CT as streaky increased density of the pericolonic fat adjacent to the diseased bowel. The abscess is initially confined by the mesentery but may break through to form a localized pelvic abscess or result in generalized peritonitis. CT can also demonstrate colo-vesical fistula formation by identifying air in the bladder.

4.23. C True
CT signs of gastrointestinal perforation include extraluminal collections of gas, fluid or bowel contrast medium. The location of these depends upon the extent to which these collections become walled off. If the patient is scanned in the supine position, the anterior peritoneal surface of the liver is the most non-dependent part of the peritoneum and should be carefully evaluated for the presence of air.

4.23. D False
CT features that are more suggestive of caecal diverticulitis than appendicitis include location of the inflammatory changes several centimetres above the base of the caecum, presence of concentric rather than eccentric caecal wall thickening, presence of intramural abscesses, and demonstration of caecal diverticula. In both cases, CT will demonstrate inflammatory changes in the pericaecal fat, which can also be seen with a perforated caecal carcinoma. This diagnosis is suggested by the presence of a considerably larger soft-tissue mass than is seen with appendicitis or diverticulitis.

4.23. E True
Approach to a pelvic abscess via the greater sciatic foramen may be made difficult by the presence of the inferior gluteal vessels and the sciatic nerve. The sacrospinous ligament marks the caudal extent of the greater sciatic foramen and can often be identified at CT. Below this level, the neurovascular structures have exited the pelvis, leaving a clear transgluteal approach to an abscess.

(Reference 38)

4.24. A False
A normal appendix is occasionally identified on routine abdominal CT as a tiny ring-like or tubular soft-tissue density structure in the right lower quadrant. It may be collapsed or filled with fluid or air. Its base is related to the posteromedial aspect of the caecal pole and it usually extends medially but may also ascend in a retrocaecal location.

4.24. B True
Periappendiceal or pericaecal inflammation is manifest on CT as higher attenuation, linear streaky densities or as poorly defined areas of increased density in the normally homogeneous pericaecal fat. This is a non-specific finding. An abnormal appendix is occasionally seen and, when acutely inflamed, CT may demonstrate wall thickening and a pus-filled distended appendix. An appendicolith may also be identified but is more commonly seen in association with an appendix mass.

4.24. C True
Gaseous distension of a thick-walled appendix in association with surrounding inflammatory changes indicates gangrenous appendicitis. Other CT findings in acute appendicitis include focal indentation or thickening of the posteromedial wall of the caecum, thickening of the distal small bowel, thickening of the anterior renal and lateroconal fasciae and enlargement of the ipsilateral psoas muscle.

4.24. D False
Perforation of the appendix occurs in approximately 25% of cases of acute appendicitis, and may result in the formation of an 'appendix mass', which may be due either to a phlegmon or an abscess. The former appears on CT as a poorly defined soft-tissue density mass that may engulf or displace the caecum or adjacent bowel loops, and is not amenable to drainage. Liquefaction of such an inflammatory mass will result in abscess formation, which typically appears as a mass of near-water density (0–20 HU) which may contain air and be surrounded by an enhancing wall on postcontrast scans.

4.24. E False
Mucocele of the appendix appears on precontrast CT as a well-defined water-density mass that indents the medial aspect of the caecal pole. Mural calcification may be identified, and when uncomplicated there are no pericaecal/periappendiceal inflammatory changes present. These features may also be seen with cystadenoma or cystadenocarcinoma of the appendix.

(Reference 38)

4.25. A False
The CT findings of primary colonic carcinoma include focal wall thickening, semi-circular or circumferential wall thickening or a discrete mass. The tumour usually has homogeneous attenuation, but areas of low attenuation due to necrosis may be seen in large tumours. A mild degree of enhancement may occur on postcontrast scans. Gas density within the mass is seen in the case of a perforated carcinoma, which may then be difficult to differentiate from diverticulitis.

4.25. B False
The CT criteria for diagnosing invasion are loss of fat planes and enlargement of the muscle. Common sites of local invasion from rectosigmoid carcinoma include the pelvic muscles, bladder, prostate, seminal vesicles and ovaries. Metastatic disease also involves lymph nodes, the liver and adrenals, and the mesentery.

4.25. C True
There are several causes for a presacral mass in a patient who has had an abdominoperineal resection (APR) for low rectal carcinoma. Normal structures that may fill this space include the uterus and seminal vesicles, as well as small bowel loops. Within 1 month following APR, a presacral mass may be due to postoperative oedema/haematoma. These usually become smaller and better defined over the next 4–9 months and the remaining fibrosis appears as minimal streaky soft-tissue density. Occasionally, a postoperative fibrotic mass will show no change.

4.25. D True
Approximately one-third of tumour recurrences occur within the first 6 months and more than 80% occur within 2 years. The presacral space is the commonest site of recurrence following APR. On CT, the tumour usually appears as a globular soft-tissue mass. CT-guided biopsy may be necessary in cases where fibrosis and tumour are indistinguishable.

4.25. E True
The differentiation between fibrosis and recurrent tumour may be made with MRI. Typically, fibrosis has low signal intensity on T1- and T2-WSE scans and does not enhance following Gd-DTPA. Recurrent tumour is typified by high signal intensity on T2-WSE scans and enhancement after Gd-DTPA. However, tumour may occasionally stimulate a marked desmoplastic response mimicking fibrosis, and small-volume tumour may not be identified within a large fibrotic mass. Furthermore, radiation damage that produces oedema, inflammation and hypervascularity will have high signal intensity on T2-WSE scans that may be seen up to 4 years after treatment. Such areas can show enhancement with Gd-DTPA

(References A, C, D, 1; B, 2; E, 40)

4.26. A False
Primary carcinomas of the anal canal appear grossly as exophytic mass lesions in just under 20% of cases, as a hard nodular mass which infiltrates local structures in 40% of cases and as an ulcerating mass in 40% of cases. On postcontrast CT, the primary tumour may appear as a mass that is either hypodense or isodense to surrounding muscle, in which case its presence can only be identified by indirect signs (see B).

4.26. B True
Indirect signs of anal canal carcinoma include symmetric or asymmetric bulging of the sphincter or levator ani muscles (with supralevator masses), asymmetry of the perirectal fat between the anorectal walls and the levator muscles and linear soft-tissue strands in the perirectal fat or upper ischiorectal fossa (an indirect sign of tumour extension). Gas in the perineum may be seen with infected tumours.

4.26. C False
Spread of anal canal carcinoma is mainly local. Anal muscles are invaded early because the mucosa is very close to the underlying sphincters. Growth is characteristically circumferential, resulting in stenosis of the anal sphincter. Invasion of the sphincter is followed by spread into the ischiorectal fossae, prostatic urethra and bladder in men, and vagina in women. However, such local invasion is difficult to assess since fat planes between the anorectal muscle ring and local structures may be very thin or absent.

4.26. D True
Approximately 15–30% of anal canal carcinomas are associated with metastases to the inguinal nodes, the superficial medial nodes being most commonly affected. Bilateral inguinal lymphadenopathy is unusual, and diffuse enlargement of all inguinal nodes is rare. Inguinal nodal metastases are most common when the primary tumour arises below the dentate line.

4.26. E True
Other sites of lymphatic metastasis include the perirectal nodes and nodes at the bifurcation of the superior rectal artery. However, in up to 40% of cases, lymphatic spread may follow the middle rectal artery, resulting in internal and common iliac lymphadenopathy, and eventually para-aortic lymphadenopathy. Haematogenous spread occurs in less than 10% of cases, with the liver being

the commonest site. These metastases are usually from tumours involving the anorectal junction.

(Reference 41)

4.27.   A True
Of the various infectious agents that produce colitis, CMV is the only one that produces distinctive radiographic findings. The colitis may be mild or fulminant with toxic dilatation. Haemorrhagic colitis and abnormality resembling pseudomembranous colitis are also reported. The whole colon may be involved, but disease isolated to the caecum is also seen. Colonic disease may coexist with ileitis.

4.27.   B True
Other findings on double-contrast barium enema in patients with CMV colitis include aphthous ulceration, caecal spasm, superficial or deep ulceration, multiple large discrete ulcers and submucosal haemorrhage. MAI and cryptosporidium may also involve the colon but have no distinguishing features.

4.27.   C True
CT in patients with CMV colitis demonstrates colonic wall thickening with a 'target' sign due to submucosal oedema. This helps to differentiate inflammatory from neoplastic wall thickening. There is usually a pancolitis, but right-sided involvement is always present. Thickening of the terminal ileum may also be seen.

4.27.   D False
As with other sites of bowel involvement, colonic KS manifests as submucosal plaques or focal areas of submucosal infiltration. Patients may present with acute appendicitis due to KS lesions obstructing the appendix. Lymphoma involves the colon more commonly in AIDS patients than in the general population. Polypoid lesions or bulky masses may be seen.

4.27.   E True
There is a high incidence of rectal involvement in AIDS-related lymphoma. This may appear as bulky perirectal masses that invade the pelvic floor. Lymphoma may also involve the anal canal. Perirectal inflammatory changes, thickening of the rectal wall and enlargement of the seminal vesicles may be seen with PGL, presumably due to chronic proctitis from repeated venereal infections.

(Reference 7)

4.28.   A True
With high-resolution transabdominal or endoscopic ultrasound, five layers to the bowel wall can be demonstrated. The first, third and fifth layers are hyperechoic and represent the probe/mucosa interface, submucosa and serosa. The second and fourth layers are relatively hypoechoic and represent the muscularis mucosa and muscularis propria.

4.28. B False
The jejunum and ileum are distinguished by the presence of valvulae conniventes in the former. These appear on ultrasound as thin, echogenic lines crossing the bowel lumen, producing a ladder appearance when the bowel is scanned parallel to its long axis. The ileal mucosa is featureless. The colon is recognized by its haustral pattern.

4.28. C True
When the bowel is well distended the maximal wall thickness is usually about 3 mm, and when non-distended the maximal thickness should not exceed 5 mm.

4.28. D False
Bowel wall thickening may be focal or diffuse, symmetrical or asymmetrical, and with or without retention of the normal bowel wall layers. Causes of bowel wall thickening include inflammatory diseases such as inflammatory bowel disease, pseudomembranous colitis, infectious colitis, ischaemia and neoplasm.

4.28. E False
As a generalization, long segments of bowel wall thickening are more commonly due to inflammatory causes, whereas neoplasms cause relatively short segments of thickened bowel. However, lymphoma may cause a long segment of thickened bowel wall, being indistinguishable from an inflammatory cause.

(References A, B, C, 11; D, E, 51)

4.29. A False
Acute appendicitis may be investigated with the use of graded compression sonography in which gradual and uniform pressure is applied to the area of interest. With this technique, normal gas and fluid-filled loops of bowel will be displaced, whereas thickened loops of bowel and distended segments will be trapped between the anterior abdominal wall and the iliopsoas muscle. An appendicolith is only occasionally identified.

4.29. B False
The sonographic criteria for the diagnosis of acute appendicitis using the above technique include the demonstration of a non-compressible blind-ending, aperistaltic tubular structure arising from the tip of the caecum with a diameter equal to or greater than 6 mm. A normal appendix can occasionally be identified but will have a diameter of less than 6 mm and will be compressible.

4.29. C False
Sonography in acute diverticulitis demonstrates bowel wall thickening due to a combination of smooth muscle hypertrophy, inflammation, oedema and muscle spasm. Oedema of the pericolic fat appears as an echogenic 'mass' that separates the inflamed segment from adjacent uninvolved loops.

4.29. D False
In patients with diverticular disease, non-inflamed diverticula are infrequently visualized, whereas inflamed diverticula appear as bright echogenic foci either within or beyond the thickened bowel wall. These echogenic foci are thought

to represent either inspissated faecoliths or microabscesses in the base of the inflamed diverticula.

4.29. E True
Crohn's disease is associated with fibro-fatty proliferation of the mesentery (creeping fat) which appears on ultrasound as an echogenic mass adjacent to a loop of hypoechoic, thickened bowel.

(Reference 11)

4.30. A False
Many bowel wall tumours ulcerate and gas within the ulcer crater is identified on ultrasound as bright echogenic foci with distal acoustic shadowing. The differential diagnosis of bowel wall masses with or without ulceration include mesenchymal tumours (especially of smooth muscle origin), lymphoma, metastases and adenocarcinoma with local extension.

4.30. B False
Ultrasound cannot demonstrate isolated superficial mucosal ulceration with no thickening of the bowel wall as occurs in ulcerative colitis. However, bowel wall thickening that occurs in the presence of transmural inflammation, as occurs in Crohn's colitis, can be demonstrated.

4.30. C True
Leiomyomas and leiomyosarcomas of the bowel characteristically produce intramural masses with a large extraluminal component. These smooth muscle tumours are often necrotic and ultrasound therefore commonly demonstrates a large solid/cystic mass.

4.30. D True
Similar to lymphoma elsewhere, bowel lymphoma characteristically results in an anechoic or hypoechoic bowel wall mass on ultrasound. Enlarged lymph nodes may have a similar appearance.

4.30. E False
The 'pseudokidney' or 'target' sign on ultrasound is non-specific and is produced by an outer hypoechoic rim due to the thickened bowel wall and a central hyperechoic region representing the narrowed bowel lumen or ulceration of the mucosa.

(References A, C, D, E, 11; B, 51)

4.31. A True
Primary rectal adenocarcinomas arise from the rectal mucosa and appear on transrectal ultrasound as superficial hypoechoic masses. Transrectal ultrasound demonstrates the various histological layers of the rectal wall as alternating echogenic and echo-poor rings. Involvement by carcinoma of deeper layers of the bowel wall is identified as thinning, thickening or destruction of these layers.

4.31. B False

Lymph nodes appear as small hypoechoic round or oval masses in the hyperechoic perirectal fat. Although they are uncommonly identified normally, their demonstration does not guarantee malignant involvement since nodes may be enlarged due to reactive changes.

4.31. C False

Microscopic tumour invasion of the rectal wall or perirectal lymph nodes cannot be reliably identified by transrectal ultrasound. *In vitro* studies have demonstrated benign nodes as being typically oval, with heterogeneous echotexture and central linear hilar echogenicity, and malignant nodes to be typically round and hypoechoic with loss of the normal internal echo texture. However, these features are not easily demonstrated *in vivo* and both benign and malignant nodes tend to have a round, hypoechoic appearance.

4.31. D False

Local recurrence of rectal adenocarcinoma following a sphincter-saving resection is usually extraluminal with secondary invasion of the suture line. Transrectal ultrasound typically demonstrates hypoechoic, exophytic growths that do not arise from the epithelial lining of the rectum.

4.31. E True

Metastases to the rectum typically appear on transrectal ultrasound as hypoechoic masses in the deeper layers of the rectal wall with sparing of the rectal mucosa. Metastases may be due to spread from primary tumours of the upper gastrointestinal tract, or from direct invasion from local tumours.

(Reference 11)

4.32. A False

The main CT criterion for determining whether a lymph node is abnormal is enlargement. CT cannot demonstrate abnormal architecture within a normal-sized node. This is the major cause of false-negative CT studies when CT is used for staging of malignant tumour. Measurement of nodes should be made using the short-axis diameter, to minimize errors due to node orientation. False-positive results may occur when benign conditions (such as sarcoidosis, tuberculosis, Crohn's disease, mastocytosis, Whipple's disease and non-tropical sprue) cause lymph node enlargement.

4.32. B False

Normal and abnormal lymph nodes are differentiated on MRI by size criteria (as for CT) rather than any change in signal characteristics. Similarly, MRI cannot differentiate benign reactive nodes from those involved by tumour, nor can it demonstrate tumour within normal-sized nodes. On T1-WSE scans, nodes are slightly hyperintense to muscle and markedly hypointense to surrounding fat. On T2-WSE scans, nodes become markedly hyperintense to muscle but the contrast between nodes and fat is reduced.

4.32. C False

Calcification within enlarged nodes may either be due to inflammatory disease, particularly tuberculosis, but can also be seen with mucinous carcinoma and sarcoma metastases, and with treated lymphoma.

4.32. D True
Lymphangiography has the advantage over CT and MRI in that it can accurately display the internal nodal architecture (thereby identifying metastases in normal sized nodes). However, nodes that are completely replaced by tumour will not be demonstrated and several groups of intra-abdominal nodes are not opacified (including portal, retrocrural, coeliac and mesenteric).

4.32. E False
Nodal metastases from non-seminomatous testicular carcinomas are a classical cause of low-attenuation intra-abdominal lymphadenopathy. Other malignancies that can cause this appearance include epidermoid genitourinary carcinomas and lymphoma. Benign conditions that result in low-density, enlarged lymph nodes include Whipple's disease (due to fat deposition within the nodes) and tuberculosis (due to necrosis).

(References A, C, D, E, 42; B, 2)

4.33. A False
Retrocrural nodes are considered enlarged when measuring greater than 6 mm in short-axis diameter. Lymphoma and lung carcinoma are the most common causes of retrocrural lymphadenopathy. The retrocrural space allows communication between the posterior mediastinum and the retroperitoneum.

4.33. B True
Retroperitoneal nodes are distributed around the aorta and IVC and are grouped into the periaortic, pericaval and interaorticocaval chains. Confluent lymphadenopathy may elevate the great vessels off the spine. Although nodes greater than 10 mm are considered enlarged, if multiple 8–10 mm sized nodes are present abnormality should be suspected. The commonest causes of retroperitoneal lymphadenopathy are lymphoma and metastases from renal, testicular, cervical and prostatic carcinomas, the latter two after involvement of pelvic nodes.

4.33. C True
Tumours which commonly result in gastrohepatic lymphadenopathy include carcinomas of the lesser curvature of the stomach, distal oesophagus, and disseminated lymphoma. Also, retrograde spread may occur from coeliac axis nodes involved by carcinoma of the pancreas, colon, breast and melanoma. The upper margin of the pancreas or transverse colon may extend into the gastrohepatic ligament, mimicking lymphadenopathy.

4.33. D False
Portal nodes lie within the porta hepatis, extending down the hepatoduodenal ligament and interconnecting with the gastrohepatic ligament. They drain centrally to the coeliac nodes. They are located anterior and posterior to the portal vein, and when enlarged may completely obliterate this structure. Portal nodes are considered enlarged if measuring greater than 6 mm in short-axis diameter.

4.33. E True
However, portal lymphadenopathy is most commonly due to metastatic involvement from a variety of primary sites, including the gallbladder and biliary tree, liver, stomach, pancreas, colon, lung and breast, as well as

lymphoma. Portal lymphadenopathy is commonly associated with liver metastases and may also result in high extrahepatic biliary obstruction.

(Reference 42)

4.34. A True
Coeliac and superior mesenteric nodes are located around the origins of the respective arteries, and together with inferior mesenteric nodes are termed pre-aortic nodes. The coeliac and superior mesenteric nodes are the terminal nodes of the gastrointestinal tract from the ligament of Treitz to the splenic flexure, and can therefore be involved by metastases from many intra-abdominal sites.

4.34. B True
Pancreaticoduodenal nodes are located anterior to the IVC, between the duodenal sweep and the pancreatic head. Enlargement of these nodes may result in distal extrahepatic biliary obstructions. Confluent nodal enlargement may produce a mass that is difficult to distinguish from a mass in the pancreatic head.

4.34. C False
Pancreaticoduodenal nodes are considered enlarged if measuring over 10 mm in short-axis diameter. Causes of enlargement include lymphoma, and carcinomas of the pancreatic head, colon, stomach, lung and breast. These nodes may communicate with nodes in the porta hepatis via the hepatoduodenal ligament, and therefore associated hepatic metastases are not infrequent.

4.34. D True
Perisplenic nodes lie in the splenic hilum and drain the spleen, tail of pancreas and the greater curvature of the stomach. They drain via the pancreaticosplenic nodes to the coeliac nodes. The commonest causes of perisplenic lymphadenopathy are lymphoma, and carcinomas of the pancreas, colon, stomach, lung and breast. The upper limit of normal size for these nodes is 10 mm.

4.34. E False
The small bowel mesentery contains a large number of lymph nodes that accompany the branches of the SMA and SMV. Multiple nodes are also present more distally adjacent to the small bowel wall. The nodes are considered enlarged if measuring over 10 mm in short-axis diameter. Non-Hodgkin's lymphoma, leukaemia, small bowel neoplasms, ovarian carcinoma, and carcinomas of the right and transverse colon are the commoner causes of malignant lymphadenopathy but metastases from lung and breast may also occur. Mesenteric lymphadenopathy may appear as multiple discretely enlarged nodes or as large masses of adenopathy sandwiching normal vessels and fat.

(Reference 42)

4.35. A True
Metastatic tumour can disseminate through the peritoneal cavity by four pathways: direct spread along mesenteric and ligamentous attachments, intraperitoneal seeding, lymphatic extension and embolic haematogenous

dissemination. Metastatic spread to the greater omentum (gastrocolic ligament) has a variable CT appearance ranging from small nodules or strands of soft tissue that increase the density of the fat anterior to the colon or small bowel, to large masses that separate the colon or small bowel from the anterior abdominal wall (omental cake). Ovarian carcinoma is the commonest cause of this appearance, but inflammatory thickening of the omentum can also occur.

4.35. B True
Carcinomas of the tail of the pancreas can spread along the phrenico-colic ligament to involve the splenic flexure of the colon. Other examples of direct tumour spread include extension of stomach, colon and pancreatic carcinomas via the transverse mesocolon and spread of biliary neoplasms along the gastrohepatic or hepatoduodenal ligaments.

4.35. C False
Intraperitoneal seeding depends upon the natural flow of fluid within the peritoneal cavity. The most common sites for pooling of ascites and subsequent fixation and growth of peritoneal tumour are the pouch of Douglas, the lower small bowel mesentery near the ileocaecal junction, the sigmoid mesocolon and the right paracolic gutter. The commonest tumours to spread via this route are adenocarcinomas of the ovary, colon, stomach and pancreas.

4.35. D False
CT frequently demonstrates ascites in association with intraperitoneal tumour implants. If a large amount of ascites is present, metastases as small as 1 cm can be identified. If metastases are very small, ascites may be the only sign of intra-abdominal carcinomatosis. If ascites is absent, soft-tissue density replacement of the mesenteric fat is all that may be seen. Diffuse thickening of the peritoneum is also a feature.

4.35. E False
Pseudomyxoma peritonei most commonly results from rupture of a mucinous cystadenoma (carcinoma) of the ovary or appendix. CT findings include multiple low-attenuation masses, that may be surrounded by discrete walls, or diffuse low-attenuation intraperitoneal material that may contain septations and causes scalloping of the surface of the liver. The walls or septations may calcify.

(Reference 2)

4.36. A False
The attenuation value of ascitic fluid generally ranges from 0 to 30 HU but may be higher with exudative ascites due to the higher protein content of the fluid. However, attenuation values of ascitic fluid are non-specific, and infected or malignant ascites cannot be differentiated reliably from transudative ascites. Acute intraperitoneal haemorrhage may be identified since it can have attenuation values significantly higher than 30 HU.

4.36. B False
The distribution of peritoneal fluid within the abdominal cavity may differ depending upon the aetiology. Patients with benign transudative ascites tend to have large greater sac collections with relatively little fluid in the lesser sac,

whereas patients with malignant ascites tend to have proportionally equal volumes of fluid in these two spaces.

4.36.   C False
When a small amount of ascites is present, fluid tends to collect in the most dependent parts of the abdominal cavity, which in the supine patient are the right subhepatic space and the pouch of Douglas. When a large volume of ascites is present, the small bowel loops tend to be located centrally within the abdomen and the ascites may appear as triangular collections of fluid in between the leaves of the small bowel mesentery and the loops of bowel.

4.36.   D True
Intraperitoneal bile accumulation (biloma) is caused by iatrogenic, traumatic or spontaneous rupture of the biliary tree. On CT, most bilomas appear as round or oval fluid collections with attenuation values of less than 20 HU (which may be higher if complicated by haemorrhage or infection). They are usually confined to the right upper quadrant but may be located in the left upper quadrant in about 30% of cases.

4.36.   E True
Transudative ascites has uniform low signal intensity on T1-WSE scans and high signal intensity on T2-WSE scans. With exudative ascites, the increased protein content results in shortening of T1 relaxation times and the fluid will have intermediate to high signal intensities on T1-WSE scans and appear hyperintense on T2-WSE scans.

(Reference 2)

4.37.   A False
The psoas major muscle is formed from fibres that arise from the transverse processes of T12–L5. In its upper extent, it lies in a paraspinal location and normally appears triangular on axial imaging. The size of the muscle increases as it passes caudally and it assumes a more rounded appearance. In the pelvis it fuses with the iliacus to form the iliopsoas, which inserts into the lesser trochanter of the femur. Between the psoas and vertebral body, several small soft-tissue structures may be identified, which represent the sympathetic chain, lumbar veins and arteries, and lymph nodes.

4.37.   B False
The psoas minor muscle originates from the sides of the bodies of T12 and L1. When visualized, it appears on CT as a small, round structure immediately anterior to the psoas major. It terminates as a long flat tendon that inserts into the iliopectineal eminence of the innominate bone. Its major significance is that it may be confused with enlarged lymph nodes.

4.37.   C False
Neoplastic involvement of the psoas muscle by lymphoma or other retroperitoneal malignancies most often appears on precontrast CT as enlargement of the muscle with attenuation values similar to normal muscle. Areas of low

attenuation may be present. The psoas may also be displaced laterally by paraspinal lymphadenopathy.

4.37. D True
The normal psoas muscle has low signal intensity on both T1-WSE and T2-WSE MRI. T1-weighted scans provide the best contrast between the muscle and retroperitoneal fat, while T2-weighted scans are more sensitive to abnormalities within the muscle. Tumour results in higher signal on both scan sequences. On T1-WSE scans, the signal intensity is still less than that of fat, but may be increased if haemorrhage has occurred.

4.37. E True
Infection within the psoas muscle may occur secondary to sepsis within the spine, kidney, bowel or pancreas. CT typically shows a diffusely enlarged muscle with central areas of low density (0–30 HU). Gas may occasionally be identified. However, in the absence of gas density, CT cannot reliably differentiate abscess from haematoma or necrotic tumour.

(Reference 2)

4.38. A True
A lymphangioma with a high lipid content can simulate a lipoma on CT. Otherwise, they appear as well-defined, near water density masses. Lipomas appear as sharply marginated, homogeneous masses with CT attenuation values equal to that of normal fat. Rarely, a very well-differentiated liposarcoma cannot be differentiated from a lipoma.

4.38. B False
Three distinct CT patterns have been described for liposarcomas which reflect the amount and distribution of fat within the tumour. The solid pattern has CT numbers greater than 20 HU; the mixed pattern has areas of fat density mixed with areas of soft-tissue density; the pseudocystic pattern is described below. The well-differentiated liposarcoma with abundant mature fat usually has a mixed pattern. The poorly differentiated liposarcoma has little fat and has a solid appearance. Areas of soft-tissue density show enhancement on postcontrast scans.

4.38. C True
The pseudocystic pattern of liposarcoma results from a homogeneous distribution of fat cells within a mucinous connective tissue stroma. CT demonstrates a well-defined homogeneous mass with attenuation values between 20 and −20 HU. It may be indistinguishable from a cystic lymphangioma.

4.38. D True
Leiomyosarcomas tend to have a greater degree of necrosis than other malignant retroperitoneal tumours, accounting for the regions of near water density observed on CT. However, this appearance can also be seen with pleomorphic liposarcoma, neurofibrosarcoma, malignant fibrous histiocytoma and haemangiopericytoma. Areas of calcification may also be identified in the latter two neoplasms.

4.38. E False
Retroperitoneal paraganglionomas are found in a para-aortic location from the renal arteries to the aortic bifurcation. Benign tumours tend to be small and of homogeneous muscle density, whereas malignant tumours are larger and contain low-density regions due to necrosis. The criterion for malignancy is the demonstration of metastases.

(References A, B, C, 2; D, E, 43)

4.39. A False
Retroperitoneal fibrosis (RPF) is idiopathic in about 70% of cases. It may be associated with hypersensitivity to methysergide, and with primary or metastatic retroperitoneal malignancy, which stimulates a desmoplastic response resulting in fibrosis that is indistinguishable from idiopathic RPF. Perianeurysmal fibrosis is another recognized form. The fibrous mass/plaque may extend from the level of the renal hila to the sacral promontory.

4.39. B False
The fibrotic mass typically causes encasement of the aorta and IVC. On ultrasound the lesion may appear as a bulky mass with ill-defined margins or a flat mass with smooth margins. It may be anechoic or hypoechoic. Hydronephrosis may be identified in up to 90% of cases. Intravenous urography classically shows unilateral or bilateral hydroureter with narrowing of the ureteric calibre and medial deviation of the middle third.

4.39. C True
The CT appearances of RPF are very variable. CT may be normal. Other cases show a well-defined sheath of soft-tissue density obscuring the aorta and IVC. Rarely, CT may demonstrate a single or multiple soft-tissue masses with irregular borders, simulating lymphadenopathy.

4.39. D True
On precontrast CT, RPF usually has similar attenuation values to that of muscle. However, focal or uniform hyperdensity is also recognized and is attributed to the high collagen content of the tissue. On postcontrast CT, marked enhancement is a characteristic feature.

4.39. E False
On MRI, retroperitoneal fibrosis usually has relatively low signal intensity on both T1- and T2-WSE scans, which may help differentiate it from tumour, which would have high signal on T2-WSE scans. However, fibrosis that is actively forming, such as may occur following recent chemotherapy or radiotherapy for tumour, may have relatively high signal intensity on T2-WSE scans, making the distinction less clear-cut.

(References A, B, 37; C, D, 2; E, 15)

4.40. A False
One of the earliest changes affecting the small bowel following pelvic radiotherapy is fixation of the small bowel loops, which may be missed on small bowel barium studies unless careful palpation is performed during the examination. Other features include thickening and straightening of the

mucosal folds, with prominence of the valvulae conniventes, and nodular mucosal filling defects due to oedema. Mucosal ulceration, although evident microscopically, is rarely seen radiologically.

4.40.   B False
Barium studies also demonstrate angulation of bowel loops with wall thickening, reduced peristalsis resulting in pooling of barium within bowel loops, and stenosis, which is usually a late feature. Traction upon the bowel loops may occur due to mesenteric fibrosis, producing appearances similar to carcinoid. Fistula formation is also seen.

4.40.   C True
Following pelvic radiotherapy, damage to the large bowel is typically confined to the rectum and sigmoid. Acute exacerbation of inflammatory bowel disease or diverticular disease may occur. Chronic symptoms occur 6 months after treatment. The commonest abnormality seen on barium studies is a stricture which is typically smooth and tapered. The mid-sigmoid and upper rectum are most commonly involved. There may be multiple strictures, which can be long or short. Mucosal changes include loss of haustration, superficial or deep ulceration and a 'cobblestone' pattern.

4.40.   D True
CT demonstrates inflammatory changes in the perirectal fascia with associated bowel wall thickening. There may be an increase in the presacral space.

4.40.   E True
This increased T2 signal intensity is initially confined to the submucosa, but with progression of disease the whole of the rectal wall shows increased signal intensity, resulting in loss of the normal differentiation between the submucosa and muscle. The rectal wall will enhance following Gd-DTPA.

(Reference 40)

5 Genitourinary tract

5.1. Which of the following are true concerning ^{99m}Tc-DMSA renal scintigraphy?
A   The isotope is accumulated by the proximal tubular cells.
B   Glomerular filtration rate can be estimated.
C   Overestimation of split renal function in obstruction is a recognized limitation.
D   It is the isotope of choice for detecting renal scarring.
E   Reduced renal uptake is a feature of renal tubular acidosis.

5.2. Concerning dynamic ^{99m}Tc-DTPA renal scintigraphy
A   the isotope is excreted by a combination of glomerular filtration and tubular excretion.
B   liver activity at any stage of the scan is abnormal.
C   activity in the renal pelvis at 30 minutes after injection is always abnormal.
D   the initial upward slope of the time/activity (T/A) curve represents renal perfusion.
E   persistent activity in the ureters is usually abnormal.

5.3. Concerning ^{99m}Tc-DTPA scintigraphy in the assessment of renal tract obstruction
A   reduced perfusion is a feature of obstructive uropathy.
B   a continuously rising T/A curve is diagnostic of obstruction.
C   a false positive result for obstruction following diuretic is a finding in the presence of impaired renal function.
D   a false negative result for obstruction following diuretic is a finding in the presence of lower ureteric obstruction.
E   multiple photopenic areas are a recognized finding with pelviureteric junction (PUJ) obstruction.

5.4. Concerning dynamic renal scintigraphy in the assessment of acute renal failure
A   uptake of MAG-3 by the liver indicates renal impairment.
B   prerenal failure and acute tubular necrosis (ATN) can be differentiated.
C   absent renal blood flow in the dynamic phase is typical of acute tubular necrosis.
D   delayed intrarenal transit is a feature of acute glomerulonephritis.
E   acute arterial and venous thrombosis can be differentiated in most cases.

5.5. Concerning ultrasound of the kidneys
A parenchymal echogenicity equal to that of the liver or spleen is always abnormal.
B diffuse increase in cortical echogenicity is a feature of lupus nephritis.
C increased echogenicity of the medullary pyramids is a feature of AIDS-related nephropathy.
D renal vein thrombosis is a recognized cause of unilateral increase in cortical echogenicity.
E oxalosis is usually associated with a decrease in cortical echogenicity.

5.6. Which of the following are true concerning renal ultrasound?
A Anechoic spaces in the medulla are a finding in renal papillary necrosis.
B Renal sinus lipomatosis is a recognized cause of a hypoechoic mass in the renal sinus.
C Medullary nephrocalcinosis is typically associated with distal acoustic shadowing.
D Demonstration of a dilated pelvicaliceal (PC) system is diagnostic of obstruction.
E Obstruction is excluded if the PC system is not dilated.

5.7. Which of the following are true concerning renal ultrasound?
A A renal infarct is a recognized cause of a hyperechoic parenchymal lesion.
B A hypertrophied column of Bertin is isoechoic to normal renal parenchyma.
C Enlargement of the medullary pyramids is a feature of acute tubular necrosis.
D A hypoechoic cortical rim is a finding in renal cortical necrosis.
E Bilateral renal cysts are a recognized finding following haemodialysis.

5.8. Concerning ultrasound in the assessment of renal tumours
A the majority of renal cell carcinomas are hypoechoic relative to normal renal parenchyma.
B hypoechoic areas within a renal mass excludes angiomyolipoma.
C renal oncocytoma characteristically appears as a uniformly hyperechoic mass.
D renal lymphoma is a cause of an anechoic mass with posterior acoustic enhancement.
E a juxtaglomerular tumour is a cause of a hyperechoic parenchymal mass.

5.9. Concerning the imaging features of cystic renal masses
A the wall of a simple cyst is usually too thin to be identified on precontrast CT.
B a simple cyst never distorts the renal parenchymal outline.
C simple cysts have high signal intensity on T2-WSE MRI.
D calcification within the septa of a cyst excludes a benign aetiology.
E cysts with density values of 60–80 HU are a recognized finding in adult polycystic kidney disease (APCKD).

5.10. Concerning imaging of cystic renal masses
A the septa within a benign cyst are characteristically less than 1 mm in thickness.
B nodularity within a cyst excludes the diagnosis of a simple cyst.
C thickening of a cyst wall excludes a benign aetiology.
D the presence of calcification differentiates a cystic renal cell carcinoma from a multilocular cystic nephroma.
E on MRI, the majority of benign and malignant cysts are differentiated by the signal intensity of the cyst fluid.

5.11. Concerning imaging of renal cell carcinoma
A on precontrast CT, a mass that is hyperdense compared to normal renal parenchyma is a recognized finding.
B CT typically demonstrates a poorly defined interface with normal renal parenchyma.
C local lymph nodes greater than 2 cm in short-axis diameter are usually involved with tumour.
D thickening of Gerota's fascia always indicates extracapsular extension.
E ipsilateral psoas muscle enlargement is a sign of recurrent disease.

5.12. Concerning CT in the assessment of renal neoplasms
A renal lymphoma is a cause of thickening of the perirenal fascia.
B renal lymphoma masses are usually homogeneous on precontrast CT.
C renal cell adenomas are a cause of bilateral renal masses.
D a central, non-enhancing, stellate scar is a feature of oncocytoma.
E the majority of angiomyolipomas display some fat density.

5.13. Concerning ultrasound in the assessment of renal infections
A anechoic areas at the corticomedullary junction are a recognized finding in acute pyelonephritis.
B acute focal bacterial nephritis typically appears as a hypoechoic mass.
C xanthogranulomatous pyelonephritis typically results in a shrunken kidney.
D xanthogranulomatous pyelonephritis is a recognized cause of a solitary hypoechoic mass.
E a renal fungus ball typically appears as an echogenic mass in the renal pelvis.

5.14. Recognized CT findings in renal infections include which of the following?
A Wedge-shaped non-enhancing areas of renal parenchyma.
B Gas in the renal collecting system.
C A low-density mass with an enhancing wall.
D Multiple hypodense masses surrounding a central staghorn calculus.
E A multiloculated cystic mass with calcified walls.

5.15. Which of the following are recognized CT features of transitional cell carcinoma of the renal pelvis?
A Expansion of the renal parenchyma.
B Eccentric thickening of the wall of the renal pelvis.
C Calcification.
D Enhancement of the tumour on postcontrast scans.
E Renal hilar lymphadenopathy.

5.16. Concerning MRI of the kidney
A   the renal capsule is visualized in the majority of patients.
B   the renal cortex and medulla are isointense on T1-WSE scans.
C   the PC system is most clearly seen on T2-WSE scans.
D   loss of corticomedullary differentiation is a feature of chronic glomerulo-nephritis.
E   reduction of renal cortical signal intensity is a feature of sickle cell nephropathy.

5.17. Which of the following are recognized MRI findings in the assessment of renal masses?
A   Increased signal intensity of the renal parenchyma on T1-WSE scans in xanthogranulomatous pyelonephritis.
B   Areas of high signal intensity on T2-WSE scans in renal cell carcinoma.
C   Areas of high signal intensity on T1-WSE scans in angiomyolipoma.
D   Multiple hyperintense masses on T1-WSE scans due to metastases.
E   High signal intensity on T1-WSE scans in a renal cyst.

5.18. Concerning imaging in the assessment of the failing renal transplant
A   in acute cell-mediated rejection, ultrasound initially demonstrates increase in AP diameter of the kidney.
B   in acute cell-mediated rejection, ultrasound typically demonstrates loss of corticomedullary differentiation.
C   on Doppler ultrasound, a resistivity index (RI) of 0.9 is abnormal.
D   loss of corticomedullary differentiation on MRI is specific for acute tubular necrosis.
E   reduced perfusion on dynamic renal scintigraphy differentiates rejection from cyclosporin toxicity.

5.19. Concerning imaging in the assessment of renal transplant complications
A   stenosis at the ureterovesical junction is a recognized finding.
B   urinoma and seroma can be differentiated by dynamic renal scintigraphy.
C   on T1-WSE MRI, a hyperintense perinephric fluid collection is a recognized finding.
D   a stenosis at the arterial anastomosis is a recognized feature of rejection.
E   renal vein occlusion is usually due to thrombosis.

5.20. Concerning scintigraphy in the assessment of the adrenal glands
A   on posterior views, the right adrenal normally appears more active than the left.
B ^{131}I-iodocholesterol is a suitable radiopharmaceutical for imaging the adrenal cortex.
C   in Cushing's syndrome, unilateral uptake with suppression of the contralateral gland is typical of adenoma.
D   in Conn's syndrome due to adenoma, uptake is usually seen in both glands.
E   activity in the liver is a normal feature of ^{131}I-MIBG scans.

5.21. Concerning CT in the assessment of the adrenal glands
A the limbs of the adrenals are normally straight or concave.
B in adrenal hyperplasia the glands appear abnormal in over 80% of cases.
C a large right adrenal mass typically causes anterior displacement of the IVC.
D adrenal enlargement is a recognized finding in hyperthyroidism.
E adrenal calcification occurs in the majority of cases of Addison's disease.

5.22. Which of the following are true of CT of the adrenal glands?
A A near water density mass is a recognized appearance of a cortisol-producing adenoma.
B Normal appearance of the glands is a recognized finding in Cushing's disease.
C An atrophic contralateral gland is a recognized finding in the presence of a functioning adenoma.
D Aldosteronomas are typically greater than 4 cm in size.
E Calcification is a recognized feature of adrenal carcinoma.

5.23. Which of the following are true concerning phaeochromocytoma?
A There is an increased incidence in von Hippel–Lindau disease.
B A mass in the bladder wall is a recognized CT finding.
C Extra-adrenal tumours are more likely to be malignant.
D The majority are greater than 3 cm at the time of diagnosis.
E Hyperechoic regions are a recognized feature on ultrasound.

5.24. Which of the following are true of the imaging features of adrenal tumours?
A Calcification is a recognized finding in a non-functioning adenoma.
B The majority of non-functioning adenomas are less than 3 cm in diameter.
C Isolated adrenal enlargement is a typical presentation of lymphoma.
D Myelolipoma is typically associated with retroperitoneal lymphadenopathy at diagnosis.
E On ultrasound, myelolipoma is a recognized cause of an echogenic adrenal mass.

5.25. Concerning MRI of the adrenal glands
A normal glands are isointense to fat on T2-WSE scans.
B inhomogeneity within a mass on T1-WSE scans is a feature of carcinoma.
C adrenal adenomas are typically hyperintense to normal glands on T1-WSE scans.
D metastases are typically hypointense to fat on T2-WSE scans.
E phaeochromocytoma is a recognized cause of a mass that is hyperintense to fat on T2-WSE scans.

5.26. Concerning the normal CT and MRI anatomy of the male pelvis

A   on CT, the spermatic cord typically appears as a homogeneous structure of soft-tissue density.

B   the seminal vesicles are always separated from the bladder by a fat plane.

C   differentiation of the bladder wall from urine is optimal on T1-WSE MRI scans.

D   the central and peripheral zones of the prostate gland can be separated on postcontrast CT.

E   on MRI, the central and transitional zones of the prostate can be distinguished from the peripheral zone on T2-WSE scans.

5.27. Concerning CT and MRI in the assessment of transitional cell carcinoma of the bladder

A   the primary tumour enhances on postcontrast CT.

B   poor definition of the outer aspect of the bladder wall is a feature of extravesical spread.

C   effacement of the fat between the seminal vesicles and the posterior bladder wall is diagnostic of invasion.

D   the primary tumour is isointense to bladder wall on T2-WSE MRI.

E   on MRI, tumour involvement of perivesical fat is most clearly identified on T1-WSE scans.

5.28. Which of the following are true of transrectal ultrasound (TRUS) of the prostate gland?

A   The prostatic urethra passes through the transitional zone.

B   The peripheral zone constitutes the largest volume of the gland in the adult.

C   Prostatic carcinomas are typically hyperechoic.

D   Granulomatous prostatitis is a cause of a hypoechoic lesion.

E   Infiltration of the seminal vesicles can be identified.

5.29. Concerning CT in the assessment of prostatic carcinoma

A   carcinoma cannot be specifically diagnosed in a normal sized gland.

B   an AP diameter of the gland of 5 cm is diagnostic of malignancy.

C   enlargement of the obturator nodes is a recognized finding.

D   enlargement of the seminal vesicles is diagnostic of invasion.

E   a presacral mass is a recognized finding.

5.30. Concerning MRI of the prostate gland

A   on T2-WSE scans, areas of high signal intensity in the central zone are diagnostic of benign prostatic hypertrophy.

B   prostatic carcinoma typically appears as a hyperintense focus on T2-WSE scans.

C   increased signal intensity in the periprostatic fat on T1-WSE scans is a feature of extracapsular spread of carcinoma.

D   reduced signal intensity of the seminal vesicles on T2-WSE scans is a feature of invasion by prostatic carcinoma.

E   increased signal intensity of pelvic lymph nodes on T2-WSE scans is a feature of metastatic spread from prostatic carcinoma.

5.31. Concerning imaging of the pelvis following radiotherapy
A   ureteric strictures usually have a smoothly tapering distal margin.
B   vesicoureteric reflux is a recognized finding.
C   thickening of the bladder wall is typically most marked anteriorly.
D   increased signal intensity of the bladder mucosa on T1-WSE MRI is a recognized finding in the acute phase.
E   homogeneous reduction in signal intensity of the peripheral zone of the prostate on T2-WSE MRI is a recognized finding.

5.32. Concerning ultrasound and MRI in the assessment of scrotal anatomy
A   on ultrasound, the epididymal body and testis are usually isoechoic.
B   demonstration of intratesticular arterial flow on colour Doppler ultrasound is usually abnormal.
C   the testis and epididymis are isointense on T2-WSE MRI scans.
D   the mediastinum testis is hypointense to the testis on T2-WSE MRI scans.
E   on MRI, the tunica albuginea is only identified if thickened.

5.33. Which of the following are true concerning imaging of the scrotum?
A   On ultrasound, a hydrocele appears as an anechoic collection completely surrounding the testis.
B   Ultrasound typically demonstrates skin thickening in association with a simple hydrocele.
C   Simple hydroceles are isointense to urine on T1-WSE MRI scans.
D   Haematocele is a recognized cause of a hyperintense peritesticular collection on T1-WSE MRI scans.
E   Chronic haematocele is a cause of scrotal calcification.

5.34. Which of the following are true concerning imaging of varicoceles?
A   The majority occur on the right side.
B   The normal pampiniform plexus should not be identified on ultrasound.
C   Ultrasound typically demonstrates varicoceles posterior to the testis.
D   Varicoceles increase in size following the Valsalva manoeuvre.
E   High signal intensity within a varicocele is a recognized finding on T2-WSE MRI scans.

5.35. Scrotal ultrasound demonstrates a cystic epididymal mass. The differential diagnosis includes which of the following?
A   Epididymal cyst.
B   Spermatocele.
C   Cystadenoma of the epididymis.
D   Non-papillary benign mesothelioma (adenomatoid tumour).
E   Sperm granuloma.

5.36. Which of the following are recognized ultrasound features of epididymo-orchitis?
A   Increased echogenicity of the epididymis.
B   Isolated enlargement of the tail of the epididymis.
C   A diffusely hypoechoic testis.
D   Hydrocele.
E   Demonstration of flow in the epididymal head on colour Doppler ultrasound.

5.37. Concerning imaging of testicular torsion
A   on scintigraphy, increased blood pool activity in the testis excludes the diagnosis.
B   on scintigraphy, demonstration of increased blood flow around the testis is a feature.
C   on ultrasound, the testis invariably appears hypoechoic within the first 6 hours.
D   Doppler ultrasound usually demonstrates complete absence of flow.
E   diffuse decrease in signal intensity of the testis on T2-WSE MRI is a feature in chronic cases.

5.38. Concerning MRI in the assessment of the scrotum
A   acute epididymitis is associated with increased signal intensity of the epididymis on T2-WSE scans.
B   chronic epididymitis is associated with reduced signal intensity of the epididymis on T2-WSE scans.
C   in acute epididymo-orchitis, the testis typically has increased signal intensity on T2-WSE scans.
D   seminoma is most clearly identified on T1-WSE scans.
E   non-seminomatous germ cell tumours are characterized by homogeneous reduction of signal intensity on T2-WSE scans.

5.39. Concerning the ultrasound features of testicular neoplasms
A   the epididymis is typically enlarged.
B   testicular infarction may result in identical appearances.
C   seminoma usually presents as a focal hypoechoic mass.
D   hyperechoic foci at the periphery of a lesion are a feature of teratoma.
E   Leydig cell tumours typically appear as hyperechoic masses.

5.40. Concerning CT in the assessment of spread from testicular neoplasms
A   left para-aortic lymphadenopathy is a recognized finding with right-sided tumours.
B   isolated internal iliac lymphadenopathy is a typical feature.
C   mediastinal lymphadenopathy is a feature of metastatic seminoma.
D   a cystic para-aortic mass is a recognized finding.
E   following treatment, a residual soft tissue density mass indicates active disease.

Genitourinary tract: Answers

5.1. A True
Static renal imaging is performed about 3 hours after an intravenous injection of ^{99m}Tc-DMSA (dimercaptosuccinic acid). DMSA is taken up by and fixed in the proximal tubular cells. Since rapid loss of isotope does not occur, multiple views of the kidney can be obtained. Static imaging with DMSA demonstrates the distribution of functioning renal tissue, and the ratio of uptake between the two kidneys allows an estimation of split renal function.

5.1. B False
Estimation of glomerular filtration rate requires the use of isotopes such as ^{99m}Tc-DTPA, ^{125}I-iothalamate or ^{51}Cr-EDTA which are excreted by glomerular filtration. ^{99m}Tc-DMSA and ^{99m}Tc-glucoheptonate are only suitable for static imaging and can assess divided renal function, the distribution of intrarenal function and sites of functioning ectopic renal tissue.

5.1. C True
In the normal situation, 10–20% of injected DMSA will be excreted by the kidney, and the bladder may be identified if the field of view is large enough. In an obstructed kidney, retention of isotope in the dilated renal pelvis will result in an overestimation of functioning renal tissue in the obstructed kidney. Estimation of renal function at 2 minutes as well as 24 hours will help to overcome this problem.

5.1. D True
If divided renal function is all that is required, anterior and posterior views are adequate. However, if renal scarring needs to be identified, then both posterior oblique views are necessary to provide good definition of the renal cortex. In this respect, imaging using single photon emission computed tomography (SPECT) is more sensitive than planar static imaging.

5.1. E True
In some cases of renal tubular dysfunction, for example renal tubular acidosis, renal uptake of DMSA may be poor in the presence of good renal function.

(Reference 5)

5.2. A False
^{99m}Tc-DTPA (diethylenetriamine penta-acetic acid) is excreted solely by glomerular filtration, whereas an isotope such as ^{131}I-hippuran is excreted by a combination of glomerular filtration and tubular secretion. Dynamic renal imaging with DTPA may be used to assess renal perfusion, total renal function, divided renal function and renal obstruction.

5.2. B False

Dynamic renal imaging with DTPA is performed continuously after the rapid intravenous injection. In the first 30 seconds, activity is seen normally in the major vascular structures, the liver, spleen and in the heart. During this phase, renal perfusion can be assessed. Imaging at 1–2 minutes shows the distribution of renal function, allowing the split function to be estimated. Sequential imaging thereafter will demonstrate the collecting systems and bladder. At this stage, renal tract obstruction can be assessed.

5.2. C False

Activity may be identified in the calices and collecting systems at later stages in the scan due to pooling of isotope, especially in an extrarenal pelvis. Imaging following a diuretic or after standing up will show that this is not due to obstruction.

5.2. D True

The T/A curve from a normal DTPA renal scan consists of three separate segments. The initial steep upward rise represents the vascular phase and is dependent upon renal blood flow. The more gradually rising second segment is the accumulation phase and reflects filtration of the isotope and entry into the renal tubules, with the point of maximum accumulation at the peak of the curve. The third, downward-sloping segment is the excretory curve and represents net loss of activity as the rate of accumulation of isotope is exceeded by the rate of drainage.

5.2. E True

The ureters may normally not be identified at all on the dynamic DTPA scan. Transient visualization of the ureter is also not abnormal, but continuous visualization of the ureter indicates an abnormal ureter.

(Reference 5)

5.3. A True

Reduced vascularity in an obstructed kidney is manifest on the DTPA scan as delayed appearance of activity in the renal cortex and on the T/A curve as a slow rising curve. The time to maximal activity, which may be represented by T_{max}, is increased.

5.3. B False

A spontaneous fall in the T/A curve before 20 minutes virtually excludes obstruction. A rising curve, while suggestive, does not prove the presence of obstruction. If the effect of gravity or a diuretic results in prompt emptying of the collecting systems, obstruction can be excluded. If there is impaired washout following diuretic then obstruction can be diagnosed, taking into account various limitations (see below).

5.3. C True

When there is significant reduction of renal function, the kidney's response to diuretic may be impaired, producing a false positive result for obstruction. This may be overcome by using a double dose of diuretic. Other situations that may result in a false positive scan include dehydration, in the neonate and elderly patient (due to a lower GFR and therefore an impaired diuretic response), if the diuretic is not given intravenously, or in the presence of gross reflux.

5.3. D True
In the presence of lower urinary tract obstruction, washout of activity from the collecting system into a grossly dilated ureter may occur, and if only activity over the renal areas is measured, a false negative result may be obtained.

5.3. E True
The scintigraphic features of a PUJ obstruction on dynamic renal scans include reduced perfusion, a rim of functioning tissue around multiple photopenic areas that represent dilated calices, gradual passage of activity into the calices and absent drainage into the bladder. Administration of diuretic may result in ballooning of the renal pelvis.

(Reference 5)

5.4. A False
^{99m}Tc-MAG-3 (benzoylmercaptoacetyltriglycerine) is a newer renal imaging agent that is excreted by tubular secretion in a similar manner to hippuran. It has an extraction efficiency three times greater than ^{99m}Tc-DTPA. However, normal uptake by the liver may make it difficult to obtain satisfactory regions of interest over the right kidney for split function analysis.

5.4. B True
In cases of prerenal failure (i.e. due to hypovolaemia), dynamic renal scanning typically demonstrates normal blood flow with good uptake of isotope. However, there will be delayed transit of isotope and minimal excretion.

5.4. C False
In patients with ATN, dynamic scanning shows some reduction in perfusion, with absent or poor uptake and no excretion into the bladder. The reduced or absent uptake is manifest on the 2-minute images, which show essentially a blood-pool image in both kidneys which is equal to that in the liver and spleen. On later images, there is progressive fading of renal blood pool activity as the isotope enters the extracellular space. Dynamic renal scanning can provide a positive diagnosis of ATN.

5.4. D True
Parenchymal diseases such as acute glomerulonephritis typically show severe impairement of renal perfusion, with poor uptake, delayed intrarenal transit and poor or absent excretion. The features allow differentiation from ATN but most other causes of acute renal failure will show similar features.

5.4. E False
In both renal vein thrombosis and renal artery occlusion, the dynamic scan will demonstrate severely impaired perfusion, poor or absent uptake and absent excretion.

(References A, 18; B, C, D, E, 5)

5.5. A False
The echogenicity of the renal cortex was originally described as being less than that of the liver or spleen. However, re-evaluation with modern, electronically focused transducers has indicated that renal cortical echogenicity may be equal to that of the liver or spleen, but should not exceed it. Diffusely increased parenchymal echogenicity is a non-specific finding that may require renal biopsy for diagnosis.

5.5. B True
Cellular infiltration within the renal parenchyma results in diffusely increased cortical echogenicity (greater than that of the liver or spleen) with preservation of corticomedullary differentiation. This pattern of abnormality is seen particularly with diffuse renal disease, including acute and chronic glomerulo-nephritis, ethylene glycol poisoning, amyloid and leukaemia.

5.5. C True
Renal abnormalities in AIDS patients include acute tubular necrosis, focal nephrocalcinosis, interstitial nephritis, intrarenal infections and proteinuria. Ultrasound may demonstrate diffusely increased parenchymal echogenicity with preservation of corticomedullary differentiation, attributed to either tubular dilatation or focal segmental glomerulosclerosis. Increased pyramidal echogenicity is attributed to medullary nephrocalcinosis.

5.5. D True
In the acute stage of renal vein thrombosis, ultrasound demonstrates a global decrease in cortical echogenicity. After approximately 7–10 days the parenchymal echogenicity increases, eventually becoming greater than that of the liver. Suppurative pyelonephritis is another cause of unilateral increase in parenchymal echogenicity.

5.5. E False
Conditions associated with bilateral cortical nephrocalcinosis can all result in increased cortical echogenicity. These include oxalosis, Alport syndrome, chronic glomerulonephritis and renal cortical necrosis.

(Reference 11)

5.6. A True
Ultrasound can demonstrate advanced cases of renal papillary necrosis when there has been separation and sloughing of the necrotic papilla with a resulting cavity in the medulla. This may appear as triangular, cystic spaces arranged around the renal sinus. At the periphery of the cyst, a bright echo representing the arcuate vessels may be identified. This finding helps to distinguish the condition from hydronephrosis.

5.6. B True
Renal sinus lipomatosis is a common condition in the elderly when it represents a response to parenchymal atrophy. It may also occur when there is parenchymal destruction or in obesity. Ultrasound may demonstrate an enlarged kidney with a thin hypoechoic rim of atrophic cortex. The fat in the

enlarged sinus usually appears hyperechoic, as does the normal renal sinus fat, but may occasionally appear as a hypoechoic mass.

5.6. C False
In cases of medullary nephrocalcinosis, the renal cortex has a normal appearance but the medullary pyramids appear hyperechoic, typically without distal acoustic shadowing. Causes include distal renal tubular acidosis, frusemide therapy (in children), prolonged ACTH therapy, Cushing's syndrome, Bartter's syndrome and primary and secondary hyperparathyroidism.

5.6. D False
Hydronephrosis results in anechoic spaces within the hyperechoic renal sinus fat caused by the dilated calices. However, this does not always correlate with obstructive uropathy. PC dilatation may be seen in the presence of a full bladder, especially in children. Also, situations in which there is increased urine flow (overhydration, diuretic administration, diabetes insipidus) can result in prominence of the calices. Finally, an extrarenal pelvis may mimic hydronephrosis.

5.6. E False
The PC systems may not be dilated in dehydrated patients, particularly in the presence of acute obstruction. With chronic renal failure, PC dilatation is to be expected, one exception being in cases of retroperitoneal fibrosis.

(References A, B, C, 37; D, E, 11)

5.7. A True
Renal infarction may appear on ultrasound as a focally increased area of parenchymal echogenicity with thinning of the involved cortex. A similar appearance has been described with chronic atrophic pyelonephritis. The echogenicity of an infarct is related to its age, with acute infarcts appearing hypoechoic.

5.7. B True
The characteristics of a hypertrophied column of Bertin include a mass that indents the renal sinus from a lateral aspect, is clearly defined from the renal sinus, has a maximum diameter of less than 3 cm, is contiguous with normal renal parenchyma, and has an echotexture that is similar to that of normal renal parenchyma. Renal scintigraphy with ^{99m}Tc-DMSA will confirm that the mass is composed of functioning renal tissue.

5.7. C True
Ultrasound features that have been reported in acute tubular necrosis include renal enlargement due to interstitial oedema and prominence and enlargement of the medullary pyramids. The renal enlargement is typically in the AP dimension and the degree of enlargement is partly related to the severity of renal damage.

5.7. D True
Renal cortical necrosis is a rare cause of acute renal failure that may complicate shock, sepsis, myocardial infarction, burns and postpartum haemorrhage. There is preservation of the medulla and a thin rim of

subcapsular cortex. Calcification may be seen at the interface of the necrotic and viable tissue as early as 6 days after the event. Ultrasound demonstrates normal sized kidneys with loss of normal corticomedullary differentiation.

5.7. E True
There is a recognized association between patients with chronic renal failure being treated with haemodialysis and the development of renal cysts. Up to 79% of patients may have cysts if treatment has been continued for more than 3 years. The cysts may be multiple and bilateral. These patients are also at increased risk of developing renal cell carcinomas, 40% of which may be multiple or bilateral.

(Reference 37)

5.8. A False
Renal cell carcinoma may have variable echogenicity on ultrasound. In one study, 44% had a similar echotexture to the normal renal parenchyma, whereas 21% appeared slightly hyperechoic and 35% were relatively hypoechoic. Approximately 5–10% of primary carcinomas are of the papillary type. A higher percentage of these are hypoechoic and calcified compared to non-papillary types of carcinoma.

5.8. B False
Angiomyolipomas are solitary, usually right-sided lesions classically occurring in middle-aged females, or bilateral masses when associated with tuberous sclerosis. Ultrasound classically demonstrates a hyperechoic mass, the echogenicity being related to its high fat content and multiple vascular interfaces. These lesions have a tendency to bleed, and areas of haemorrhage or necrosis result in hypoechoic regions.

5.8. C False
Renal oncocytomas are uncommon benign tumours that are usually single but may be multiple and bilateral. Ultrasound usually demonstrates a solid, hypoechoic mass which is characterized by a central, hyperechoic scar. Other lesions may have areas of calcification and cystic degeneration, then being indistinguishable from a renal cell carcinoma.

5.8. D True
The kidney is involved with lymphoma either by haematogenous spread or direct extension from retroperitoneal nodal disease. Multiple masses, diffuse enlargement, a single mass or extension of disease from local nodes may all be observed on ultrasound. The masses are typically anechoic or hypoechoic, and may be associated with distal acoustic enhancement.

5.8. E True
Juxtaglomerular tumour is a rare cause of renin-mediated hypertension occurring most commonly in young women. The tumours are usually small, solitary and confined to the kidney. They have a mean diameter of 2–3 cm and are usually subcapsular in location. The tumour is usually echogenic but may be hypoechoic if haemorrhage has occurred.

(Reference 37)

5.9.   A True
On precontrast CT, simple renal cysts are well-defined, homogeneous, round masses typically with attenuation values in the range of water (-10 to $20\,HU$). The wall is thin and usually imperceptible. There is no immediate change in attenuation values after contrast enhancement. CT has difficulty in the diagnosis of small cysts (less than $1\,cm$) due to partial volume averaging.

5.9.   B False
Most renal cysts are unilocular and arise in the cortex. Therefore they tend to distort the parenchymal outline. Less commonly, they originate within and are confined to the medulla. Ultrasound criteria for the diagnosis of a simple cyst include absence of internal echoes, a thin sharply defined wall, far wall enhancement and increased through transmission of sound.

5.9.   C True
On MRI, simple renal cysts have the signal characteristics of water, being uniformly hypointense on T1-WSE scans and hyperintense on T2-WSE scans. The cyst wall is usually not perceptible.

5.9.   D False
Calcification has been reported in 1–3% of cysts. A calcified cyst can be considered benign if there is only a small amount of calcium in the wall or septa, there is no associated soft-tissue mass, the centre of the mass has water density, and no portion of the mass enhances on postcontrast CT. If the calcification is extensive, irregular and thick, and especially if it is not peripheral, it cannot be considered benign.

5.9.   E True
In one series, almost 70% of patients with APCKD had one or more high-density cysts (58–$84\,HU$), usually due to infection or bleeding. Most other pathologically proven cases of hyperdense cysts have been reported as isolated cases. On CT, if these cysts are smooth, well-defined and homogeneous, do not enhance on postcontrast scans and show no change in size on consecutive scans, they may be considered benign.

(Reference 15)

5.10.   A True
Septations within a cyst may result from two adjacent cysts sharing a common wall, or may be a manifestation of healing or organization of a cyst that has haemorrhaged or been infected. Septations are often partial. If septa are thin (less than $1\,mm$), smooth, and attached to the cyst wall without associated thickening, they may be considered to be of no clinical significance.

5.10.   B True
It is impossible to distinguish benign from malignant nodularity on radiological grounds alone. Malignant nodules are frequently in the base of a cyst near the renal parenchyma. This nodularity may be the result of neoplasm arising in the cyst epithelium or due to asymmetric cystic necrosis of a solid tumour.

5.10.   C False
Thickening of a cyst wall excludes the diagnosis of a simple cyst, but the cyst may still be benign. A benign cystic mass with thick walls may be due to an infected simple cyst, abscess or organizing haematoma. However, such an appearance may be the only manifestation of a cystic renal cell carcinoma.

5.10.   D False
The two commonest multiloculated renal masses in adults are renal cell carcinoma and multilocular cystic nephroma. The latter condition is most common in childhood. In adults, there is a strong female predominance. Calcification may be present in both conditions, but extensive calcifications favour renal cell carcinoma.

5.10.   E False
On MRI, the signal intensity of cyst fluid is an unreliable sign in the distinction of benign from malignant cystic renal masses. Studies have shown that cyst fluid signal intensities of proven cystic renal cell carcinomas can be identical to those of simple renal cysts. The only indicators of malignancy are thick septations and nodularity, as is the case with other imaging modalities.

(Reference 15)

5.11.   A True
Renal cell carcinomas are usually heterogeneous and hypodense compared to renal parenchyma on precontrast CT scans but may occasionally be isodense or hyperdense. On postcontrast CT, they typically enhance to a lesser degree than normal parenchyma, although during the arterial phase of a dynamic postcontrast scan they may transiently become hyperdense.

5.11.   B True
An unsharp interface with the renal parenchyma is typical but the mass may appear encapsulated (due to a pseudocapsule). If present, the wall of a cystic tumour is usually thick and irregular. Other CT findings include renal vein or IVC invasion (manifest as an intraluminal filling defect or as enlargement of the vein), collateral circulation, nodular areas of soft-tissue attenuation in the perinephric space, and tumour calcification.

5.11.   C True
Local lymph nodes less than 1 cm in diameter are considered normal, those 1–2 cm in size are considered indeterminate while those greater than 2 cm in diameter are almost always enlarged due to tumour. Local hilar, periaortic and paracaval nodes are typically involved. Nodal enlargement may also be due to reactive hyperplasia. Direct extension into the liver, spleen and adrenal gland are also features.

5.11.   D False
CT features of extracapsular extension into the perinephric fat include an indistinct tumour margin with strands of thickened connective tissue septa extending posteriorly and laterally where the fat is most abundant. Direct anterior tumour spread is limited by the anterior pararenal fascia. However, connective tissue septa and fascial planes may also be thickened due to oedema, collateral vessels or haemorrhage.

5.11. E True
Patients at high risk of recurrence include those with large tumours at presentation, incomplete resection and lymph node or adrenal metastases. CT features suggestive of local recurrence include a soft-tissue mass in the renal fossa, asymmetric or ipsilateral psoas muscle enlargement, or involvement of local organs.

(Reference 2)

5.12. A True
Renal lymphoma is a late manifestation of systemic lymphoma, with about 5% of patients having renal abnormalities at imaging, while 33–52% of patients have postmortem evidence of renal involvement. CT findings are variable and include perinephric involvement with compression and displacement of the kidney, fascial thickening and/or obliteration of the perinephric spaces.

5.12. B True
Other CT findings include diffuse renal enlargement and bilateral renal masses that are typically homogeneous and isodense, but may show areas of low attenuation following treatment. On postcontrast CT, the masses show less enhancement than normal renal parenchyma. Direct extension from retroperitoneal disease or a solitary renal mass may also be seen.

5.12. C True
A renal adenoma has been defined as a mass confined to the cortex, less than 2 cm in size and with no evidence of haemorrhage, necrosis, mitoses, cellular atypia or metastasis. CT may demonstrate a solitary or multiple bilateral solid masses that may be calcified. They cannot be distinguished from small renal cell carcinomas.

5.12. D True
An oncocytoma is a subtype of renal adenoma. A central, non-enhancing stellate scar is a characteristic but not pathognomonic CT feature. Such an appearance may also be seen with renal cell carcinoma. Also, those tumours without a scar cannot be differentiated from renal adenomas or a papillary renal cell carcinoma.

5.12. E True
Angiomyolipomas are composed of blood vessels, smooth muscle and fat and their CT appearance will depend upon the varying amounts of these tissues, and also on the presence of haemorrhage. Some fat is seen in almost all cases. Although considered benign, they may be locally invasive and regional lymphadenopathy has been described. If the tumour is predominantly extrarenal, the differential diagnosis will include liposarcoma.

(Reference 2)

5.13. A True
In acute pylonephritis, ultrasound may demonstrate an enlarged kidney with anechoic areas at the corticomedullary junction due to the presence of microabscesses and necrosis in the outer part of the medulla. These result from

ascending bacterial infection. Renal enlargement is related to an increase in water content of the affected kidney.

5.13.   B True
Acute focal bacterial nephritis (acute lobar nephronia) represents an inflammatory mass without drainable pus. Ultrasound characteristically shows a poorly defined hypoechoic mass that may disrupt the corticomedullary junction. With antibiotic therapy, anechoic areas develop due to liquefaction. Eventually, the mass should resolve completely.

5.13.   C False
Xanthogranulomatous pyelonephritis is a rare form of inflammatory disease classically in patients with obstructive uropathy secondary to long-standing calculi (seen in 50–80% of cases) and chronic infection. The disease process may be diffuse or segmental (focal). In the diffuse form, the kidney is enlarged but maintains its reniform shape. The parenchyma is replaced by multiple circular masses that surround the central echo complex. These represent the dilated pus-filled calices. Their echogenicity depends upon the amount of debris and necrosis that is present.

5.13.   D True
In the rarer form of segmental xanthogranulomatous pyelonephritis, ultrasound may demonstrate a single or multiple hypoechoic masses surrounding a single calix that contains a calculus.

5.13.   E True
Renal fungus balls are most commonly due to *Candida* in patients who are immunocompromised. Renal involvement is usually secondary to systemic infection. Ultrasound characteristically shows an echogenic mass in the renal pelvis that does not shadow. The differential diagnosis includes tumour, blood clot, or pyogenic debris.

(Reference 37)

5.14.   A True
Acute pyelonephritis is associated with interstitial oedema and severe vasoconstriction that may be in a lobar distribution (acute lobar nephronia). CT findings include diffuse renal enlargement and areas of low attenuation due to oedema on precontrast scans. Postcontrast CT shows wedge-shaped poorly enhancing or non-enhancing areas that may become hyperdense on delayed scans. Loss of corticomedullary differentiation, obliteration of renal sinus fat and caliceal distortion are additional findings which may disappear after successful treatment.

5.14.   B True
Gas may be seen in the renal collecting system, parenchyma, in a subcapsular location or in the perinephric space in cases of emphysematous pyelonephritis. Gas may also be seen in cases of xanthogranulomatous pyelonephritis.

5.14.   C True
Renal abscess may complicate acute pyelonephritis or result from infection of a renal cyst. CT typically demonstrates a well-defined, hypodense mass, with a thick, slightly irregular wall that may show enhancement on postcontrast scans.

Gas is seen within a minority of renal abscesses. Thickening of the perirenal fascia is also seen.

5.14. D True
CT features of xanthogranulomatous pyelonephritis include a central staghorn calculus that may fill the whole of the renal pelvis, poor or absent contrast excretion, discrete solid masses and extension into the perinephric space. Multiple non-enhancing round areas within the renal contour correspond to dilated calices. Their attenuation values are greater than that for urine. Pericaliceal enhancement is also a feature.

5.14. E True
Renal hydatid is a cause of a calcified, cystic renal mass. Hydatid may be focal or diffuse. The calcification occurs in the cyst walls, which may be thick or thin, and may enhance on postcontrast scans. Larger cysts may contain daughter cysts (brood capsules). The density values of the cyst contents may be 10–20 HU higher than for simple cyst fluid.

(Reference 2)

5.15. A True
Transitional cell carcinoma of the upper ureter, renal pelvis and collecting system accounts for 5–10% of all renal tumours. Tumours may be solitary or multiple. Three CT patterns have been described. Large infiltrating tumours may expand the renal parenchyma, invade the renal pelvis and obliterate renal sinus fat. The kidney usually retains its reniform shape.

5.15. B True
Other CT appearances of transitional cell carcinoma include a sessile intraluminal mass (the commonest finding) and concentric or eccentric ureteric or pelvic wall thickening. CT can easily differentiate soft-tissue masses from other causes of radio-opaque renal pelvic filling defects such as blood clot or calculi.

5.15. C True
Surface calcification is a rare finding with transitional cell tumours and will result in increased density of the tumour on precontrast CT. Adherent blood may also increase the attenuation values. Intratumoral calcification is suggestive of a squamous component to the tumour.

5.15. D True
On precontrast CT, the tumour is usually slightly hypodense to renal parenchyma. It is typically avascular or hypovascular at angiography but, rarely, may show slight transient enhancement on postcontrast scans.

5.15. E True
Extrarenal spread of transitional cell carcinoma is usually central, with extension out through the renal hilum, with or without local lymphadenopathy. CT cannot differentiate between a stage I tumour, confined to the urothelial mucosa, and a stage II tumour, that has invaded the muscularis.

(Reference 2)

5.16. A False
The renal capsule cannot be visualized by MRI. However, chemical shift artefact can occur around the lateral borders of the kidney at the interface between water-containing renal parenchyma and perinephric fat. The perirenal fascia can sometimes be identified on T1-WSE scans as thin, low-intensity bands contrasted against the hyperintense fat.

5.16. B False
Corticomedullary differentiation is apparent on T1-WSE scans when the renal cortex appears relatively hyperintense to the medulla. This differentiation becomes more apparent if the patient is well hydrated and on STIR sequences, but is lost on T2-WSE scans.

5.16. C False
The renal collecting system is best visualized on T1-weighted images, when the low signal intensity of urine is contrasted well against the hyperintense renal sinus fat. On T2-WSE scans, urine in the PC system appears hyperintense to renal sinus fat. However, the non-dilated PC system is usually not distinguishable from central renal sinus fat.

5.16. D True
On MRI, loss of corticomedullary differentiation is a sensitive sign of renal parenchymal disease but is very non-specific, having been demonstrated in cases of chronic glomerulonephritis, end-stage renal failure and chronic ischaemia.

5.16. E True
Reduction of renal cortical signal intensity in patients with sickle cell nephropathy is most evident on T2-WSE scans. This may be due to the presence of abnormal iron metabolism in the renal cortex, rather than simply due to the paramagnetic effect of the iron. The renal medulla appears normal in these patients. Reduced renal cortical signal intensity has also been described in paroxysmal nocturnal haemoglobinuria.

(Reference 3)

5.17. A True
Lipid within the renal parenchyma in cases of xanthogranulomatous pyelonephritis can result in increased signal intensity of the renal parenchyma. The signal intensity, however, is not the same as that of surrounding fat. Purulent urine within obstructed calices also has higher signal intensity than normal urine on T1-WSE scans, but renal calculi may not be identified.

5.17. B True
Regardless of the presence of haemorrhage or necrosis, renal cell carcinomas can demonstrate areas of high signal on both T1- and T2-WSE scans. The tumour generally shows prolongation of T1 relaxation times, appearing hypointense to normal renal parenchyma. Lymphadenopathy and venous invasion are also well demonstrated.

5.17. C True
The signal intensity of angiomyolipoma will depend upon the relative amounts of fatty and soft-tissue components present. Areas of fat will appear hyperintense on T1-WSE scans and show moderately increased signal intensity on T2-WSE scans.

5.17. D False
Renal metastases have been demonstrated by MRI and typically appear as multiple masses with low signal intensity on T1-WSE scans and large flip-angle GE sequences. Renal lymphoma can also produce homogeneous low signal intensity masses on T1-WSE scans with associated loss of corticomedullary differentiation. Adjacent lymphadenopathy may have a similar appearance.

5.17. E True
Simple renal cysts appear as homogeneous, well-defined masses of low signal intensity on T1-WSE scans and high signal intensity on T2-WSE scans. The wall is not visible. Infection or haemorrhage within a cyst can cause reduction of T1 relaxation times and increased signal intensity. In APCKD, MRI may demonstrate multiple cysts of varying signal intensities due to haemorrhage at different stages.

(Reference 3)

5.18. A True
Early cellular-mediated rejection is usually seen within the first week after transplantation. On ultrasound, the kidney is enlarged secondary to oedema. Initially there is an increase in the AP diameter of the transplant kidney (normal AP diameter is 3–5 cm). At this stage, an increase in renal length is rare.

5.18. B False
The renal parenchyma becomes hyperechoic and there is accentuation of corticomedullary differentiation, with clearer identification of the enlarged medullary pyramids. The collecting system may be mildly dilated due to atony. Mucosal oedema may result in thickening of the urothelial epithelium.

5.18. C True
The RI is the ratio of the peak systolic velocity minus the peak diastolic velocity divided by the peak systolic velocity. Values of 0.9 or greater are considered abnormal, indicating a reduction in diastolic flow. The most likely cause of the abnormal vascular impedance depends upon the time after surgery. In the early postoperative period, increased RI may be due to renal vein occlusion or rarely due to hyperacute rejection. Loss of diastolic flow within 24 hours is suggestive of severe ATN. During the second postoperative week, the commonest cause is acute rejection. Rarer causes include acute hydronephrosis, extrarenal compression and acute pyelonephritis.

5.18. D False
Acute tubular necrosis is the third diagnosis that needs consideration in the assessment of 'medical' causes of graft failure (the other two being acute rejection and cyclosporin toxicity). It typically complicates transplantation with a cadaver kidney. Imaging will show features that may be seen in both

rejection and cyclosporin toxicity, including loss of corticomedullary differentiation on MRI.

5.18. E False
Reduced perfusion of the kidney on dynamic renal scintigraphy is also a non-specific feature that cannot definitely differentiate between ATN, rejection or cyclosporin toxicity. However, the disparity between perfusion and excretion may be important, with ATN typically showing mild loss of perfusion with severe impairment of excretion, while the reverse is true in cases of rejection. Renal biopsy may also be misleading since rejection can be focal and missed if the biopsy site is inappropriate.

(References A, B, D, E, 15; C, 11)

5.19. A True
'Surgical' causes of renal transplant failure may be divided into urologic (obstructive uropathy, perinephric/perivesical collections) or vascular. Obstructive uropathy is usually due to stenosis at the ureterovesical junction, which may either be a result of ischaemia of the donor ureter due to devascularization, or due to an improper ureteroneocystostomy. Later in the post-transplant course, such a stenosis may be a manifestation of rejection.

5.19. B True
The causes of perirenal or perivesical fluid collections include seroma, lymphocele, urinoma, abscess or haematoma. Urinoma may appear on renal scintigraphy as an area of extraluminal activity typically between the transplant site and the bladder. Other causes of a fluid collection may be 'identified' as photopenic areas or due to their mass effect on the bladder.

5.19. C True
MRI can demonstrate perinephric and perivesical fluid collections as regions with the signal characteristics of water on all pulse sequences. However, a heamatoma may be differentiated since it can appear hyperintense on both T1- and T2-WSE scans depending upon its age.

5.19. D True
Causes of arterial narrowing/stenosis include kinking of the renal artery, a rejection phenomenon in the wall of the donor renal artery with subsequent fibrosis, and also due to narrowing of the anastomosis during surgery. Dynamic renal scintigraphy with ^{99m}Tc-DTPA may show delayed uptake of isotope (there should be renal uptake simultaneous with the visualization of the iliac arteries). However, this is not diagnostic and arteriography must be performed.

5.19. E False
Renal vein occlusion is an uncommon finding and may be suspected in the presence of massive proteinuria. In most cases the occlusion is from extrinsic compression (e.g. due to lymphocele) rather than thrombosis.

(Reference 15)

5.20. A True
On adrenal scintigraphy, regardless of the radiopharmaceutical used, the right adrenal gland may appear more active than the left on posterior views since it is more posterior in position. The right adrenal also usually appears higher than the left. The normal adrenal uptake is less than or equal to 0.2% of the injected dose.

5.20. B True
Adrenal cortical imaging may be performed with ^{131}I-iodocholesterol or ^{75}Se-selenocholesterol. The adrenal medulla is imaged with ^{131}I-MIBG (*meta*-iodobenzylguanidine).

5.20. C True
Cushing's syndrome may be due to an adenoma, carcinoma or bilateral hyperplasia secondary to an ACTH-secreting pituitary adenoma or ectopic ACTH secretion. Uptake by carcinoma is variable and depends upon the metabolic activity of the tumour. Usually there is no uptake but occasionally uptake is faint or marked.

5.20. D True
The typical scan appearance in Conn's syndrome due to adenoma is of moderately increased activity in one gland with normal activity in the other gland. Administration of dexamethasone will result in suppression of activity in the normal gland but not in the gland with the functioning adenoma. When the cause is bilateral hyperplasia, suppression of activity may only occur with high-dose dexamethasone.

5.20. E True
Normal sites of uptake of ^{131}I-MIBG include the liver, the bladder and the colon. MIBG scintigraphy may be of use in the detection of extra-adrenal phaeochromocytomas (approximately 20% of cases) and may also be taken up in sites of metastatic disease from malignant tumours (approximately 10% of cases).

(Reference 5)

5.21. A True
The adrenal glands may have varying shapes depending upon the level of the CT slice and the orientation of the gland. The limbs may be up to 4 cm in length and have a uniform thickness, except at the apex of the gland where the limbs converge. The thickness of the limbs is usually less than 5–7 mm, and any measurement above 10 mm is suggestive of disease. A small adrenal mass may only be identified as a focal enlargement that causes a convex margin to the gland.

5.21. B False
In up to 50% of patients with biochemical evidence of hyperplasia, CT may demonstrate a normal appearance to the adrenals. When abnormal, CT demonstrates bilateral enlargement of the glands, which usually retain their shape. Hyperplastic adrenal glands may contain microscopic or macroscopic

nodules, which may result in a nodular appearance to the glands. The nodules may be up to 2 cm in size.

5.21.   C True
In the presence of a large mass in the right posterior upper abdomen, differentiation between a primary adrenal, hepatic or renal mass may be difficult. Anterior displacement of the IVC is more likely to be due to a retroperitoneal mass than a hepatic mass.

5.21.   D True
Bilateral adrenal enlargement may occur in acromegaly (in almost 100% of cases), in hyperthyroidism (40%), hypertension with atherosclerosis (16%), in diabetes mellitus (3%) and in a variety of malignancies. However, these findings are rarely of any clinical significance.

5.21.   E False
Adrenal hypofunction, either due to autoimmune processes or ACTH deficiency, results in bilateral small glands without calcification. Adrenal calcification may be seen in cases of granulomatous disease or following adrenal haemorrhage. Adrenal hypofunction with bilaterally enlarged glands may be due to haemorrhage, active granulomatous disease, including sarcoid, or, less commonly, metastases. Patients with haemochromatosis may have normal sized or small glands that are slightly hyperdense.

(Reference 2)

5.22.   A True
Cushing's syndrome is due to bilateral adrenal hyperplasia in 70% of cases, an adrenal adenoma in 20% and an adrenal cortical carcinoma in 10%. Adrenal adenomas appear on CT as well-defined, homogeneous masses of soft-tissue or near-water density (if there is a high lipid content) that may occasionally be calcified. Cushing's adenomas are usually at least 2 cm in diameter.

5.22.   B True
If the adrenal glands appear normal at CT, adrenal hyperplasia is the cause. The glands may also appear bilaterally enlarged, although the enlargement may be unilateral.

5.22.   C True
In cases of adrenal hyperfunction, the demonstration of bilaterally enlarged glands with nodularity is suggestive of hyperplasia rather than adenoma since, in the latter case, the contralateral gland usually appears normal or atrophic. However, if a single hyperplastic nodule is identified and the contralateral gland appears normal, an adenoma cannot be reliably excluded.

5.22.   D False
Hyperaldosteronism is due to an adrenal adenoma in 75% of cases and due to hyperplasia in the majority of the rest. Adrenal carcinoma is a rare cause. CT identifies approximately 70% of aldosteronomas, which may only be 5 mm in diameter, and are typically less than 3 cm in size. Most patients with hyperplasia as the cause have normal CT appearances. Conn's adenomas are characteristically hypodense.

5.22. E True
Approximately 50% of adrenal carcinomas are functioning. CT typically demonstrates a large mass (usually at least 4 cm and commonly greater than 10 cm), with areas of low density due to necrosis, patchy enhancement on postcontrast scans, and occasionally calcification. Other findings include lymphadenopathy, renal vein and IVC invasion and metastases (not uncommon at presentation).

(Reference 2)

5.23. A True
Phaeochromocytomas also occur with increased frequency in patients with neurofibromatosis, multiple cutaneous neuromas and multiple endocrine neoplasia type II (MEN-II). In the latter case, the tumours may be asymptomatic with normal urinary biochemistry. When the tumour is symptomatic, urinary catecholamine, vanillylmandelic acid (VMA) and metanephrine levels are almost always raised.

5.23. B True
Approximately 10% of phaeochromocytomas are extra-adrenal in location, most frequently in the paracaval or para-aortic regions along the course of the sympathetic ganglia, or near the organ of Zuckerkandl (located adjacent to the distal aorta at the site of the inferior mesenteric artery origin). Rarer locations are in the mediastinum or in the wall of the urinary bladder (1% of cases).

5.23. C True
Extra-adrenal phaeochromocytomas are malignant in up to 40% of cases, compared to 10% of adrenal lesions. They may be single or multiple. Tumours associated with MEN-II are rarely extra-adrenal and about 75% are bilateral, compared to only 10% of tumours otherwise.

5.23. D True
On precontrast CT, phaeochromocytomas may appear as homogeneous soft-tissue density masses. Larger tumours may show central hypodense areas due to necrosis and, rarely, extensive necrosis may result in a cystic appearance. Calcification is a recognized feature. On postcontrast CT, solid areas of the tumour enhance.

5.23. E True
On ultrasound, a phaeochromocytoma may appear as a purely solid mass of homogeneous or heterogeneous echotexture, with anechoic or hypoechoic regions due to necrosis and hyperechoic regions due to fresh haemorrhage. The masses are usually well-defined.

(References A, B, C, D, 2; E, 37)

5.24. A True
Non-functioning adrenal adenomas have been identified in 2–8% of postmortems and appear to be seen with increased incidence in elderly obese diabetics, elderly women, hypertensives and in association with malignant tumours of the

kidney, bladder and endometrium. The masses are of homogeneous soft-tissue or near-water density and may be calcified.

5.24. B True
Non-functioning adenomas may range in size from 1 to 6 cm. They are well defined with an imperceptible wall and may show mild enhancement on postcontrast CT. They have no distinguishing CT features from functioning tumours and cannot be differentiated by MRI.

5.24. C False
The adrenals are occasionally involved with lymphoma, most frequently of the non-Hodgkin type. Isolated involvement is unusual, with associated retroperitoneal lymphadenopathy being present in most cases. There may be unilateral or bilateral adrenal enlargement. The MRI features are similar to those for adrenal metastases.

5.24. D False
Adrenal myelolipomas are benign tumours with a postmortem incidence of 0.2–0.4%. They are composed of myeloid and erythroid elements as well as fat. They are usually unilateral and asymptomatic but may cause pain due to haemorrhage. CT demonstrates a well-defined mass which may be up to 12 cm in diameter. The density of the mass depends upon the relative amounts of fat and myeloid/erythroid elements present.

5.24. E True
The ultrasound appearance of an adrenal myelolipoma depends upon its variable tissue components and also on the presence of haemorrhage or calcification. Predominantly fatty tumours are echogenic and may result in discontinuity and posterior displacement of the echogenic line of the diaphragm behind the mass.

(References A, B, C, D, 2; E, 37)

5.25. A False
On T1- and T2-WSE MRI pulse sequences, normal adrenal glands have a signal intensity that is greater than that for the diaphragmatic crura and less than that of fat. The signal intensity is usually homogeneous but occasionally the adrenal cortex and medulla can be differentiated due to the lower signal intensity of the latter.

5.25. B True
Adrenal carcinomas have no particular distinguishing features on MRI. On T1-WSE scans, they may have heterogeneous signal intensity which is less than that of retroperitoneal fat. On T2-WSE scans, they may appear hyperintense to surrounding fat. In general, all adrenal masses are most clearly identified on T1-weighted scans.

5.25. C False
Both functioning and non-functioning adrenal adenomas appear on T1- and T2-WSE scans as well-defined masses that have similar signal intensity to the normal adrenal glands. High signal may be seen in the event of haemorrhage,

which is an uncommon complication. Adrenal hyperplasia is not associated with changes in signal intensity.

5.25. D False
Adrenal metastases may appear similar to adenomas on T1-WSE scans but can usually be differentiated on T2-WSE scans since they are commonly isointense to surrounding fat (whereas adenomas are almost always hypointense). Calculation of mass/fat signal intensity ratios from T2-WSE scans may help in the differentiation, but even then up to 33% of cases may be indeterminate.

5.25. E True
Phaeochromocytomas may produce a mass that is markedly hyperintense to surrounding organs and also hyperintense to fat on T2-WSE scans. They have low signal intensity on T1-WSE scans. Large tumours may show heterogeneous signal intensities.

(Reference 2)

5.26. A False
The spermatic cord can be identified anterolateral to the symphysis pubis and medial to the femoral vein. It may appear as a small, oval soft-tissue structure or more usually as a thin-walled 'ring-like' structure containing multiple soft-tissue densities due to the vas deferens and spermatic vessels.

5.26. B False
The seminal vesicles are identified posterior to the bladder, cephalad to the prostate and anterior to the rectum. They are oval or pear-shaped structures of soft-tissue density. There is usually a small triangular fat plane between the bladder wall and the seminal vesicles but this may be effaced if the rectum is distended or the patient is scanned prone. On T1-WSE MRI scans, the seminal vesicles have low to intermediate signal intensity appearing hyperintense to fat on T2-WSE scans.

5.26. C False
On T1-WSE scans, the bladder has homogeneous low signal intensity due to its urine content, and the hypointense bladder wall is difficult to identify. On T2-WSE scans, the urine is hyperintense and the bladder wall remains hypointense, allowing easy differentiation. Chemical shift artefact (due to the fat/urine interface) may obscure the thin bladder wall. It appears on axial views as a hypointense band along one lateral wall and a hyperintense band along the other lateral wall.

5.26. D False
On precontrast CT, the prostate appears as a homogeneous soft-tissue density structure located just posterior to the symphysis pubis and anterior to the rectum. The zonal anatomy cannot be differentiated on CT.

5.26. E True
On T1-WSE MRI, the prostate gland has homogeneous low signal intensity similar to surrounding muscle. On T2-WSE scans, the central and transitional zones can be differentiated from the peripheral zone due to the higher signal

intensity of the latter. This differentiation is consistently made in men under the age of 35 years, but with increasing age there is variation in the size and signal intensity of the peripheral zone.

(Reference 2)

5.27. A True
Primary bladder carcinoma may be either polypoid or infiltrative. The former appears on CT as an irregular filling defect (occasionally with subtle surface calcification) arising from the inner bladder and projecting into the bladder lumen. Infiltration of the adjacent wall is suggested by the presence of contiguous bladder wall thickening (in the absence of biopsy or radiotherapy). Purely infiltrative carcinomas thicken the bladder wall without evidence of an intraluminal mass. Contrast enhancement may help to distinguish tumour from fibrosis.

5.27. B True
CT cannot differentiate between superficial and deep extension into the bladder wall muscle. The earliest feature of extravesical spread is poor delineation of the outer wall of the bladder adjacent to an area of wall thickening. However, this finding is not specific and can also occur due to inflammation, infection, radiotherapy or previous biopsy/surgery.

5.27. C False
Loss of the normal fat plane between the seminal vesicle or rectum and the posterior bladder wall as an isolated sign is not a reliable indicator of invasion. MRI is more sensitive in the diagnosis of invasion into both the seminal vesicles and prostate due to the associated signal changes.

5.27. D False
On MRI, polypoid tumours appear as intermediate signal intensity masses contrasted against the hypointense urine on T1-WSE scans, while any intramural tumour appears as a region of increased signal intensity relative to the hypointense bladder wall on T2-WSE scans. Replacement of a segment of bladder wall by high signal intensity with normal perivesical fat characterizes tumour invasion of deep muscle (stage T3b according to the TNM classification).

5.27. E True
Tumour extension into the perivesical fat is manifest on T1-WSE scans as reduced signal intensity within the fat. Lymphadenopathy is also best evaluated on T1-WSE scans whereas direct extension to the pelvic side wall muscles, seminal vesicles and prostate are optimally assessed on a combination of T2-WSE scans.

(Reference 24)

5.28. A True
Transrectal ultrasound of the prostate can identify the various glandular zones of the prostate (transitional, central and peripheral) as well as the non-glandular anterior fibromuscular stroma. In the young adult, the transitional zone constitutes about 5% of total prostatic volume and is located either side

of the prostatic urethra. It is in this zone that benign hyperplasia develops. The transitional zone is relatively hypoechoic compared to the peripheral zone.

5.28. B True
The peripheral zone constitutes approximately 70% of prostatic volume and is situated on the posterolateral aspect of the gland. The vast majority of carcinomas develop in this region. The central zone is situated at the base of the gland, accounting for approximately 25% of prostatic volume in the young adult. The ejaculatory ducts pass through this zone to reach the verumontanum. The central zone is relatively resistent to disease processes.

5.28. C False
The commonest ultrasound appearance of prostatic adenocarcinoma is of a poorly defined hypoechoic area, typically situated in the peripheral zone. Prostatic carcinomas may extend into the transitional zone, resulting in loss of clear demarcation between the peripheral and transitional zones. It is also possible that up to 30% of carcinomas originate in the transitional zone.

5.28. D True
Unfortunately, not all hypoechoic lesions demonstrated by TRUS are carcinomas. Other biopsy-proven causes of hypoechoic lesions include atypical glandular hyperplasia, prostatic atrophy, ductal dilatation and muscle around the ejaculatory ducts. Therefore, ultrasound-guided biopsy is required when a hypoechoic area is demonstrated in the prostate.

5.28. E True
TRUS can demonstrate capsular breaches by tumour as well as seminal vesicle invasion. The latter is very likely in the presence of a carcinoma in the base of the gland that measures greater than 2 cm in diameter. Ultrasound findings in seminal vesical and ejaculatory duct involvement include a 'halo' sign around the ejaculatory duct due to surrounding tumour, obliteration or displacement of the ejaculatory duct, tumour spreading directly into the seminal vesicle, or an extraprostatic mass at the entrance of the seminal vesicle that obliterates the seminal vesicle–prostate angle.

(Reference 44)

5.29. A True
The CT attenuation value of prostatic carcinomatous tissue is the same as that of normal prostatic parenchyma, and therefore intraglandular carcinomas cannot be identified. Occasionally, small carcinomas may be associated with a low-density focus, but this is not a specific feature and is probably the result of haemorrhage or necrosis within the tumour.

5.29. B False
The prostate is usually enlarged in advanced disease, the normal size being 3–5 cm for transverse diameter and 2.5–4.5 cm for AP diameter. Enlargement of the gland can also be due to benign hypertrophy and CT cannot distinguish this from enlargement due to carcinoma. Spread into the periprostatic fat is manifest as streaky soft-tissue density in the fat.

5.29.   C True
Lymphatic spread from prostatic carcinoma is primarily to the external and
internal iliac node groups then to common iliac nodes. Obturator nodes are
part of the external iliac group and are located posterior to the external iliac
artery and just superior to the obturator internus muscle. Lymph nodes that
are greater than 2 cm in short-axis diameter are virtually always malignant.

5.29.   D False
The size of the seminal vesicles is so variable that enlargement is not a reliable
criterion for diagnosing seminal vesicle invasion. If there is marked asymmetry
between the sizes of the seminal vesicles, invasion of the larger one is
suggestive but is again not totally specific. Also, absence of the normal triangle
of fat anterior to the seminal vesicles may be normal and is unreliable for
diagnosing invasion.

5.29.   E True
Advanced prostatic carcinomas may extend into the bladder neck, distal
ureters, and pelvic side walls. Rectal involvement is unusual due to an effective
barrier formed by the fascia of Denonvilliers. However, spread to the presacral
space may occur.

(References A, B, 1; C, D, E, 24)

5.30.   A False
Benign prostatic hypertrophy (BPH) includes several pathological types
(mixed, sclerotic, fibromuscular, and glandular) that occur in the central zone
of the gland. MRI may demonstrate generalized glandular enlargement and a
wide variation of signal intensity, including areas of both increased and
decreased signal on T2-WSE scans. Calcification is either not identified, or if
large enough results in areas of signal void. Because of the wide range of
abnormalities demonstrated, MRI is of limited value in differentiating nodules
of BPH from other conditions such as carcinoma and chronic prostatitis.

5.30.   B False
On T2-WSE MRI, prostatic carcinoma usually appears as a focus of low signal
intensity in the high intensity peripheral zone. Other findings include areas of
inhomogeneous signal intensity. These features are not specific and can also be
produced by chronic prostatitis.

5.30.   C False
On SE MRI, periprostatic spread is best evaluated on T1-WSE scans and
appears as areas of reduced signal intensity in the normally hyperintense fat.
Obliteration of the normal high signal rim around the anterolateral aspect of
the gland on T2-WSE scans (representing the periprostatic venous plexus) is
also a feature of extracapsular tumour spread.

5.30.   D True
Invasion of the seminal vesicles, bladder neck and local muscle is best assessed
on T2-WSE sequences. Direct extension into the bladder neck or levator ani is
manifest as increased signal intensity whereas invasion of the seminal vesicles
results in reduced signal intensity. The differential diagnosis of hypointensity in
the seminal vesicles includes atrophy and fibrosis.

5.30. E True
T2-WSE scans are useful for distinguishing between lymph nodes and muscle, since nodes involved with tumour are relatively hyperintense to muscle. However, metastatic and reactive lymphadenopathy cannot be distinguished.

(References A, 3; B, C, D, E, 24)

5.31. A True
Ureteric damage following pelvic radiotherapy occurs in approximately 1–4% of cases. The commonest manifestation is stricture formation usually in the distal ureter just above the vesicoureteric junction and extending proximally for a variable length. However, strictures may also occur higher where the ureters cross anterior to the iliac vessels. A smooth tapering distal margin may also be seen with recurrent tumour, which can be assessed with CT.

5.31. B True
Incompetence of the intravesical portion of the ureter may occur as a result of bladder wall fibrosis, leading to vesicoureteric reflux. Ureteric fistula is another recognized though rare complication of pelvic radiotherapy.

5.31. C False
The bladder is the most radiosensitive organ in the urinary tract, being involved in up to 20% of patients following pelvic radiotherapy. Acute symptoms are due to mucosal oedema, haemorrhage and necrosis. With time, fibrosis results in a small-volume contracted bladder with symptoms of frequency and incontinence. In the chronic stage, CT will demonstrate bladder wall thickening, which is most marked posteriorly, and inflammatory changes in the perivesical fat. Vesicovaginal and vesicoenteric fistulae are other complications.

5.31. D True
The earliest MRI abnormality identified is increased signal intensity of the bladder mucosa on T2-WSE scans, initially seen in the posterior bladder wall and trigone area, but eventually involving the whole bladder. With more severe changes, the whole of the bladder wall shows increased signal intensity and becomes thickened. Focal thickening may also be seen, affecting the posterior wall and trigone. Increased signal intensity on T1-WSE scans is thought to be due to haemorrhage. Enhancement of the bladder wall is seen following Gd-DTPA.

5.31. E True
Following radiotherapy, the prostate atrophies and calcification can be identified on plain films and CT. The seminal vesicles also shrink and become of uniformly low signal intensity on T2-WSE scans.

(Reference 40)

5.32. A False
On ultrasound, the epididymal head may appear relatively echogenic but the body and tail are normally slightly hypoechoic compared to the testis. The testis itself has medium-level homogeneous echogenicity and the mediastinum testis can be identified as a hyperechoic band orientated in a longitudinal direction.

5.32.   B False
The major vascular supply to the testis is via the testicular artery, with some supply via arterial anastomoses from the cremasteric and deferential arteries. The testicular artery runs along the posterior aspect of the testis to form one or more capsular arteries that run beneath the tunica albuginea. The capsular artery gives rise to centripetal arteries that enter the testis and branch into recurrent rami. Occasionally, a major branch of the testicular artery penetrates the mediastinum to cross the testis and anastomose with capsular arteries on the other side. Capsular, centripetal and transtesticular arteries can all be demonstrated by colour Doppler ultrasound.

5.32.   C False
The testis is of homogeneous signal intensity. On T1-WSE scans, it has signal intensity between that of water and fat, while on T2-WSE scans it becomes hyperintense to fat, with a signal intensity just less than that of water. On T2-WSE scans, the normal epididymis has intermediate signal intensity, appearing hypointense to the testis. Intratesticular vessels are usually not visualized.

5.32.   D True
The mediastinum testis represents an invagination of the tunica albuginea along the bare area of the testis, through which the testis receives its blood supply, nerves, lymphatics and tubules. The mediastinum testis is recognized as a region of low signal intensity, ranging from 1 to 3 cm in length, on both PDW and T2-WSE scans.

5.32.   E False
The tunica albuginea is a thin layer of dense fibrous tissue that completely surrounds the testis. It is normally identified as a thin layer of very low signal intensity on T2-WSE scans. Because it is well visualized, its integrity in cases of trauma and tumour can be easily demonstrated.

(References A, B, 11; C, D, E, 3)

5.33.   A False
A hydrocele represents a collection of fluid between the visceral and parietal layers of the tunica vaginalis. This fluid completely surrounds the testis except posterolaterally, where the epididymis is closely related to the testis. A hydrocele will not separate these two structures. The hydrocele is confined to the hemiscrotum by the median raphe.

5.33.   B False
Simple hydroceles are not associated with scrotal skin thickening. In chronic hydroceles, or those complicated by haemorrhage or infection, skin thickening, low-level internal echoes, layering debris or septa may be seen.

5.33.   C True
Simple hydroceles have the same signal characteristics to urine on MRI being uniformly hypointense on T1-WSE scans, hyperintense on T2-WSE scans and of intermediate signal intensity on PDW scans. Hydroceles complicated by infection may have inhomogeneous intermediate signal intensity on T1-WSE and PDW scans.

5.33. D True
Hydroceles that are complicated by haemorrhage (haematoceles) may have high signal intensity on all pulse sequences depending upon the age of the haematoma. Septations, if present, appear as thin hypointense bands.

5.33. E True
The ultrasound appearance of a haematocele is related to its age, ranging from an anechoic fluid collection to a complex solid/cystic mass that may eventually form a septate hydrocele. Scrotal calcifications may be present peripherally in a chronic hydrocele. These may not be seen on MRI.

(Reference 24)

5.34. A False
A varicocele is an abnormal tortuosity and dilatation of the veins of the pampiniform plexus in the spermatic cord. Approximately 90% occur on the left side, the majority of the remainder being bilateral. A solitary right-sided varicocele should prompt an investigation for situs inversus or causes of venous obstruction such as pelvic or renal neoplasm.

5.34. B False
Normal veins in the pampiniform plexus are identified on high-resolution real-time ultrasound as multiple tubular anechoic structures surrounding the spermatic cord. The vessels are usually less than 1.5 mm in diameter. One main draining vein measuring up to 2 mm in diameter may be seen around the spermatic cord or posterior to the testis.

5.34. C True
On ultrasound, a varicocele typically appears as a collection of irregular vessels in the region of the spermatic cord or posterior to the testis. The vessels can usually be traced to the inguinal canal, are compressible and usually have diameters exceeding 6 mm.

5.34. D True
Varicoceles will increase in calibre following a Valsalva manoeuvre or on erect positioning. Small reducible varicoceles may not be seen at all when the patient is supine. Venous flow can be detected by Doppler ultrasound and will be augmented by the Valsalva manoeuvre.

5.34. E True
MRI findings in cases of varicocele include widening of the spermatic canal and prominence of the intrascrotal spermatic cord accompanied by prominent serpiginous structures representing dilated veins. The intravascular signal may be high, particularly on T2-WSE scans, due to slow flow.

(Reference 24)

5.35. A True
Epididymal cysts are common masses of unclear aetiology. Ultrasound typically demonstrates a well-defined anechoic mass with through transmission of sound. They occur anywhere along the course of the epididymis. They may contain low-level echoes due to debris from previous haemorrhage. They may

indent the testis, making differentiation from a tunica albuginea cyst difficult. This may also result in heterogeneity of the adjacent testicular parenchyma, mimicking neoplasm.

5.35. B True
Spermatoceles usually originate from the efferent ducts in the head of the epididymis or more proximal rete testis. Multiple spermatoceles may be seen following vasectomy. Ultrasound demonstrates a simple cystic fluid collection that is usually unilocular and most commonly seen in the epididymal head. The differential diagnosis includes a hydrocele of the spermatic cord or an epididymal cyst. They have similar signal characteristics to water on MRI.

5.35. C True
Papillary cystadenomas of the epididymis are uncommon benign tumours that usually occur in the epididymal head. They may be unilateral or bilateral, and are associated with von Hippel–Lindau disease. On ultrasound, they typically appear as multiseptate cystic masses measuring 1–6 cm in diameter.

5.35. D False
Adenomatoid tumours are uncommon benign mesothelial cell tumours that present as an enlarging painless mass. They are most commonly located at the poles of the epididymis, in the tunica vaginalis of the testis, or involving the spermatic cord. Ultrasound demonstrates a solid, homogeneous extratesticular mass that is typically more echogenic than the testis. Other causes of solid epididymal masses include leiomyoma, fibroma, hamartoma, primary or metastatic carcinoma, sarcoma, seminoma and teratoma.

5.35. E False
Sperm granulomata may occur anywhere in the epididymis or vas deferens and are usually small (averaging less than 1 cm in diameter), solid nodules. Ultrasound typically demonstrates a solid mass that is hypoechoic relative to the adjacent testis.

(Reference 24)

5.36. A True
In acute epididymo-orchitis, the epididymis is usually enlarged and hypoechoic but may occasionally appear hyperechoic due to early fibrosis or haemorrhage. Involvement may be either focal or diffuse. The epididymis is usually enlarged and hyperechoic in cases of chronic epididymitis.

5.36. B True
In cases of tuberculous infection the epididymal head is usually involved if infection is blood-borne, whereas retrograde spread from the prostate usually affects the vas deferens. Central necrosis may produce a complex mass on ultrasound. Isolated enlargement of the epididymal tail is suggestive of tuberculous infection.

5.36. C True
Associated orchitis usually manifests as decreased echogenicity and enlarge-ment of the testis. Focal orchitis is less common and usually appears as a peripheral elongated hypoechoic region adjacent to the epididymis. In the

latter situation, the differential diagnosis will include a primary testicular neoplasm exending into the epididymis. Approximately 10% of patients with an intratesticular neoplasm present with an acute scrotum.

5.36. D True
Reactive hydroceles are a frequent finding and may be very large, typically being larger than those associated with tumour. Scrotal skin thickening also occurs with time. Adhesions may be seen between the tunica vaginalis and the epididymis, indicating severe infection or tuberculous epididymitis.

5.36. E True
There is normally no detectable flow in the epididymal head on colour Doppler ultrasound. Therefore, demonstration of flow should be considered abnormal. In one study, of 51 patients with epididymo-orchitis, hypervascularity was demonstrated in the epididymis alone or in the epididymis and testis in all cases. In one case, hypervascularity was limited to the testis.

(References A, E, 11; B, C, D, 24)

5.37. A True
Assessment of testicular torsion can be performed using first pass and equilibrium imaging following intravenous administration of ^{99m}Tc-pertechnetate. Testicular torsion can be diagnosed if there is reduced scrotal blood flow on the dynamic phase with a central photopenic area on the equilibrium phase, indicating an avascular testis.

5.37. B True
In cases of delayed torsion, peritesticular inflammatory changes may result in increased blood flow and a central photopenic area due to the avascular testis. These appearances can also be seen with epididymo-orchitis complicated by testicular ischaemia. The typical findings in epididymo-orchitis are increased blood flow with diffuse increase in blood pool and no photopenic central area.

5.37. C False
On ultrasound, the testis commonly appears normal within the first 4–10 hours although some cases show slight increase in size of the testis and epididymis, with slight decrease in testicular echogenicity. With time the testis becomes significantly hypoechoic and intense peritesticular inflammatory tissue develops, which manifests as peritesticular hypervascularity on colour flow Doppler.

5.37. D True
By the time the patient is scanned, vessels are usually not detected in the torted testis. In one study, of 33 patients with torsion, no flow was demonstrated in 30 cases, decreased flow in 2 cases and normal flow in 1 case. Spontaneous detorsion may result in normal flow on Doppler ultrasound.

5.37. E True
MRI features of subacute torsion (3–5 days post-event) have been described and include demonstration of the site of the twist, swelling and areas of increased signal intensity (due to haemorrhage) in the epididymis, thickening and reduced vascularity of the proximal spermatic cord, and testicular changes. The latter include reduction in size of the affected testis with inhomogeneous

reduction of signal intensity on T2-WSE scans. With time, the testis becomes markedly hypointense and small, and thickening of the tunica albuginea can also be seen.

(References A, B, 5; C, D, 11; E, 3)

5.38.   A True
In acute epididymo-orchitis, the signal intensity of the epididymis on T2-WSE scans is increased and the epididymis appears enlarged. In some cases, heterogeneity of signal intensity is present, likely due to various stages of infection and haemorrhage. Increased flow to the inflamed epididymis is manifest as serpiginous tubular structures with absent signal coursing through the spermatic cord.

5.38.   B True
In chronic epididymitis, the epididymis may be either focally or diffusely enlarged and has decreased signal intensity on T2-WSE scans. Acute exacerbation of chronic epididymitis may not result in increased signal intensity, and therefore cannot be excluded if MRI demonstrates an enlarged, hypointense epididymis.

5.38.   C False
On T2-WSE MRI, orchitis is manifest as focal or diffuse reduction in signal intensity of the testis. Frank abscess formation may result and is identified as a focal area of increased signal intensity. Increased vascularity to the testis may be identified as areas of tubular signal void within the testicular parenchyma.

5.38.   D False
Seminoma has been reported to be almost isointense to normal testis on T1-WSE scans, whereas it produces mildly inhomogeneous reduction of signal intensity on T2-WSE scans. A small rim of normal testicular tissue is usually identified around the tumour.

5.38.   E False
Non-seminomatous germ cell tumours (embryonal cell carcinoma, teratoma and choriocarcinoma) are characterized pathologically by the presence of haemorrhage and necrosis which results in a greater heterogeneity of signal intensity compared to seminoma. On T2-WSE scans, areas of high and low signal intensity will be seen. A hypointense rim due to a fibrous capsule is also a characteristic finding on T2-WSE scans.

(Reference 3)

5.39.   A False
Enlargement of the epididymis due to extension of a testicular tumour is an unusual feature and is more suggestive of inflammatory disease or torsion rather than neoplasm. Other features favouring inflammation include the presence of a large hydrocele and scrotal skin thickening. Also, purely echogenic masses are often benign.

5.39. B True
Testicular infarcts usually appear as hypoechoic areas and may be sonographically indistinguishable from malignancies. Infarcts may become partially hyperechoic due to fibrosis. Global infarction may result in a shrunken testis that is either diffusely hypoechoic or hyperechoic depending upon the degree of fibrosis.

5.39. C True
Seminoma usually appears as a poorly defined homogeneous hypoechoic mass and less commonly as a well-defined focal mass or a focal mass with areas of increased echogenicity. A characteristic but uncommon ultrasound appearance is that of a diffusely hypoechoic enlarged testis. A similar appearance can be seen with acute lymphocytic leukaemia. Seminoma is most common in the fourth and fifth decades of life.

5.39. D True
Hyperechoic shadowing foci within a teratoma are due to calcification, bone or cartilage. Another recognized ultrasound pattern is of an enlarged testis with diffuse alteration of echotexture and broad bands of dense shadowing echoes. This pattern, however, has also been described with testicular carcinoid. Other germ cell tumours include embryonal cell carcinoma and choriocarcinoma. Both are characterized by haemorrhage and necrosis, resulting in complex testicular masses on ultrasound.

5.39. E False
Gonadal stromal/sex cord tumours account for approximately 6% of testicular neoplasms and include Leydig and Sertoli cell tumours. Ultrasound demonstrates a round homogeneous well-defined hypoechoic mass. About 20% are malignant and may be associated with retroperitoneal lymphadenopathy. Gynaecomastia occurs with 20% of benign Sertoli cell tumours and in the majority of malignant tumours.

(Reference 24)

5.40. A True
Initial spread from testicular carcinomas is lymphatic, following the gonadal veins. Left-sided tumours spread to para-aortic and pre-aortic nodes at the level of the left renal vien (L1–2), while right-sided tumours spread to interaorticocaval and paracaval nodes at the L2–3 level (at the right renal hilum and junction of the right gonadal vein with the IVC). Crossover, particularly from right to left, can occur, while crossover from left to right is atypical.

5.40. B False
Internal iliac and femoral nodes are usually not involved unless there has been trans-scrotal orchidectomy or other scrotal/inguinal surgery. High external iliac nodes may be seen, usually only in the presence of renal hilar nodes. Atypical patterns of spread may occur if the tumour has spread beyond the capsule.

5.40. C True
From the renal hilar nodes, spread occurs to the para-aortic/paracaval nodes and then retrocrural nodes from T11–L4. This is followed by extension to

posterior mediastinal, subcarinal, hilar, anterior mediastinal and supraclavicular nodes. Haematogenous spread to lung, liver, bone and brain also occurs. Thoracic metastases from seminoma are usually to mediastinal nodes whereas parenchymal lung metastases are commoner with teratoma.

5.40.   D True
Lymph nodes or other metastases from testicular primaries can appear cystic with CT numbers between −10 and 20 HU. This phenomenon is more common following chemotherapy. Solitary para-aortic/paracaval nodes greater than 1.5 cm or a group of nodes larger than 1 cm should be considered abnormal, as should retrocrural nodes larger than 6 mm.

5.40.   E False
Absence of active disease can be assumed when lymph nodes have returned to normal size, no new masses have developed and serum markers are normal. Raised serum markers indicate residual active disease. Residual masses on CT (either solid or cystic) with normal markers may represent malignancy, teratomatous transformation, or fibrosis. Uniform low-density masses may represent necrosis or mature cystic teratoma, but the presence of active foci cannot be ruled out from the scan appearances.

(Reference 24)

6 Musculoskeletal system

6.1. Which of the following are true concerning ^{99m}Tc-MDP (methylene diphosphonate) bone scintigraphy?
A Bone uptake is primarily dependent upon osteoclastic activity.
B Focal areas of increased activity are a recognized feature of primary hyperparathyroidism.
C Reduced renal activity is a feature of osteomalacia.
D Disuse osteoporosis is characterized by focal reduction in activity.
E Periarticular activity is a feature of Sudek's atrophy.

6.2. Concerning ^{99m}Tc-MDP bone scintigraphy in the assessment of skeletal trauma
A rib fractures are indistinguishable from rib metastases.
B stress fractures typically result in a linear region of increased activity parallel to the cortex.
C shin splints can be differentiated from stress fracture.
D following a fracture, activity is maximal during the first 24 hours.
E fracture non-union can be demonstrated.

6.3. Concerning ^{99m}Tc-MDP bone scintigraphy in the assessment of benign bone conditions
A osteoid osteoma typically shows increased activity on the dynamic phase of a triple-phase bone scan.
B uncomplicated simple bone cyst is not associated with increased activity.
C a bone island is a recognized cause of a focal 'hot' spot.
D increased activity limited to the diaphysis of a long bone is a characteristic finding in Paget's disease.
E the scan appearances are typically normal in the lytic phase of Paget's disease.

6.4. Concerning ^{99m}Tc-MDP scintigraphy in the assessment of musculoskeletal infection
A acute osteomyelitis is a recognized cause of focally reduced bone activity.
B a normal scan appearance excludes the diagnosis of acute osteomyelitis.
C cellulitis is associated with increased blood-pool activity on a triple-phase bone scan.
D acute osteomyelitis is associated with increased activity on the dynamic phase of a triple-phase bone scan.
E increased activity in cases of chronic osteomyelitis indicates active infection.

6.5. Concerning ⁹⁹ᵐTc-MDP bone scintigraphy in the assessment of skeletal metastases

A   the majority of lytic metastases on radiography are associated with a normal scan appearance.
B   worsening of the scan appearance after treatment indicates progressive disease.
C   a 'superscan' is associated with reduced renal activity.
D   in a patient with known malignancy, a solitary rib lesion is most likely benign.
E   metastasis is a recognized cause of focally reduced activity.

6.6. Which of the following are recognized findings on ⁹⁹ᵐTc-MDP bone scintigraphy?

A   Diffuse activity in the lungs in hyperparathyroidism.
B   Gastric activity in the milk-alkali syndrome.
C   Increased renal activity in patients taking chemotherapy.
D   Thyroid activity in patients with amyloidosis.
E   Muscle activity in cases of rhabdomyolysis.

6.7. Concerning ⁹⁹ᵐTc-MDP scintigraphy in the assessment of joints

A   asymmetrical activity in the shoulders may be a normal finding.
B   in the normal hand, activity in the proximal interphalangeal joints is greater than that in the distal interphalangeal joints.
C   in rheumatoid arthritis, activity in the knee is typically equal in the medial and lateral femorotibial compartments.
D   increased activity in the superoposterior aspect of the calcaneus is a recognized finding in ankylosing spondylitis.
E   a sacroiliac index of 2.0 is diagnostic of ankylosing spondylitis.

6.8. Concerning scintigraphy in the assessment of the bone marrow

A   ⁹⁹ᵐTc-sulphur colloid is a suitable radiopharmaceutical.
B   extension of activity into the distal femoral shaft is diagnostic of polycythaemia rubra vera.
C   reduction of activity is a feature of myelofibrosis.
D   focally reduced activity is a feature of sickle cell disease.
E   paravertebral activity is a recognized finding in thalassaemia.

6.9. Concerning imaging of lumbar disc disease

A   CT is as sensitive as MRI in differentiating normal from degenerate disc.
B   approximately 30% of disc herniations are identified at the L3–4 level.
C   on CT, gas density in the spinal canal is a recognized sign of disc herniation.
D   lateral herniation of the L4–5 disc causes compression of the L4 nerve root.
E   at myelography, deformity of the dural sac by disc herniation is most easily identified at the L5–S1 level.

6.10. Which of the following are true concerning imaging of the lumbar spine?
A The posterior margin of the L3–4 disc is normally convex.
B Sequestrated disc fragments are typically located posterior to the theca.
C CT can demonstrate Schmorl's nodes.
D A synovial cyst is a sign of a degenerate posterior facet joint.
E Conjoined root sheaths are typically bilateral.

6.11. Which of the following are true concerning lumbar spinal stenosis?
A Achondroplasia is a recognized cause of developmental stenosis.
B An AP canal diameter of 18 mm is a typical finding in developmental stenosis.
C Hypertrophy of the ligamentum flavum is a recognized finding.
D Hypertrophy of the base of a superior articular facet is a cause of lateral recess stenosis.
E Facet joint hypertrophy is a cause of foraminal stenosis.

6.12. Which of the following are true concerning spondylolisthesis?
A Paget's disease is a recognized cause.
B In lytic spondylolisthesis, the AP diameter of the spinal canal is typically reduced at that level.
C Lytic spondylolisthesis is a cause of lateral recess stenosis.
D On axial CT, a pars defect is located at the level of the pedicle.
E In degenerative spondylolisthesis, the AP dimensions of the spinal canal are increased at that level.

6.13. Concerning CT imaging of the postoperative lumbar spine
A epidural scar is typically hyperdense to disc on precontrast CT.
B epidural scar typically causes indentation on the thecal sac.
C epidural scar enhances uniformly on postcontrast CT in the majority of cases.
D false meningocele is a recognized finding after laminectomy.
E arachnoiditis is a recognized cause of an intrathecal mass.

6.14. Concerning MRI of the spine
A the annulus fibrosus is hypointense on all pulse sequences.
B the posterior longitudinal ligament (PLL) cannot be distinguished from the annulus fibrosus.
C on T1-WSE scans, the spinal cord and CSF are isointense.
D on GE sequences, the CSF appears hyperintense.
E parasagittal scans demonstrate the neural foramina better in the cervical spine than in the lumbar spine.

6.15. Concerning MRI in the assessment of degenerative disease of the lumbar spine
A on T2-WSE scans, disc degeneration is associated with reduced signal intensity of the nucleus pulposus.
B disc bulge and herniated nucleus pulposus can be differentiated.
C sequestrated disc fragments are typically hyperintense to normal nucleus pulposus on T2-WSE scans.
D tears of the annulus fibrosus can be identified.
E enhancement of postoperative scar following Gd-DTPA only occurs within 1 year of operation.

6.16. Concerning imaging of degenerative diseases of the cervical spine
A   the majority of cervical disc herniations occur at the C4–5 level.
B   on precontrast CT, herniated disc appears hyperdense to the thecal sac.
C   in cervical spondylosis, bony impingement on the thecal sac is typically from apophyseal joint osteophytes.
D   the cervical cord normally has a flat ventral surface.
E   ossification of the PLL is a recognized finding.

6.17. Concerning CT in the assessment of the thoracic spine
A   the majority of thoracic disc herniations are seen at the T4–5 level.
B   calcification in thoracic disc herniations is identified in less than 10% of cases.
C   ossification of the PLL (OPLL) is most common at the T4–7 levels.
D   OPLL cannot be differentiated from a calcified disc.
E   epidural lipomatosis typically compresses the thecal sac from the ventral aspect.

6.18. Concerning the imaging features of vertebral haemangiomas
A   they are most commonly identified in the cervical spine.
B   extradural cord compression is a recognized feature.
C   CT demonstrates thickening of the bony trabeculae.
D   high signal intensity on T1-WSE MRI is a recognized finding.
E   they are invariably associated with increased activity on scintigraphy.

6.19. Concerning the imaging features of primary tumours of the vertebra
A   osteoid osteoma is typically located in the pedicle.
B   a paraspinal soft tissue mass is a recognized finding with osteoblastoma.
C   extension into the vertebral body excludes the diagnosis of aneurysmal bone cyst (ABC).
D   on MRI, fluid–fluid levels are diagnostic of ABC.
E   giant cell tumours are a recognized cause of a destructive sacral mass.

6.20. Concerning ultrasound of the knee joint and popliteal fossa
A   a popliteal cyst appears as an anechoic structure posterior to the medial femoral condyle.
B   the posterior surface of the quadriceps tendon can be visualized only in the presence of a joint effusion.
C   the menisci are homogeneously echogenic.
D   synovial hypertrophy can be identified.
E   the patella plica syndrome can be diagnosed.

6.21. Concerning MRI in the assessment of the cruciate ligaments and knee menisci
A   on T1-WSE scans, the menisci appear isointense to hyaline cartilage.
B   a meniscal tear results in a focus of increased signal intensity on T1-WSE scans.
C   the posterior cruciate ligament (PCL) normally has higher signal intensity than the anterior cruciate ligament (ACL).
D   absence of the ACL on both sagittal and coronal images is a sign of ACL disruption.
E   a chronic tear of the PCL appears as a region of increased signal intensity on T1-WSE scans.

6.22. Concerning MRI in the assessment of the knee joint

A   complete and partial tears of the medial collateral ligament (MCL) can be differentiated.

B   the lateral collateral ligament (LCL) is optimally imaged on posterior coronal images.

C   subchondral low signal intensity on T1-WSE images is a feature of chondromalacia patellae.

D   meniscal cysts have high signal intensity on T2-WSE scans.

E   intra-articular loose bodies may have central high signal intensity on T1-WSE scans.

6.23. Which of the following are true of MRI of the knee joint?

A   Synovial hypertrophy typically appears hyperintense on T2-WSE scans.

B   Synovial masses in pigmented villonodular synovitis (PVNS) characteristically show areas of low signal intensity on all pulse sequences.

C   Irregularity of the posterior surface of the infrapatellar fat pad is a sign of early synovitis.

D   An area of low signal intensity on T2-WSE scans in the lateral femoral condyle is a finding in osteochondritis dissecans.

E   A popliteal cyst typically has low signal intensity on T1-WSE scans.

6.24. Concerning conventional and CT arthrography of the shoulder joint

A   double-contrast arthrography can demonstrate partial tears of the inferior surface of the rotator cuff.

B   on arthrography, demonstration of contrast medium superolateral to the greater tuberosity is a feature of a full-thickness rotator cuff tear.

C   on arthrography, diminished size of the axillary recess is a feature of adhesive capsulitis.

D   on CT arthrography, the posterior aspect of the glenoid labrum is typically rounded.

E   on CT arthrography, a Hill–Sachs deformity, if present, is identified on the same axial slice as the coracoid process.

6.25. Concerning ultrasound in the assessment of the shoulder joint

A   in the normal shoulder, a hyperechoic zone separates the deltoid muscle and the rotator cuff tendon.

B   in the normal shoulder, the rotator cuff is thickest posteriorly.

C   inability to visualize the rotator cuff is a sign of a full-thickness tear.

D   thinning of the rotator cuff is a feature of impingement syndrome.

E   identification of focal echogenic areas within an intact rotator cuff tendon is diagnostic of a partial tear.

6.26. Concerning MRI of the shoulder joint

A   the glenoid labrum has intermediate signal intensity on T2-WSE scans.

B   the rotator cuff is optimally imaged in the sagittal plane.

C   the subscapularis bursa is usually only identified in the presence of a joint effusion.

D   increased signal intensity in the rotator cuff tendon on T1-WSE scans is a feature of a partial tear.

E   high signal intensity within the subacromial/subdeltoid bursa on T2-WSE scans is a feature of full-thickness rotator cuff tear.

6.27. Concerning imaging of the hip joint
A   thickening of the medial cortex of the femoral neck is a feature of osteoarthritis.
B   osteoid osteoma is a recognized cause of increased joint space.
C   scalloping of the femoral neck is a feature of pigmented villonodular synovitis.
D   on ^{99m}Tc-MDP scintigraphy, increased activity in the femoral head is a feature of transient osteoporosis.
E   on T2-WSE MRI, increased femoral head marrow signal intensity is a feature of transient osteoporosis.

6.28. Concerning the imaging features of osteonecrosis of the femoral head
A   loss of hip joint space is the earliest radiographic abnormality.
B   increased activity is a recognized finding on ^{99m}Tc-MDP scintigraphy.
C   a normal MRI scan excludes the diagnosis.
D   on T1-WSE MRI, a lesion with a hypointense rim is a recognized finding.
E   on T2-WSE MRI, a lesion with a hyperintense centre is a recognized finding.

6.29. Concerning imaging following hip replacement
A   on plain radiographs, presence of a radiolucent line at the cement–bone interface invariably indicates loosening.
B   heterotopic new bone formation is seen in less than 5% of cases.
C   on arthrography, failure of contrast medium to enter a radiolucent space at the prosthesis–cement interface excludes loosening.
D   on ^{99m}Tc-MDP scintigraphy, increased activity around the acetabulum at 1 year postoperation indicates loosening.
E   on ^{99m}Tc-MDP scintigraphy, focally increased activity at the tip of the femoral prosthesis is a typical finding in infection.

6.30. Which of the following are true concerning the sacroiliac joints?
A   The anteroinferior portion of the joint is cartilaginous.
B   On CT, the width of the synovial portion of the joint is normally 6 mm.
C   Inflammatory sacroiliitis initially results in erosions of the iliac side of the joint.
D   Intra-articular bony spurs are a recognized CT finding in ankylosing spondylitis.
E   Osteitis condensans ilii is typically bilateral.

6.31. Concerning imaging of the temporomandibular joints
A   anterior displacement of the disc is the commonest form of internal derangement.
B   on arthrography, the anterior recesses are filled with contrast medium on the closed mouth view.
C   in anterior displacement with reduction, arthrography is normal.
D   on arthrography, injection of contrast medium into the inferior joint space normally opacifies the superior joint space.
E   on T2-WSE MRI, the disc and posterior ligament (bilaminar zone) are isointense.

6.32. Concerning MRI in the assessment of musculoskeletal infection

A MRI is more sensitive than scintigraphy in the detection of cellulitis.

B acute osteomyelitis is a cause of decreased marrow signal intensity on T2-WSE scans.

C cellulitis and abscess can be differentiated.

D vertebral osteomyelitis is usually associated with abnormal signal in the adjacent disc.

E increased marrow signal intensity on T1-WSE scans is a recognized finding in treated osteomyelitis.

6.33. Concerning MRI in the assessment of the bone marrow

A a bone infarct is characterized by a hyperintense margin on T2-WSE scans.

B a bone island can be differentiated from a metastasis on T2-WSE scans.

C increased haematopoiesis is a recognized cause of focal hypointense lesions on T1-WSE scans.

D aplastic anaemia is a cause of increased marrow signal on T1-WSE scans.

E marrow replacement in Gaucher's disease cannot be differentiated from that due to tumour.

6.34. Concerning CT in the assessment of benign bone lesions

A peripherally located matrix calcification is typical of enchondroma.

B an osteocartilaginous exostosis contains fat density.

C a bone cyst typically has soft-tissue density centrally.

D the nidus of an osteoid osteoma characteristically shows enhancement on postcontrast CT.

E fibrous dysplasia typically has a well-defined, continuous, sclerotic margin.

6.35. Concerning CT in the assessment of malignant bone and soft-tissue tumours

A increased attenuation of the medullary fat is a feature of osteosarcoma.

B the soft-tissue component of osteosarcoma typically becomes hyperdense relative to muscle on postcontrast scans.

C invasion of the medullary cavity is not a feature of parosteal osteosarcoma.

D calcifications in the cartilaginous cap of an exostosis indicate chondrosarcoma rather than osteochondroma.

E any soft-tissue attenuation with a fat-density mass is diagnostic of liposarcoma.

6.36. Concerning MRI in the assessment of neoplasms of the musculoskeletal system

A contrast between tumour and fat is optimal on T1-WSE scans.

B increased T2 relaxation time in the muscle adjacent to a tumour is diagnostic of invasion.

C benign and malignant tumours can be differentiated by relaxation times in the majority of cases.

D lipoma usually has high signal intensity on both T1- and T2-WSE scans.

E leukaemia typically results in focal areas of reduced signal intensity on T1-WSE scans.

6.37. Concerning MRI in the assessment of tumours and tumour-like conditions of the musculoskeletal system

A   areas of low signal intensity on T2-WSE scans are a finding in haemophilic pseudotumour.

B   chondrosarcoma typically appears hypointense to muscle on T2-WSE scans.

C   a soft-tissue haemangioma typically demonstrates signal void on all pulse sequences.

D   residual tumour and postradiotherapy marrow changes cannot be distinguished.

E   invasion of cortical bone can be identified.

6.38. Concerning MRI in the assessment of musculoskeletal trauma

A   a bone bruise results in an area of increased marrow signal intensity on T1-WSE scans.

B   a linear area of low signal surrounded by increased signal intensity on T2-WSE scans is a finding in stress fracture.

C   pseudoarthrosis following a scaphoid fracture can be identified.

D   muscle tears are optimally imaged using T1-WSE sequences.

E   diffuse increase in muscle signal intensity on T2-WSE scans is a finding in compartment syndromes.

6.39. Concerning ultrasound in the assessment of tendons

A   a normal tendon may appear hypoechoic if it lies obliquely to the ultrasound beam.

B   the Achilles tendon is round in its mid-portion on transverse scans.

C   a partial tear of the patellar tendon is a cause of a well-defined, hypoechoic lesion within the tendon.

D   in acute patellar tendinitis, the tendon appears enlarged and hypoechoic.

E   hyperechoic foci within a tendon are a finding in chronic tendinitis.

6.40. Which of the following are true concerning intraoperative ultrasound of the spine?

A   Scanning is typically performed by direct application of the transducer to the cord.

B   Extramedullary tumours are usually more echogenic than the cord.

C   Obliteration of the central canal echo by a mass is diagnostic of an intramedullary lesion.

D   Anterior compression by disc and bony bar cannot be distinguished.

E   Post-traumatic intramedullary haematoma can be identified.

Musculoskeletal system: Answers

6.1. A False
^{99m}Tc-MDP is currently the radiopharmaceutical of choice for bone scintigraphy. Skeletal activity is dependent upon both osteoblastic activity and vascularity. Static imaging is performed 2–4 hours after intravenous injection. Any ^{99m}Tc-MDP that is not taken up by bone is extracted by the kidneys, and therefore the kidneys and bladder should be visualized on a normal bone scan.

6.1. B True
Metabolic bone diseases such as hyperparathyroidism and osteomalacia may result in an overall increase in skeletal activity resulting in a 'superscan'. However, both conditions can also show focal areas of increased activity, due to brown tumours in the former case and pseudofractures (Looser's zones) in the latter case.

6.1. C True
Increased skeletal uptake of radiopharmaceutical leaves less for renal excretion. The characteristic findings with metabolic bone disease include increased uptake of tracer in the long bones, wrists, calvaria, mandible, costochondral junctions and sternum ('tie sternum' sign). Similar features are also seen in renal osteodystrophy.

6.1. D False
Disuse osteoporosis is associated with high rates of local bone formation, and scintigraphy shows increased activity in the involved bones. Paralysed limbs may also show increased activity initially, possibly due to increased blood flow.

6.1. E True
The reflex sympathetic dystrophy syndrome (Sudek's atrophy) is characterized by increased predominantly periarticular activity on the static bone scan. During the dynamic phase of a triple-phase bone scan increased activity may also be seen, reflecting increased vascularity.

(References A, 5; B, C, D, E, 49)

6.2. A False
On bone scintigraphy, rib fractures typically appear as focal regions of increased activity that are aligned so that two or more ribs in the same location are involved. Their activity decreases over a period of 3–6 months. Conversely, rib metastases characteristically appear as more diffuse areas of increased activity that are randomly distributed.

6.2. B False
Stress fractures are classified as fatigue fractures or insufficiency fractures. The former are the result of repetitive abnormal forces on normal bone, whereas the latter result from normal forces on abnormal bones, as may occur in osteoporosis, osteomalacia, Paget's disease, osteopetrosis, rheumatoid arthritis,

fibrous dysplasia, irradiation, and hyperparathyroidism. Scintigraphy typically shows linear increased activity that is orientated perpendicularly to the cortex of the bone.

6.2.   C True
'Shin splints' refers to a condition which is thought to be due to a periostitis in the tibia at the insertion of the soleus muscle. It is typically seen in long-distance runners and dancers. Scintigraphy shows a linear area of increased uptake in the cortex of the tibia on static imaging with normal vascular and blood-pool phases during a triple-phase scan.

6.2.   D False
Following a fracture, repair usually begins within 24 hours and the bone scan may show increased activity at this stage. Abnormal activity is usually present at 3 days and gradually increases, reaching a maximum within several weeks. Following this, activity subsides and eventually reaches normal levels, although increased activity at the fracture site can be seen several years later.

6.2.   E True
In patients with fracture non-union, scintigraphy may demonstrate two patterns. In the first, there is intense homogeneous activity at the fracture site. In the second, there is a line of decreased activity at the fracture site with increased activity on either side.

(References A, B, D, E, 49; C, 34)

6.3.   A True
Osteoid osteoma is a vascular tumour that will show increased activity on all phases of a triple-phase bone scan. The static images may show a characteristic finding with a central zone of markedly increased activity representing the nidus, and a surrounding zone of lower but still increased activity due to the associated sclerosis.

6.3.   B False
Uncomplicated simple bone cysts show either normal or slightly increased activity on bone scintigraphy. In the latter situation, an area of decreased activity may be present in the centre of the lesion. When complicated by a fracture, the lesion will appear 'hot'.

6.3.   C True
Generally, bone islands that are less than 3 cm in diameter show normal ^{99m}Tc-MDP activity on bone scintigraphy. However, lesions that are larger may be associated with increased activity. In the latter situation, absence of activity on the dynamic phase of a triple-phase scan helps differentiate them from a more aggressive lesion. Other benign causes of increased activity include fibrous dysplasia, fibrous cortical defect, eosinophilic granuloma, enchondroma, aneurysmal bone cyst and osteochondroma.

6.3.   D False
In long bones, Paget's disease is characterized by extension from an articular surface into the shaft of the bone. Pathologically, there is increased vascularity and osteogenesis in the active phase of the disease that results in intense

activity in the involved bones. The scan may also demonstrate enlargement and deformity of the bones. However, sarcomatous change cannot be reliably identified.

6.3. E False
During the early osteoporotic (lytic) phase of Paget's disease, most typically seen in the cranial vault as osteoporosis circumscripta, the bone scan shows markedly increased activity, with greater activity in the advancing margins of the lesion and less increased activity in its central portion. Radiographically, disease progression is manifest by a mixed lytic/sclerotic phase which may be followed by a predominantly sclerotic phase. In the late stages of disease, the bone scan may be normal in the presence of marked radiographic abnormality. Response to treatment may also result in a reduction of activity.

(References A, D, 5; B, C, E, 49)

6.4. A True
In acute osteomyelitis, ^{99m}Tc-MDP scintigraphy may show abnormal activity as early as 24 hours after the onset of symptoms. The scan is usually positive until the lesion has healed. Occasionally, very early in the course of the disease, the scan shows a 'cold' area, probably due to interruption of the blood supply. This finding is particularly seen in neonates and young children.

6.4. B False
In the situation described above, the abnormal area changes from one of decreased activity to one of increased activity, as the infarctive process is replaced by one of hyperaemia and repair. In between, there will be a time at which the scan appearances are normal, resulting in false negative scintigraphy. False negative scans are commonest in children, another reason being the difficulty in identifying foci of infection adjacent to the growth plate. Gallium citrate or labelled-WBC scans are more sensitive in these situations but are associated with a higher radiation dose.

6.4. C True
Triple-phase bone scans are performed by obtaining images dynamically during the intravenous injection of radiopharmaceutical, followed by a blood-pool phase after several minutes, followed by delayed static images at 2–4 hours. Cellulitis is characterized by increased activity on the vascular and blood-pool phases followed by a normal appearance on the delayed images.

6.4. D True
In cases of acute osteomyelitis, there is increased activity on all phases of the triple-phase bone scan. Increased activity on the static images differentiates osteomyelitis from cellulitis.

6.4. E False
In chronic osteomyelitis, increased activity on the ^{99m}Tc-MDP scan may be due to active infection, or due to bone remodelling in the absence of infection. Gallium citrate is less sensitive to the process of bone remodelling and is therefore a better indicator of active infection.

(References A, B, E, 49; C, D, 5)

6.5.   A False
The majority of radiographically lytic lesions will show increased activity on ^{99m}Tc-MDP scintigraphy, since at a cellular level both osteoclastic and osteoblastic activity will be present, the latter at a level which is greater than that in the surrounding normal bone. However, approximately 5% of cases of metastatic disease will show a normal scan appearance in the presence of radiographic abnormality. The commonest causes are anaplastic carcinomas and multiple myeloma.

6.5.   B False
Following radiotherapy or chemotherapy, the bone scan may show resolution of abnormalities, findings interpreted as representing a favourable therapeutic response. However, conversion of a scan to a normal appearance does not exclude the presence of viable tumour. Transient worsening of the scan appearance ('flare phenomenon') is also recognized and is attributed to bone healing rather than progressive disease.

6.5.   C True
In cases of advanced, diffuse metastatic involvement of the skeleton, greater than 50% bone uptake of ^{99m}Tc-MDP may occur, resulting in less being available for renal excretion. Consequently, the kidneys may not be visualized and the bone-to-background ratio of counts may be unusually high, resulting in a 'superscan'. This type of scan appearance can be differentiated from that due to metabolic bone disease by the non-uniformity of uptake and since uptake in the calvarium and mandible is much less striking.

6.5.   D True
In 6–8% of cases of metastatic disease, the bone scan shows a solitary lesion, of which approximately 50% will be malignant. However, in one study, solitary rib lesions were identified in 1.4% of cancer patients and 90% of these were due to a benign cause, such as fracture or radiation therapy.

6.5.   E True
Metastatic disease may manifest as one or more areas of focally reduced activity on the bone scan. The cause of this may either be due to total interruption of blood supply to the involved bone by the tumour, or due to total replacement of the bone by tumour, leaving no viable osteoblasts to produce bone that can accumulate the tracer.

(Reference 49)

6.6.   A True
Lung uptake of ^{99m}Tc-MDP can occur in both hyperparathyroidism and mitral stenosis and is possibly related to the diffuse pulmonary microcalcification seen in these conditions. Other recognized causes of thoracic activity include calcified metastases, sarcoidosis, radiotherapy, fibrothorax, berylliosis, bronchogenic carcinoma, breast carcinoma and malignant pleural effusion.

6.6.   B True
The hypercalcaemia associated with the milk-alkali syndrome can result in diffuse gastric uptake of ^{99m}Tc-MDP. Gastric activity also occurs in cases of gastric calcification, mucinous adenocarcinoma of the stomach and in the

presence of significant free pertechnetate. Other recognized causes of non-skeletal uptake in the abdomen include aortic aneurysm, intestinal infarction, metastatic calcification, malignant ascites and metastases from other mucinous adenocarcinomas.

6.6. C True

The kidneys may show an increase in the level of activity in cases of acute tubular necrosis, due to chemotherapy and radiotherapy, in cases of nephrocalcinosis, in sickle cell anaemia and in renal carcinoma. Other causes of increased uptake in the genitourinary tract include thalassaemia major, leiomyoma of the uterus, ovarian carcinoma and metastatic seminoma.

6.6. D True

Other causes of thyroid activity following ^{99m}Tc-MDP include uptake in calcified thyroid cartilage and uptake due to free pertechnetate. Medullary carcinoma of the thyroid, calcified thyroid carcinomas and some thyroid nodules may also accumulate the tracer.

6.6. E True

Any situation associated with muscle necrosis can result in muscle activity on ^{99m}Tc-MDP scintigraphy, and forms the basis of myocardial infarction imaging with pyrophosphates. Inflammatory conditions such as polymyositis are also associated with muscle uptake. Soft-tissue activity may be due to calcification and can be seen in myositis ossificans, calcific tendinitis, gouty tophi and dermatomyositis.

(Reference 49)

6.7. A True

Usually, normal joints show symmetrical activity, although the shoulder joint on the side of dominant handedness can occasionally show greater activity than the other shoulder. On ^{99m}Tc-MDP scans, activity is greatest in the periarticular regions due to the greater bone mass at this site. Also, for this reason, activity will be slightly greater in the medial femoral condyle than in the lateral femoral condyle, since the former is slightly larger.

6.7. B True

In the hands and wrists, the greatest activity is seen around the wrists and the metacarpophalangeal joints. The level of activity diminishes gradually from the thumb to the little finger. Also, within a digit, activity diminishes from the proximal interphalangeal joint to the distal interphalangeal joint.

6.7. C True

Increased activity in inflamed joints is related to the increased degree of osteoblastic activity in the periarticular bone and also to increased vascularity of the synovium and adjacent bone. Therefore, joint inflammation results in focally increased activity in the periarticular bone. The typical pattern in rheumatoid arthritis is one of symmetrically increased activity in the peripheral joints. In the knee, activity is typically equally increased in both femorotibial compartments, unlike the scan appearances in osteoarthritis.

6.7.   D True

Increased activity may also occur at sites of subligamentous erosion, such as occurs at the insertion of the Achilles tendon into the calcaneus in the seronegative arthropathies. Activity may also be identified at the site of eroded calcaneal spurs. The typical pattern of activity in the seronegative 'rheumatoid variants' is one of central skeletal involvement and asymmetrical peripheral joint uptake.

6.7.   E False

The sacroiliac joints (SIJ) are difficult to assess scintigraphically since they normally demonstrate a relatively high level of activity. Cases of unilateral sacroiliitis may be easily identified. In a situation when bilateral disease is suspected, quantitative analysis can be performed using the sacroiliac index, which represents the ratio of counts in an ROI over the SIJ compared to the counts in an ROI over the adjacent sacrum. The normal ratio varies but is typically in the region of 1.4. Values above this support a diagnosis of sacroiliitis but without being able to differentiate different causes.

(References A, B, C, D, 49; E, 5)

6.8.   A True

Following the intravenous injection of ^{99m}Tc-sulphur colloid, about 80–90% of the substance is phagocytosed by the liver and spleen, with the remainder being accumulated in the bone marrow. In normal adults, uptake occurs in the axial skeleton and proximal portions of the femora and humeri. Sulphur colloid uptake also assesses erythroblastic activity, since this corresponds to reticuloendothelial activity, except in cases of aplastic anaemia.

6.8.   B False

Normally, long-bone activity is seen in the humeral heads and the proximal one-third of the femoral shafts. Extension of marrow activity down the humeri and into the distal femoral shafts is seen in any case of marrow hyperplasia. Causes include polycythaemia, haemolytic anaemia, chronic anaemia due to blood loss, megaloblastic anaemia, and also leukaemia and lymphoma. The increased activity may be uniform or patchy.

6.8.   C True

Myelofibrosis may be a long-term complication of polycythaemia. In the early stages of the disease, there may be a relative reduction in the axial uptake of sulphur colloid with extension of activity into the long bone shafts. In late disease, there is a virtual absence of all marrow activity. The scan may also demonstrate massive splenomegaly. Other causes of reduced marrow activity are metastases and radiation.

6.8.   D True

In sickle cell disease, the scan may show focally reduced uptake in asymptomatic patients due to old medullary infarcts. In acute crises, additional areas of reduced activity may be seen. Following a crisis, the infarcted marrow may return to normal in 1–12 months, or progress to permanent fibrosis. The combination of a bone marrow scan with a ^{99m}Tc-MDP bone scan can help distinguish osteonecrosis from osteomyelitis. Osteonecrosis is suggested by a large defect on the marrow scan and a smaller defect on the bone scan.

6.8. E True
Bone marrow scanning with ^{99m}Tc-sulphur colloid may also outline sites of extramedullary haematopoiesis in cases of chronic anaemia. Thus, the accumulation of activity in a paravertebral mass differentiates extramedullary haematopoiesis from a neurogenic or other tumour and from an abscess.

(References A, D, E, 49; B, C, 5)

6.9. A False
The earliest manifestation of disc degeneration is a reduction in water content, which cannot be identified by CT. However, further disc degeneration results in the accumulation of gas in the disc space (vacuum phenomenon) which is well identified. CT will also identify disc calcification, which occurs in approximately 10% of lumbar disc degenerations.

6.9. B False
Approximately 43% of disc herniations occur at the L5–S1 level, 47% at L4–5 and the majority of the remainder at L3–4. Less than 3% of all herniations are at L1–2 and L2–3. More than 60% of herniations are paracentral in location, 30% are central and 10% are lateral.

6.9. C True
The CT features of disc herniation include a focal disruption of the normal posterior disc margin with extension of relatively hyperdense disc material into the spinal canal. This may obliterate the epidural fat, cause an impression on the thecal sac or displace the nerve root. Other findings include demonstration of gas density within the spinal canal if a gas-containing disc fragment is herniated. Gas within the spinal canal may also be due to a vacuum effect in a synovial cyst arising from a degenerate facet joint.

6.9. D True
The location of the disc herniation will determine the clinical presentation. A paracentral disc herniation will compress the ipsilateral traversing nerve root designated by the lower vertebral level (i.e. a right L4–5 paracentral disc herniation will compress the right L5 nerve root). However, a lateral disc herniation will compress the ipsilateral nerve root designated by the upper disc level (i.e. a right lateral L4–5 disc herniation will compress the right L4 nerve root).

6.9. E False
The degree of deformity of the thecal sac by herniated disc material depends upon the relative size of the thecal sac to the spinal canal. At L5–S1, the spinal canal is typically large compared to the thecal sac and deformity will be least at this level.

(References A, B, C, 46; D, E, 2)

6.10. A False
The posterior margin of the intervertebral disc parallels the margin of the vertebral end-plate. At the L5–S1 level, the posterior disc margin may be slightly convex, but at higher levels the posterior disc margin is slightly

concave. The attenuation value of normal disc is 50–100 HU, approximately twice that of the thecal sac.

6.10. B False
A sequestrated disc fragment refers to herniated disc material that has lost continuity from the native disc and is lying free within the epidural space. Such free fragments may migrate in either a caudal or cephalad direction and, rarely, may rupture into the thecal sac. Sequestrated fragments appear on CT as material of disc density within the lateral recess of the spinal canal.

6.10. C True
A Schmorl's node represents a herniation of the nucleus pulposus through the end-plate into the vertebral body. The lesion appears on CT as a hypodense area at the upper or lower border of the vertebral body, with a dense surrounding margin due to the displaced rim of end-plate. The lesions are often multiple. Their dense margin and location away from the centre of the vertebral body help to distinguish them from metastases.

6.10. D True
A synovial cyst appears on CT as a mass in the lateral recess of the spinal canal and may mimic a disc herniation. However, it is more closely related to the medial margin of the facet joint than the posterior margin of the disc, allowing differentiation. The adjacent facet joint will show degenerative changes including loss of joint space, sclerosis, osteophytosis and a 'vacuum' phenomenon.

6.10. E False
A conjoined root sheath is usually unilateral and typically involves the L5 and S1 roots. The affected nerve root sheaths have their own subarachnoid space but share the same dural sheath. CT demonstrates an asymmetric appearance to the root sheaths, the conjoined sheath being larger and resembling a 'mass' in the lateral recess. However, it has the same density as the thecal sac and is continuous with it, allowing differentiation from a lateral disc herniation.

(References A, 1 and 46; B, C, D, E, 2)

6.11. A True
Lumbar spinal stenosis may be classified as developmental or acquired. The former is usually idiopathic, secondary to inadequate development of the canal or foramina (i.e. a congenitally small canal). Other causes include Morquio's syndrome, hypochondroplasia and Down's syndrome. Patients with congenitally small canals usually develop symptoms of spinal stenosis in later life due to superimposed degenerative changes. Stenosis is also classified as being central canal, lateral recess or foraminal, depending upon the location of the abnormality.

6.11. B False
Features of developmental spinal stenosis include a decrease in the AP diameter of the spinal canal (typically to less than 12 mm), a reduced interpedicular distance (less than 20 mm), thickening of the laminae (to above 14 mm), shortening and thickening of the pedicles and enlargement of the

inferior articular processes. It has also been suggested that the lower limit of canal cross-sectional area is $1.45\,cm^2$.

6.11. C True
The commonest cause of acquired spinal stenosis is degenerative disease involving the discovertebral and posterior facet joints. Disc bulging/herniation combined with spur formation at the margin of the vertebral end-plate causes anterior encroachment upon the thecal sac. Posterolateral encroachment occurs due to facet hypertrophy and thickening and/or buckling of the ligamentum flavum, which normally measures 2–3 mm in thickness. This combination of abnormalities results in central canal stenosis.

6.11. D True
The lateral recess is that part of the spinal canal that is bounded anteriorly by the posterolateral aspect of the vertebral body, laterally by the medial and inferomedial aspect of the pedicle, and posteriorly by the base of the superior articular facet. The lateral recess should measure at least 3 mm in AP dimension. Hypertrophy of the base of the superior articular process may compress the nerve root against the posterior border of the vertebral end-plate.

6.11. E True
The intervertebral foramen contains the exiting nerve root in its upper aspect. It is bounded anteriorly by the posterolateral aspect of the vertebral end-plate and disc, superiorly by the pedicle and posteriorly by the ligamentum flavum and superior articular process of the posterior facet joint. The neural foramen should measure at least 5 mm in its AP dimension. Hypertrophy and upward elongation of the tip of the superior articular facet can compress the nerve root, causing the foraminal type of spinal stenosis.

(References A, 46; B, C, D, E, 46 and 2)

6.12. A True
Spondylolisthesis refers to the situation in which one vertebral body slips anteriorly in relation to the vertebral body below. It may be divided into five categories: (1) congenital; (2) isthmic; (3) degenerative; (4) traumatic; (5) pathological. Disorders of bone such as hyperparathyroidism, Paget's disease, metastatic disease and infection can all result in pathological spondylolisthesis.

6.12. B False
Lytic spondylolisthesis is the result of bilateral defects of the pars interarticularis. The body of the involved vertebra slips forwards, while the neural arch remains aligned with that below, since the facet joints are intact. Consequently, the AP canal diameter is usually increased at the level of the lysis.

6.12. C True
Hypertrophic bone formation on the proximal side of the pars defect in cases of lytic spondylolisthesis can compress the nerve root against the posterior lateral border of the vertebra below, resulting in lateral recess stenosis at the level of the slip. Degenerative spur formation from the end-plate can worsen the situation, as can associated disc herniation.

6.12.   D True
In spondylolysis, axial CT demonstrates bilateral defects in the pars interarticularis that are differentiated from the facet joints by their location, being seen on slices at or just cephalad to the pedicle (facet joints are located at the level of the neural foramen). The defect may have irregular, sclerotic margins. Associated spondylolisthesis is manifest by an increase in AP canal diameter, and the 'pseudobulging disc' sign.

6.12.   E False
In degenerative spondylolisthesis, CT demonstrates erosion of the medial aspects of the superior articular facets, allowing forward slip of the inferior articular facet and neural arch toward the posterior aspect of the upper border of the vertebral body below. This results in a reduction of the AP canal diameter at the affected level(s) and central canal stenosis. Also, forward movement of the inferior articular facet toward the posterior border of the vertebral body below can trap the nerve root in the lateral recess.

(References A, 46; B, C, D, E, 2)

6.13.   A False
Recurrent disc herniation and epidural scar formation are two causes of postoperative symptoms following hemilaminectomy and discectomy. Features of epidural scar tissue on precontrast CT include a linear or curvilinear band of tissue anterior, anterolateral and/or lateral to the dural sac and following its contour. Scar tissue typically has higher density than the dural sac but is hypodense relative to disc material.

6.13.   B False
The thecal sac is characteristically retracted toward the site of scar tissue. However, occasionally scar tissue may appear nodular and cause indentation of the thecal sac, simulating recurrent disc.

6.13.   C True
On postcontrast CT, epidural scar enhances uniformly in over 67% of cases, allowing differentiation from recurrent disc herniation, which does not enhance. The trapped nerve root may be identified within the enhancing scar as a hypodense filling defect. Rim enhancement of herniated disc may also be seen, presumably due to surrounding scar tissue. Using the above criteria (answers A–C), differentiation of disc and scar has been possible in up to 83% of cases.

6.13.   D True
The inadvertent opening of the dural sac or a dural sheath at the time of surgery can lead to the accumulation of CSF in the epidural and deep subcutaneous tissues at the operation site, resulting in a false meningocele. CT demonstrates a CSF-density collection that may show rim enhancement on postcontrast scans. It is not a cause of back pain.

6.13.   E True
Arachnoiditis can be demonstrated on myelography, at CT following intrathecal contrast medium and on MRI. Three patterns have been described: (1) central adhesion of the nerve roots into a clump of soft tissue; (2) adhesion

of the nerve roots to the meninges, giving rise to the 'empty thecal sac' sign; (3) the final stage, an inflammatory mass that fills the thecal sac and which can produce a complete block at myelography. Such a mass displays little enhancement after Gd-DTPA, a feature that helps to distinguish it from neoplasm.

(References A, B, C, D, 2; E, 46)

6.14. A True
The annulus fibrosus has a high collagen content, accounting for its low signal intensity on all MRI pulse sequences. It is distinguished from the nucleus pulposus, which has an 80–85% water content and appears of intermediate signal intensity on T1-WSE scans and hyperintense on T2-WSE and GE scans. The high signal intensity is maximal at the centre of the disc.

6.14. B True
The PLL appears as a thin (1 mm) hypointense band that merges with the outer fibres of the annulus fibrosus and cannot be distinguished from this structure. It may be seen separate from the posterior cortex of the vertebral body in mid-sagittal sections, particularly on GE sequences. The ligamentum flavum is also identified on T1-WSE scans as a thicker band of low signal intensity between adjacent laminae.

6.14. C False
On T1-WSE scans, the spinal cord has intermediate signal intensity and is contrasted against the low signal intensity of the CSF (which has the signal characteristics of water). Cortical bone, ligaments and the dura all appear hypointense, while the intramedullary and epidural fat are hyperintense.

6.14. D True
GE sequences produce a 'myelographic' effect allowing the greatest contrast between the medium signal intensity spinal cord and the markedly hyperintense CSF. The nucleus pulposus appears moderately hyperintense due to its high water content. Vertebral bone marrow and epidural fat are relatively hypointense in comparison.

6.14. E False
In the cervical spine, the neural foramina are angled approximately 45° to the sagittal plane and are therefore not optimally demonstrated on parasagittal scans, as in the lumbar spine. T1-WSE parasagittal scans demonstrate the intermediate signal nerve root as it exits just beneath the pedicle, surrounded by hyperintense epidural fat. A low signal intensity 'dot' may be identified anterosuperior to the nerve root and represents a radicular vein.

(References A, C, D, E, 3; B, 46)

6.15. A True
Intervertebral disc degeneration is associated with loss of water from the nucleus pulposus, resulting in reduction of signal intensity on T2-WSE and GE scans. A normal finding with increasing age is a linear area of reduced signal

within an otherwise normal appearing disc. This represents an intranuclear cleft due to invagination of the annulus and should not be confused with degenerative disc disease. Lengthening of the annular fibres may occur, resulting in a diffuse disc bulge.

6.15. B True
Localized weakening of annular fibres may result in a focal disc bulge or protrusion. Progressive degeneration or trauma can produce further interruption of the fibres, resulting in herniation of the nucleus pulposus (HNP). In the former case, MRI shows the disc bulge to be outlined posteriorly by a hypointense rim representing the intact annulus and PLL. HNP may be identified by discontinuity of this low signal intensity rim and the presence of disc material in the spinal epidural space.

6.15. C False
The signal intensity of sequestrated disc fragments depends upon their degree of hydration. Acute disc herniations may retain some water and have similar signal to the native disc on T2-WSE scans, whereas chronic disc herniations appear hypointense due to loss of water. Rarely, a sequestrated fragment may appear hyperintense.

6.15. D True
Tears of the annulus fibrosus without associated HNP may be a cause of discogenic low back pain. These tears appear on T2-WSE scans as areas of increased signal intensity in the normally hypointense annulus. They may show enhancement on postgadolinium T1-WSE scans, due to the formation of reparative granulation tissue. Such tears are also identified by discography, although with this technique reproduction of the patient's symptoms is thought to be the most important finding.

6.15. E False
Postoperative scar tissue characteristically enhances immediately after administration of Gd-DTPA, allowing differentiation from recurrent disc herniation. Scar enhancement is maximal in the first 9 months following surgery, gradually decreasing thereafter. Disc can enhance if imaging is performed 20–30 minutes following injection of Gd-DTPA, due to passive diffusion of contrast medium into the disc.

(References A, C, D, E, 46; B, 3)

6.16. A False
Cervical disc herniation is less common than lumbar disc herniation. Of cases that come to surgery, the C5–6 level is most frequently involved, followed by the C6–7 level. C4–5, C3–4 and C7–T1 are involved in decreasing order of frequency. Acute disc herniation is usually post-traumatic, most commonly occurring in young adults.

6.16. B True
On precontrast CT, a cervical disc herniation appears as a soft-tissue mass that is slightly hyperdense to the thecal sac. It may project centrally or paracentrally into the spinal canal, or posterolaterally into the neural foramen. On postcontrast scans, an enhancing rim may outline the disc herniation

(retrocorporeal enhancement). Such a finding is not normally seen below the C2–3 level and may be the only sign of a subtle disc herniation.

6.16. C False
The major causes of encroachment upon the spinal canal and neural foramina are osteophytes from the discovertebral junction and uncinate processes. These are a consequence of disc degeneration and loss of disc height. Disc herniation and thickening/buckling of the ligamentum flavum may also contribute to the resulting spinal canal stenosis.

6.16. D False
The cervical cord is best assessed in the presence of intrathecal contrast medium (postmyelography CT). The normal cervical cord is elliptical in cross-section. Disc herniation or osteophytes may result in cord compression and rotation. Initially, the ventral surface of cord becomes flattened. This may progress to central infolding and widening of the anterior median fissure. Marked cord atrophy may be associated with poor recovery after decompression.

6.16. E True
Ossification of the PLL is most commonly found in the cervical region and can cause radiculopathy or myelopathy. Calcification/ossification in the ligamenta flava is also recognized. The abnormality typically extends across several vertebral levels.

(Reference 35)

6.17. A False
Intervertebral disc herniation is rare in the thoracic spine, but when present most frequently occurs at levels below T8. The incidence is greatest in the fourth to sixth decades. Degenerative change within the disc is the major causative factor, but trauma may play a part.

6.17. B False
Calcification is identified in thoracic disc herniations in up to 55% of cases. CT without intrathecal contrast medium may demonstrate a calcified mass in the spinal canal. Disc herniation is usually central or paracentral but lateral herniations do occur.

6.17. C True
OPLL can also occur in the thoracic region, in which case it is frequently associated with OPLL in the cervical canal. A minority of patients with cervical OPLL will have involvement in the thoracic region. Ligamentum flavum ossification also occurs and the two may produce radiculopathy or myelopathy.

6.17. D False
On CT, OPLL appears as calcific density separated from the posterior vertebral body margin by a thin hypodense zone. This finding, and the extension over several levels, allows it to be distinguished from an osteophyte. Calcification may also be laminated. Ligamentum flavum calcification is thickest adjacent to the superior articular process, in the capsular portion of the ligament.

6.17.   E False
Epidural lipomatosis is a rare condition that results in excessive accumulation
of fat in the epidural space. It is associated with steroid therapy, Cushing's
syndrome and obesity. It is a rare cause of cord compression. CT typically
shows the fat to be located dorsal to the thecal sac.

(Reference 35)

6.18.   A False
Haemangioma is the commonest benign bone neoplasm of the spine, being
identified in up to 11% of postmortem studies. It occurs most often in the
lower thoracic and upper lumbar vertebrae. Pathologically, it consists of
multiple abnormal vascular channels within a fatty matrix.

6.18.   B True
The vast majority of haemangiomas are confined to, and involve, a portion of
a vertebral body. They may extend into the posterior elements and break
through the cortex to involve the epidural and/or paraspinal spaces.
Haemangiomas may produce cord compression.

6.18.   C True
Haemangiomas are associated with thickening of and reduction in number of
bony trabeculae, giving a 'stippled' appearance on CT. The thick trabeculae
appear as dense nodules in a background of fat. Similar appearances may
occur with plasmacytoma and lymphoma. Cortical thickening and enlargement
of the vertebral body distinguishes Paget's disease from haemangioma.
Enhancement occurs on postcontrast scans.

6.18.   D True
The MRI features of haemangioma include areas of high signal intensity on
T1-WSE scans due to the fat content of the lesion, and areas of high signal on
T2-WSE scans due to the cellular/vascular component of the mass. The signal
characteristics of any extravertebral component differ due to the absence of fat
in this portion of the mass. Signal void due to blood flow is not a prominent
feature.

6.18.   E False
Vertebral haemangioma may result in either increased or reduced activity on
bone scintigraphy, or in a normal scan.

(References A, B, C, D, 35; E, 5)

6.19.   A True
The typical location of a spinal osteoid osteoma is the pedicle of a thoracic or
upper lumbar vertebra. Plain films and CT demonstrate a sclerotic focus of
bone, usually not larger than 1–2 cm. CT may also identify a small hypodense
central nidus.

6.19.   B True
Osteoblastoma is characterized by diffuse lytic expansion of a portion of the
neural arch with cortical thinning. The cervical region is most commonly
affected. Intraspinal and paraspinal extension by a soft-tissue mass may be

present and result in compression of the cord and/or nerve roots. The differential diagnosis includes aneurysmal bone cyst.

6.19. C False
ABC is characterized by tremendous expansion of the bone with marked thinning of the cortex. Approximately 20% occur in the spine, particularly the posterior elements. The lesion may extend into the vertebral body and into the epidural and paraspinal spaces. Bony expansion can result in cord compression. Postcontrast CT may demonstrate enhancement within the mass.

6.19. D False
MRI in cases of ABC typically shows multiple well-defined cystic spaces with low signal rims. The signal intensities of the contained fluid vary due to the presence of paramagnetic blood breakdown products. Fluid–fluid levels are very suggestive of ABC but are not pathognomonic. Other (rare) causes include fibrous dysplasia, simple bone cyst, malignant fibrous histiocytoma, telangiectatic osteosarcoma and synovial sarcoma.

6.19. E True
Giant cell tumours in the spine typically involve the sacrum, involvement elsewhere being uncommon. CT shows a poorly defined expanded lesion, occasionally with cortical destruction. The differential diagnosis includes chordoma, metastasis and lymphoma.

(References A, B, C, E, 35; D, 46)

6.20. A True
A popliteal (Baker's) cyst represents a synovial cyst of the semimembranosus–gastrocnemius bursa and results from the presence of a joint effusion. Ultrasound can identify cysts as small as 3 mm in diameter. They are well defined, anechoic and may contain thin septa. A poorly defined inferior cyst margin and fluid in the calf soft tissues indicate cyst rupture. Complete decompression of the cyst following rupture may result in a false negative ultrasound diagnosis.

6.20. B False
Both anterior and posterior surfaces of the quadriceps tendon are well demonstrated by ultrasound since they are bordered by fat and soft tissue. The posterior surface of the tendon becomes more clearly visualized in the presence of an effusion.

6.20. C True
On ultrasound, the menisci have a triangular shape, with the apex of the triangle pointing into the joint. There is no visible interface between the meniscus and the joint capsule at the point of attachment. The posterior horn of the medial meniscus is commonly the only part that can be imaged. A normal finding is a hypoechoic peripheral portion of the mid-to-posterior lateral meniscus, due to the popliteus tendon. Meniscal tears also cause reduced echogenicity and heterogeneous echotexture in the affected meniscus.

6.20. D True
In patients with suprapatellar swelling, ultrasound differentiates joint effusion from synovial hypertrophy, which results in irregular echogenic masses projecting into the suprapatellar bursa from its anterior and posterior surfaces. Causes include ankylosing spondylitis, rheumatoid arthritis, haemophilic arthropathy and pigmented villonodular synovitis. Synovial chondromatosis causes multiple, highly echogenic intra-articular bodies with acoustic shadowing.

6.20. E True
The patella plica is an embryological synovial remnant which usually disappears during fetal life but persists into adulthood in 20% of the population. Thickening and impingement of the plica between the patella and medial femoral condyle may cause knee pain. The plica forms an echogenic band that moves into the patellofemoral joint as the knee is flexed.

(Reference 45)

6.21. A False
The menisci have a low water content and therefore give little or no signal on any magnetic resonance pulse sequence (similar to cortical bone). The femoral and tibial hyaline cartilage has a higher water content and appears as a uniform 2 mm band of intermediate signal intensity on T1- and T2-WSE scans, having higher signal intensity on GE techniques. GE techniques allow detection of degenerative change in the articular cartilage.

6.21. B True
T1-WSE, 3D-GRASS and GE sequences are most sensitive for detecting meniscal tears and degeneration. These have been graded into three categories. Mucoid degeneration (grade 1) appears as a focus of high signal within the meniscus that does not reach the meniscal surface. This may progress to a linear increase in signal that does not reach the surface (grade 2) and finally to a focus of high signal extending to the meniscal surface (grade 3), representing a macroscopic tear.

6.21. C False
The PCL appears homogeneously hypointense on T1-WSE images, whereas the normal ACL may have intermediate signal intensity. The ACL is orientated oblique to the sagittal plane and is imaged in its full length by scanning the patient supine, with the knee externally rotated 15°.

6.21. D True
Complete tears of the ACL may be manifest as absence or discontinuity of the ligament with or without loss of its normally straight parallel margins. Strains result in a focus of high signal on T2-WSE images. Sagittal scans may also demonstrate anterior displacement of the tibia relative to the femur, and a high arched PCL. Tears are most common in the mid-portion of the ACL, or near its femoral insertion.

6.21. E True
On T2-WSE scans, chronic PCL tears with fibrous scarring may show no increase in signal intensity. However, acute tears associated with haemorrhage and oedema appear as areas of increased signal intensity. Complete disruption

of the PCL is imaged as a discontinuity of the ligament. PCL tears may be associated with detachment of the posterior horn of the medial meniscus, tears of the medial and lateral collateral ligaments, capsular tears and avulsion of its tibial insertion.

(References A, C, 3; B, D, E, 47)

6.22. A True
On T2-WSE scans, partial tears or sprains of the MCL appear as thickening of the ligament and increased signal intensity (due to haemorrhage and oedema) around the low-signal-intensity ligamentous fibres. Complete tears are manifest as discontinuity of the ligament and are associated with haemorrhagic joint effusion and extravasation of joint fluid along the ligament.

6.22. B True
The LCL consists of several structures and is also identified on the most lateral sagittal images at its insertion into the fibular head. Tears of the LCL are usually not associated with as much signal increase as in the case of MCL tears, since it is separate from the joint capsule and is not associated with extravasation of joint fluid.

6.22. C True
In cases of chondromalacia patellae, subchondral low signal intensity is due to subchondral sclerosis. Other MRI findings include thinning and irregularity of the patellar cartilage, typically over the medial facet. Inhomogeneous signal intensity on T2-WSE scans, due to oedema, may also be seen.

6.22. D True
Meniscal cysts are commoner in the lateral meniscus and are always associated with horizontal tears. On MRI, they typically have low signal intensity on T1-WSE scans and high signal on T2-WSE and GE scans. The signal may vary depending upon the protein content of the cyst, and the meniscal tear may also be seen.

6.22. E True
Bony loose bodies can be identified by the high-signal fatty marrow within their centre. This is surrounded by a low signal intensity rim. Non-calcified loose bodies are better assessed with MRI than plain radiography.

(References A, B, C, 3; D, E, 34)

6.23. A False
Synovial hypertrophy and pannus generally show intermediate signal intensities on both T1- and T2-WSE scans but become hyperintense on T1-WSE scans following intravenous administration of Gd-DTPA. Synovial hypertrophy and masses are best contrasted with associated effusions on T2-WSE or GE scans, where the fluid appears hyperintense.

6.23. B True
PVNS is characterized pathologically by the deposition of haemosiderin-laden macrophages in hyperplastic synovial tissue. The paramagnetic effect of the iron results in signal loss which is most marked on GE sequences. Similar

features may be seen in other conditions associated with recurrent haemor-rhage, such as haemophilic arthropathy and intra-articular haemangioma.

6.23.   C True
The infrapatellar fat pad appears as an area of high signal intensity on T1-WSE scans immediately anterior to the synovial membrane of the knee joint. Its posterior surface is normally smooth. Synovial inflammation produces the 'irregular infrapatellar fat pad' sign. Causes include haemophilia, rheumatoid arthritis, PVNS, and reactive synovitis due to post-traumatic haemorrhagic effusions.

6.23.   D True
Osteochondritis dissecans is a disorder of adolescents in which a segment of articular cartilage and subchondral bone becomes partially or totally separated from the underlying bone. It typically involves the non-weight bearing lateral aspect of the medial femoral condyle. The detection of fluid between the lesion and the underlying bone is a reliable sign of an unstable fragment.

6.23.   E True
Popliteal (Baker's) cysts are located between the tendons of the medial head of the gastrocnemius and the semimembranosus muscle, posterior to the medial femoral condyle. Their signal characteristics are those of synovial fluid, appearing hypointense on T1-WSE scans and hyperintense on T2-WSE and GE images.

(Reference 3)

6.24.   A True
Rotator cuff tears are usually post-traumatic, either isolated or in association with anterior dislocations, dislocation with avulsion of the greater tuberosity, or fractures of the greater tuberosity without dislocation. Chronic impingement also plays a role. Tears may be complete or incomplete. Partial tears may involve the superior or inferior surfaces of the tendon, or the substance of the tendon. Only partial inferior surface tears can be identified by arthrography, which demonstrates an irregular collection of contrast medium above the opacified joint cavity, near the anatomical neck of the humerus.

6.24.   B True
A full-thickness rotator cuff tear results in communication between the glenohumeral joint cavity and the subacromial/subdeltoid bursa, which is manifest arthrographically by the identification of contrast medium adjacent to the undersurface of the acromion and the superolateral aspect of the greater tuberosity.

6.24.   C True
Adhesive capsulitis may be post-traumatic. Arthrography demonstrates reduced capsular volume (5–10 ml) and size of the axillary recess and subscapular bursa, and irregularity of the capsular insertions. Arthrography may also be used therapeutically in a procedure called brisement. This involves the gradual distension of the capsule by a combination of contrast medium, saline and local anaesthetic.

6.24. D True
On CT arthrography, the normal glenoid labrum has a triangular anterior portion that is slightly larger than the posterior portion. The type of capsular insertion can also be identified. In type I attachments, the capsule inserts close to the glenoid labrum. In type II and III attachments, the capsular insertion is located further medially along the scapular neck. Type III capsular insertions are thought to predispose to glenohumeral instability.

6.24. E True
CT findings with recurrent anterior dislocation include stripping of the anterior capsule, tears of the glenohumeral ligaments and subscapularis tendon, tears and detachments of the anterior labrum with or without a fracture of the anterior glenoid (Bankart lesion) and the Hill–Sachs deformity. The latter represents a compression fracture of the posterolateral aspect of the humeral head due to impaction against the anterior glenoid rim at the time of dislocation.

(Reference 34)

6.25. A True
Ultrasound of the shoulder demonstrates several recognizable layers. Most superficially are the skin and echogenic subcutaneous fat. Next is the relatively hypoechoic deltoid muscle, which is separated from the rotator cuff tendon by a hyperechoic layer made up of the subdeltoid bursa and its surrounding fat. The lateral insertion of the rotator cuff tendon into the greater tuberosity should always be identified.

6.25. B False
The rotator cuff tendon thins posteriorly and at its insertion into the greater tuberosity. Therefore measurements of tendon thickness should be made from its anterior portion. The normal tendon is approximately 6 mm thick, equal to the thickness of the overlying deltoid muscle. Deep to the tendon is the humeral head and greater tuberosity. The acromion is located medially.

6.25. C True
Abnormal ultrasound findings include changes in echogenicity or thickness of the tendon and tendon disruption. The latter is the most reliable sign of a rotator cuff tear. Full-thickness tears may be associated with retraction of the cuff beneath the acromion, with consequent inability to identify the tendon.

6.25. D True
Impingement may result in thinning and increased echogenicity of the tendon. A perforation may be present in the thinned tendon but be impossible to identify by ultrasound. Thickening of the tendon with areas of mixed echogenicity is a less common finding and may be due either to impingement, or oedema and haemorrhage from an acute tear.

6.25. E False
A focal hyperechoic region within an otherwise normal tendon may be due to a small complete or partial tear. Ultrasound cannot differentiate the two. Identical ultrasound features may be due to healing partial tears, scars or calcific tendinitis. Ultrasound has the advantage over arthrography in that it

can identify tears limited to the superior surface of the tendon or those within
the substance of the tendon.

(Reference 34)

6.26.   A False
On T1-WSE MRI scans, fat appears hyperintense, muscle and hyaline cartilage
have intermediate signal intensity while cortex, ligaments, tendons, the capsule
and the fibrocartilaginous glenoid labrum produce no signal due to their lack of
water. However, small foci of increased signal intensity may occur in the
labrum in asymptomatic individuals, a finding that has been attributed to
mucoid degeneration.

6.26.   B False
The rotator cuff is formed by the supraspinatus, infraspinatus, teres minor and
subscapularis muscles and tendons. The first three muscles insert into the
greater tuberosity and the latter into the lesser tuberosity. The supraspinatus
tendon is most commonly involved by impingement, tendinitis and tears. It is
best imaged in the coronal oblique plane which can demonstrate the entire
length of the muscle and tendon. This plane also demonstrates the relationship
of the acromioclavicular joint and the rotator cuff to advantage.

6.26.   C True
The axial plane is best for assessing the glenoid labrum and the anterior
capsular structures. The presence of a joint effusion also allows identification
of the type of capsular insertion. The assessment of shoulder instability is best
in the axial plane.

6.26.   D True
On MRI, a partial tear of the rotator cuff appears as a region of increased
signal intensity in the tendon on T1-WSE scans that does not extend
completely through the tendon. T2-WSE scans demonstrate a further increase
in signal intensity. A complete tear shows abnormal signal that occupies the
full thickness of the tendon. Other findings include tendon retraction, muscle
atrophy and subacromial spurs.

6.26.   E True
High signal intensity on T2-WSE scans within the capsule or bursa indicates
fluid. Fluid in the subacromial/subdeltoid bursa may occur with bursitis or in
association with a partial tear of the superior surface of the tendon. Fluid in
both this bursa and within the joint occurs with a full-thickness tear. A partial
tear of the inferior surface of the tendon may cause a joint effusion.

(Reference 50)

6.27.   A True
Osteoarthritis of the hip is typically associated with non-uniform cartilage loss
and usually superolateral femoral head displacement. This causes a shift of
weight-bearing from the centre of the femoral neck to the medial cortex of the

femoral neck, resulting in cortical thickening and a characteristic osteophyte from the medial aspect of the femoral head. Osteoarthritis secondary to processes that cause uniform cartilage loss, such as rheumatoid arthritis and crystal deposition diseases, is not associated with such changes.

6.27.   B True
Widening of the joint space indicates substance within the joint distending it (e.g. blood, pus, inflammatory fluid, cartilage in acromegaly, intra-articular bodies or synovium). About 25% of osteoid osteomas are located in the femoral neck, where they may produce a joint effusion or synovitis distending the joint. Osteopenia will also be evident.

6.27.   C True
Scalloping defects are also seen in cases of synovial chondromatosis, usually at the junction of the head and neck. Both this condition and PVNS may cause widening of the joint (or narrowing in the late stages) and osteoporosis. Ossification of the chondral bodies indicates synovial osteochondromatosis, whereas the identification of cysts in the acetabulum and femoral head are seen with PVNS.

6.27.   D True
Transient (regional migratory) osteoporosis manifests as osteoporosis around the hip joint but is most marked in the femoral head. The joint space is normal. If left alone, mineralization within the femoral head returns to normal in 9–12 months. Bone scintigraphy demonstrates increased activity that is limited to the femoral head and proximal neck, allowing differentiation from osteonecrosis.

6.27.   E True
In transient osteoporosis, MRI shows diffusely decreased marrow signal intensity on T1-WSE scans and increased signal intensity on T2-WSE scans due to oedema. High signal intensity within the joint on T2-WSE scans is also seen and represents an associated joint effusion.

(Reference 47)

6.28.   A False
In early osteonecrosis of the femoral head the joint space is maintained and the acetabulum appears normal. The first radiographic finding is poor definition of the trabeculae in the femoral head. The reparative stage is manifest by both lysis and sclerosis. Subchondral lucency indicates a late stage of necrosis and impending collapse of the femoral head. This may be followed by secondary osteoarthritis.

6.28.   B True
Bone scintigraphy is more sensitive than conventional radiography in detecting osteonecrosis. In the acute stage, there will be no activity in the femoral head due to the loss of blood supply. The reparative stage is associated with revascularization and increased osteoblastic activity, manifesting as focal

increased tracer uptake. The inactive stage is associated with normal scan appearances.

6.28. C False
MRI is the most sensitive test for osteonecrosis but a normal scan does not exclude the diagnosis since histological changes of necrosis precede the MRI changes. The time interval between an insult resulting in osteonecrosis and abnormal MRI marrow signal is unclear.

6.28. D True
Several MRI patterns of osteonecrosis have been described. A common finding is a focal lesion in the anterosuperior aspect of the femoral head that frequently has a hypointense rim on both T1- and T2-WSE scans. Acute lesions have central fat intensity whereas chronic lesions have central low signal consistent with fibrosis. Signal intensities compatible with blood or fluid may be seen in intermediate stages.

6.28. E True
A 'double-line' sign has been described on T2-WSE MRI scans that is possibly pathognomic for osteonecrosis and was identified in 80% of cases in one series. It consists of central hyperintensity (likely due to hyperaemia or inflammation within granulation tissue) surrounded by a hypointense rim (due to chemical shift artefact).

(Reference 47)

6.29. A False
A radiolucent line at the cement–bone interface of either the femoral or acetabular component is not always abnormal. It has been attributed to a fibrous membrane and has been considered insignificant if less than 2 mm thick and not progressive. Others suggest that it is only insignificant if it surrounds less than 50% of the femoral component or less than two-thirds of the acetabular component. Features that are suggestive of loosening include migration of either component, focal areas of lysis or endosteal scalloping, or a lucent line at the prosthesis–cement interface.

6.29. B False
Heterotopic new bone formation of some degree is seen in up to 50% of cases and may take up to 2 years to mature. It is only important if large enough to cause pain or limit movement. Scintigraphy shows a characteristic picture of a band of activity bridging the acetabulum and greater trochanter.

6.29. C False
On arthrography, loosening is identified if contrast medium injected into the joint pseudocapsule enters any spaces between the prosthesis–cement interface or cement–bone interface. A change in the thickness or distribution of contrast within a space with application of force on the component is diagnostic of loosening. False negative results may occur if there is a membrane filling the potential space, if the volume of injected contrast medium is inadequate, or if intracapsular pressure is inadequate, usually due to the presence of a large bursa.

6.29. D False
On ^{99m}Tc-MDP bone scintigraphy, the level of activity depends upon the time after surgery. Femoral shaft and lesser trochanter activity usually reduces by 6 months after surgery, while activity around the acetabulum, greater trochanter and the bone adjacent to the femoral prosthesis tip may be increased for up to 2 years. Persistent activity at theses sites can be seen in 10% of patients.

6.29. E False
On ^{99m}Tc-MDP scintigraphy, infection usually results in diffusely increased activity along the prosthesis, while in loosening focal activity at the tip of the femoral prosthesis is typical. Increased activity may also be due to heterotopic new bone, ununited osteotomy or fractures. Infection can be further assessed by gallium-67 citrate or labelled WBC scans.

(References A, B, C, D, 48; E, 5)

6.30. A False
The sacroiliac joint (SIJ) consists of a synovial anteroinferior third and a cartilaginous superoposterior two-thirds. The normal ligamentous portion of the joint is V-shaped, being narrowest anteriorly. It has deep concavities for the insertion of the interosseous ligaments. The CT criteria for a normal ligamentous portion include a uniformly thin cortex, absence of erosion, sclerosis or ligamentous mineralization and symmetry between the two sides.

6.30. B False
The width of the synovial portion of the SIJ ranges from 2.5 to 4.0 mm. CT criteria for a normal synovial portion include uniformly thin, parallel cortices, absence of erosion, sclerosis or fusion, and symmetry between the two sides.

6.30. C True
The sacral side of the SIJ is covered by relatively thick hyaline cartilage whereas the iliac side is covered only by a thin layer of fibrocartilage. The CT features of inflammatory sacroiliitis include loss of definition of the subchondral bone, widening of the joint space due to erosions, and reactive sclerosis. All these features are more severe on the iliac side of the joint.

6.30. D True
In ankylosing spondylitis, CT will demonstrate inflammatory changes earlier than plain radiography. In the later stages of the disease, intra-articular bony spurs and transarticular bony bridges may be seen prior to complete bony ankylosis. CT may also incidentally demonstrate ossification of the annulus fibrosus.

6.30. E True
Osteitis condensans ilii is a benign reactive condition that typically affects postpartum women and resolves after the menopause. It is rare in men and may be unilateral, usually due to abnormal stress on one SIJ because of contralateral hip disease. Reactive sclerosis is limited to the iliac side of the joint and erosions are not a feature.

(References A, B, 2; C, D, E, 47)

6.31.   A True
The temporomandibular joint (TMJ) is a diarthrodial joint separated into superior and inferior compartments by a biconcave fibrous disc. Normally, the central thin zone of the disc is located between the mandibular condyle and the articular eminence of the temporal bone. The disc maintains this relationship during mouth opening. Internal derangements are related to abnormalities of position and motion of the disc. Anterior displacement with or without reduction is the commonest disorder.

6.31.   B True
TMJ arthrography is performed by injecting contrast medium into the inferior joint space or into both the inferior and superior joint spaces. In the latter case, the disc is identified as a filling defect between the two contrast-filled cavities. With the mouth closed, the disc is located between the anterosuperior aspect of the mandibular condyle and the posterior slope of the articular eminence of the temporal bone. As the mouth is opened, the mandibular condyle glides anteriorly over the articular eminence and the anterior recesses are emptied of contrast medium, which now fills the posterior recesses.

6.31.   C False
In anterior displacement with reduction, the arthrogram shows the disc to be anterior to its normal position with the mouth closed. The posterior rim of the disc may cause an impression on the anterior recess of the inferior joint space. As the mouth is opened, the condyle moves forward and over the posterior margin of the disc, producing a click. The disc is relocated and the arthrogram becomes normal until the mouth is closed again. In anterior displacement without reduction, the arthrogram is always abnormal.

6.31.   D False
Perforations of the disc or ligaments usually occur in cases of long-standing irreducible anterior displacement. Perforations are usually located in the posterior ligament, where it is trapped between the articulating surfaces of the condyle and temporal fossa. Acute disc perforations may occur with trauma to the jaw. The arthrographic diagnosis of perforation is made if contrast medium injected into the inferior joint space enters the superior joint space.

6.31.   E False
MRI also identifies anterior displacements of the disc, and by direct coronal scanning medial and lateral displacements are also seen. The disc is hypointense on all pulse sequences. On T2-WSE scans, the posterior ligament has higher signal intensity than the disc. MRI also demonstrates disc morphology, which is usually normal in anterior displacement with reduction, but may assume a globular appearance when displacement is irreducible.

(References A, B, C, D, 47; E, 35)

6.32.   A True
MRI is at least as sensitive as gallium-67 citrate and ^{99m}Tc-MDP bone scintigraphy in the detection of osteomyelitis but may not be as specific as labelled WBC scanning. Disadvantages of MRI may include difficulty in the detection of small air bubbles within an abscess, which are better identified by CT. MRI has the advantage in patients being treated with antibiotics or those who are neutropenic, both possible causes of false negative labelled WBC scans.

6.32. B False
Oedema and inflammation associated with acute osteomyelitis cause reduced bone marrow signal intensity on T1-WSE scans and increased signal intensity on T2-WSE scans. Soft-tissue infection without bone involvement can be identified by the normal marrow signal.

6.32. C True
Both cellulitis and abscess result in prolonged T1 and T2 relaxation times due to an increased tissue water content. Cellulitis is characterized by a poorly defined area of signal alteration, whereas an abscess is well-defined. Chronic infection may have relatively low signal intensity on both T1- and T2-WSE scans due to fibrosis.

6.32. D True
Up to 96% of cases of vertebral osteomyelitis have been associated with abnormal signal intensity within the adjacent disc, a feature that helps to distinguish infection from tumour, since the latter rarely destroys disc. Osteomyelitis initially causes increased T1 and T2 relaxation times in the disc due to oedema, blurring of the disc/vertebral body interface and signal changes in the adjacent vertebral body marrow (reduced signal intensity on T1-WSE and increased signal intensity on T2-WSE scans). The latter findings have also been seen in vertebral bodies adjacent to discs following chemonucleolysis.

6.32. E True
After successful treatment of osteomyelitis, the marrow signal usually returns to normal over several months. However, in some cases absence of haemopoietic tissue in the recovered fatty marrow results in higher signal intensity on T1-WSE scans.

(Reference 3)

6.33. A False
A bone infarct appears on MRI as a serpiginous outer rim of low signal intensity and a central region of higher signal intensity representing yellow marrow. These features differentiate it from enchondroma, which has low central signal intensity on T1-WSE scans and increased signal intensity on T2-WSE scans but may have a hypointense margin.

6.33. B True
Bone islands, healed fractures and areas of fibrosis or sclerosis have short T2 relaxation times appearing relatively hypointense compared to tumour, oedema and infection, which have prolonged T2 relaxation times due to their increased water content. The latter conditions appear hyperintense to marrow on T2-WSE scans. On T1-WSE scans, primary and metastatic bone tumours typically appear as focal areas of reduced marrow signal intensity.

6.33. C True
The signal intensity of bone marrow depends upon the relative amounts of red (haematopoietic) and yellow (fatty) marrow present. Expansion of the haematopoietic system in response to chronic anaemia may be diffuse throughout the skeleton, or may be focal, resulting in areas of reduced signal

intensity on T1-WSE scans. The differential diagnosis includes neoplasm, infection and metabolic bone disease.

6.33.   D True
Aplastic anaemias, by definition, have a paucity of haematopoietic tissue, and replacement by fat results in a higher marrow signal than normal on T1-WSE scans. In myelofibrosis, areas of low signal intensity are seen on both T1- and T2-WSE scans. These are due to replacement of marrow by fibrous tissue, as well as fat.

6.33.   E False
In Gaucher's disease the abnormal marrow typically has reduced signal intensity on both T1- and T2-WSE scans, differentiating it from tumour and infection, which appear hyperintense due to associated marrow oedema. Bone infarcts may also be identified in Gaucher's disease.

(Reference 3)

6.34.   A False
Benign cartilaginous tumours typically have matrix calcification that is evenly distributed throughout, without sizeable areas of uncalcified soft-tissue matrix. This feature distinguishes them from bone infarcts, which are characterized by peripherally located calcification. Also, benign cartilaginous tumours tend not to cause any erosion of the endosteal surface.

6.34.   B False
A benign osteocartilaginous exostosis appears on CT as a bony mass with a sharply defined margin, a soft-tissue density central matrix, a medullary cavity that is continous with that of the bone from which it arose, and a thin cartilaginous cap. CT is particularly useful in evaluating such lesions arising from the pelvis or spine.

6.34.   C False
In uncomplicated benign bone cysts, CT demonstrates a well-defined lesion with a thin sclerotic rim and central near water density. There is no enhancement on postcontrast scans. Areas of high attenuation may be seen if haemorrhage has occurred and may result in the formation of fluid–fluid levels. High attenuation is also a feature of other blood-containing lesions such as aneurysmal bone cyst and haemophilic pseudotumour.

6.34.   D True
The CT features of an osteoid osteoma include a soft-tissue density lesion within a region of surrounding sclerosis. In some cases, a densely calcified centre may be seen. These lesions are very vascular, which results in their enhancement on postcontrast scans. It is important for CT to identify the nidus and the extent of the tumour since incomplete excision may result in recurrence.

6.34.   E True
Also, on CT, fibrous dysplasia has a matrix that may be of uniform density or contain thick, dense bands of increased attenuation resulting in a mixed lytic/sclerotic appearance. On MRI, T1-WSE scans show a lesion of uniform or

inhomogeneous low signal intensity, while T2-WSE scans show mixed signal intensities with areas of high signal.

(Reference 2)

6.35. A True
The CT findings with osteosarcoma reflect the different pathological subtypes (lytic, sclerotic or mixed) and include cortical destruction, new bone formation, periosteal reaction, and extension of tumour into the surrounding soft tissues. CT may also identify medullary skip lesions, which are a characteristic feature of osteosarcoma. On MRI, predominantly sclerotic tumour is hypointense on all pulse sequences.

6.35. B False
The soft-tissue extent of tumour may be easily assessed in patients with prominent fat planes but may be unclear otherwise. Postcontrast scans increase the contrast between normal muscle and tumour since muscle enhances to a greater degree. The relationship of tumour to the neurovascular bundle is also better assessed.

6.35. C False
Parosteal osteosarcoma appears on CT as an irregular bone-density mass arising from the cortex of the bone. Extension of tumour into the medullary cavity and the surrounding soft tissues is a recognized feature. This subtype of osteosarcoma has a less aggressive clinical course than medullary (central) osteosarcoma and may be effectively treated by local resection.

6.35. D True
Other CT features that suggest malignant transformation of a benign osteocartilaginous exostosis include thickening of the cartilaginous cap with the development of a soft-tissue density mass that invades surrounding structures. CT features suggestive of a chondrosarcoma rather than an enchondroma include destruction of surrounding cortical bone, matrix necrosis, and large areas of uncalcified matrix.

6.35. E False
Lipoma appears on CT as a well-defined, homogeneous, fat-density (-40 to -100 HU) mass that may contain areas of soft-tissue density due to septa or blood vessels. Liposarcomas vary in their CT appearance depending upon the amount of fat they contain. Poorly defined tumours are indistinguishable from other soft-tissue sarcomas (e.g. malignant fibrous histiocytoma (MFH), fibrosarcoma) appearing as infiltrative inhomogeneous soft-tissue density masses. Peripheral calcification is a feature of MFH.

(Reference 2)

6.36. A True
The T1 relaxation times of most tumours is long (due to increased water content) compared to that of fat, so that neoplasms contrast strongly with adipose tissue and fatty bone marrow on T1-WSE scans. T2-WSE scans improve the contrast between tumour and muscle, since the T2 relaxation times of most neoplasms tends to be long compared to that of muscle.

Neoplastic tissue therefore appears hypointense on T1-WSE scans and hyperintense on T2-WSE scans.

6.36. B False
High signal in a muscle adjacent to a malignant tumour on T2-WSE scans occasionally occurs without invasion, due to reactive oedema. However, thickening of the muscle and loss of fat planes associated with increased muscle signal intensity implies invasion.

6.36. C False
Prolonged T2 relaxation times are suggestive of malignancy. However, there is a great deal of overlap in T1 and T2 relaxation times between benign and malignant tumours, so that differentiation is not reliably made. Features that suggest malignancy include soft-tissue invasion, permeation of cortex, heterogeneity of signal intensity, ill-defined borders and enhancement following Gd-DTPA.

6.36. D True
Lipomas are typically well-defined masses with uniform internal signal intensity apart from fibrous septa. They have fat signal intensity on all pulse sequences (hyperintense on T1-WSE scans and moderate increase in signal intensity on T2-WSE scans). Liposarcomas typically have heterogeneous signal intensity, and very poorly differentiated tumour may contain no fat intensity, being indistinguishable from any other soft-tissue malignancy.

6.36. E False
Replacement of the normal fatty marrow by tumour cells results in low signal intensity on T1-WSE scans. Leukaemia is typically infiltrative, producing patchy or diffuse signal abnormality, whereas metastases and primary bone tumours tend to be focal. Myeloma and lymphoma may be either focal or diffuse.

(Reference 3)

6.37. A True
Haemophilic pseudotumour is characterized by regions of haemorrhage of various ages. MRI may therefore demonstrate areas with short T1 relaxation times (due to methaemoglobin) and short T2 relaxation times (due to haemosiderin). Areas of long T1 and T2 relaxation times may also be present due to the water content of the mass.

6.37. B False
Primary neoplasms arising from hyaline cartilage have a characteristic appearance on T2-WSE images. They appear as lobulated lesions of moderate to high signal intensity. On T1-WSE scans, they have intermediate signal intensity.

6.37. C False
Soft-tissue haemangiomas show no flow voids on MRI. They have high signal intensity on T2-WSE scans and may have a septate or striate appearance. The high signal is related to slow-flowing blood. Peripheral areas of high signal

intensity may be seen on T1-WSE scans, corresponding histologically to areas of fat.

6.37. D False
Radiotherapy typically results in shortening of the T1 relaxation times of bone marrow 2 months to 10 years following treatment due to the fatty replacement of haemopoietic marrow. Similar changes may occur following chemotherapy. Residual tumour will be differentiated by its prolonged T1 relaxation times.

6.37. E True
Cortical bone appears on all MRI pulse sequences as a sharply defined region of signal void due to its lack of mobile protons. Tumour invasion into the cortex is identified as a region of increased signal intensity within the bone.

(Reference 3)

6.38. A False
Bone bruises are common in knee injuries. Those associated with medial collateral ligament and anterior cruciate ligament (ACL) tears typically affect the lateral femoral condyle. The lesion is confined to medulla, with the cortex and cartilage being normal. The basic pathology is trabecular disruption with oedema and haemorrhage. T1-WSE scans show a poorly defined area of decreased marrow signal that becomes hyperintense on T2-WSE scans.

6.38. B True
MRI has demonstrated two abnormal patterns in stress fracture. The first shows a linear area of low signal intensity orientated perpendicular to the cortex and surrounded by a hypointense zone on T1-WSE scan (due to oedema) which becomes hyperintense on T2-WSE scans. The second type shows a geographic area of decreased signal on T1-WSE scans without a linear component. Some areas of this region of altered signal show increased signal on T2-WSE scans.

6.38. C True
Osteonecrosis and non-union are complications of proximal pole scaphoid fractures. In non-union, MRI may show low signal intensity at the fracture site on both T1- and T2-WSE scans, indicating fibrous union. High signal intensity at the fracture site on T2-WSE scans (due to fluid) is seen with pseudarthrosis.

6.38. D False
Normal muscle has intermediate to low signal intensity on T1-WSE scans and low signal intensity on T2-WSE scans. Muscle tears cause oedema and haemorrhage, and are best shown on T2-WSE scans, appearing as areas of poorly defined increased signal intensity.

6.38. E True
Compartment syndromes are caused by increased pressure within a confined fascial space, resulting in muscle ischaemia. Fractures are the commonest cause, particularly minimally displaced tibial shaft fractures. On MRI, compartment syndromes are characterized by an increase in size of the

compartment and a mild diffuse increase in muscle signal intensity on T2-WSE scans due to oedema.

(Reference 34)

6.39.   A True
On ultrasound, a normal tendon appears echogenic and has a fibrillar texture when imaged perpendicular to the ultrasound beam. The peritendon is sometimes identified as two highly echogenic lines surrounding the tendon. If the tendon is imaged obliquely, such as at a bony attachment, that portion may appear relatively hypoechoic, mimicking a tear.

6.39.   B False
The Achilles tendon is normally oval in cross-section at its mid-point, with its axis lying obliquely in an anteromedial direction. It is normally 4–6 mm thick and 12–15 mm wide. Rounding of the tendon, reduced echogenicity and enlargement are features of acute tendinitis.

6.39.   C True
Partial tears of the patellar tendon may result in a small encysted haematoma, producing the ultrasound findings. The tendon itself may be of normal thickness, shape and echotexture. Complete tendon ruptures may initially be identified as total discontinuity of fibres with surrounding haematoma. Later, the space between the two ends of the tendon may be filled with echogenic granulation tissue.

6.39.   D True
Acute inflammation of a tendon typically results in enlargement and reduced echogenicity of the tendon. Other ultrasound features include the presence of multiple small hypoechoic and anechoic areas within the tendon, due to mucoid degeneration.

6.39.   E True
Ultrasound findings in cases of chronic tendinitis include enlargement and inhomogeneous reduction in echogenicity. Chronic inflammation is characterized by calcification, which is demonstrated on ultrasound as highly echogenic foci within the tendon, with or without associated acoustic shadowing. Calcification is more commonly identified in the patellar and rotator cuff tendons than in the Achilles tendon.

(Reference 45)

6.40.   A False
Intraoperative ultrasound of the spine is performed following laminectomy. The laminectomy space is filled with warm saline, which acts as an acoustic window through which the cord can be imaged. A 7.5 or 10 MHz sector scanner is typically used.

6.40.   B True
The cord is a hypoechoic structure and most extramedullary tumours, including meningiomas, Schwannomas and metastases, are relatively hyper-echoic. Extramedullary tumours typically appear as echogenic masses

displacing the spinal cord laterally and either anteriorly or posteriorly. Metastatic tumour may invade and surround the thecal sac, compressing the cord from all directions.

6.40. C False
Typically, the central echo complex is obliterated by an intramedullary tumour but usually spared by an extramedullary tumour. However, if the extramedullary mass has caused severe spinal cord trauma or contusion, the central canal echo can be obliterated.

6.40. D False
Anterior compression by disc can be differentiated from that due to osteophyte when disc allows some through transmission of sound. This feature is not always present. Sonography is of use in assessing the adequacy of any decompression procedure.

6.40. E True
The ultrasound appearances of an intramedullary haematoma are dependent upon the age of the lesion. Acute haematomas appear relatively echogenic whereas chronic haematomas appear either hypoechoic or anechoic. Ultrasound is also of value in the identification of post-traumatic intramedullary cysts and can assess if decompression techniques have resulted in adequate drainage of the cyst.

(Reference 45)

7 Central nervous system

7.1. Concerning CT of intracranial neoplasms
A the majority of intracranial lipomas occur in the occipital lobe.
B CNS lymphomas typically show no enhancement on postcontrast scans.
C anterior displacement of the fourth ventricle is a feature of medulloblastoma.
D calcification is an atypical feature of supratentorial primitive neuroectodermal tumour (PNET).
E chordomas of the clivus are usually hyperdense on precontrast scans.

7.2. Which of the following are true concerning cerebral gliomas?
A The majority are low grade (I and II).
B Low-grade gliomas invariably show enhancement on postcontrast CT.
C Giant cell astrocytomas are differentiated from tubers of tuberous sclerosis by their enhancement on postcontrast CT.
D Multiple ring lesions with associated white matter oedema is a recognized CT appearance of glioblastoma multiforme.
E Enhancement occurs in less than 50% of glioblastomas on postcontrast CT.

7.3. Concerning the CT features of intracranial tumours
A oligodendrogliomas characteristically show calcification.
B enlargement of the optic canal is a feature of optic nerve gliomas.
C there is a recognized association between enlargement of the optic nerve and absence of the sphenoid wing.
D supratentorial ependymomas are usually seen in the frontal horn of the lateral ventricle.
E choroid plexus papilloma typically causes compression of the lateral ventricle.

7.4. CT demonstrates a homogeneous, hyperdense mass in the pineal region. The differential diagnosis includes which of the following?
A Colloid cyst.
B Germinoma.
C Teratoma.
D Pineoblastoma.
E Vein of Galen aneurysm.

7.5. Which of the following are true of intracranial meningiomas?
A Less than 10% are infratentorial.
B Invasion of adjacent cerebral parenchyma is not a recognized feature.
C The majority are hyperdense on precontrast CT.
D Meningioma-en-plaque usually arises adjacent to the sphenoid wing.
E Intraventricular meningiomas usually arise within the choroid plexus of the lateral ventricle.

7.6. In the differential diagnosis of a suprasellar mass
A high intensity on T1- and T2-WSE MRI is a recognized feature of craniopharyngioma.
B a calcified mass on CT is most likely a pituitary adenoma.
C a mass with irregular, lobulated borders is typical of meningioma.
D an aneurysm of the internal carotid artery is typically midline.
E arachnoid cyst does not enhance on postcontrast CT.

7.7. Which of the following statements are true of infratentorial tumours?
A The majority of cerebellar astrocytomas in children are cystic.
B Spread throughout the CSF is a feature of medulloblastoma.
C Ependymoma of the 4th ventricle is typically hyperdense on precontrast CT.
D Absence of calcification distinguishes haemangioblastoma from cystic astrocytoma.
E Brain stem gliomas most commonly occur in the pons.

7.8. In the differential diagnosis of masses in the cerebellopontine angle
A intratumoral calcification suggests meningioma rather than acoustic neuroma.
B extension of tumour into the internal auditory meatus favours acoustic neuroma rather than meningioma.
C neuroma of the trigeminal nerve must be considered.
D epidermoid tumours typically have negative CT numbers.
E dynamic postcontrast CT shows a characteristic enhancement pattern with glomus jugulare tumours.

7.9. Concerning the CT features of pituitary adenomas
A the majority of macroadenomas are hyperdense on precontrast scans.
B microadenomas cause no abnormality on precontrast scans.
C the majority of adenomas calcify.
D microprolactinomas are typically hypodense compared to the remainder of the gland on immediate postcontrast scans.
E hydrocephalus is a recognized finding.

7.10. Concerning the CT features of intracranial metastases
A extensive white matter oedema is a characteristic finding.
B central necrosis is an atypical feature.
C metastases from melanoma are usually hyperdense on precontrast CT.
D CT has a sensitivity of 90% in demonstrating leptomeningeal carcinomatosis.
E cranial vault metastases are usually associated with intracerebral metastases.

7.11. Concerning the MRI features of intracranial neoplasms
A gliomas are typically well-defined on precontrast scans.
B flow void is a recognized finding adjacent to a cerebellar haemangioblastoma.
C primary CNS lymphoma may appear isointense on T2-WSE images.
D meningioma is typically hypointense on all pulse sequences.
E pituitary microadenomas are best demonstrated on T1-WSE scans immediately following Gd-DTPA.

7.12. Which of the following are true of MRI in the assessment of cranial metastases?
A   Skull vault metastases cannot be demonstrated.
B   Peritumoral oedema may obscure lesions on T2-WSE scans.
C   Enhancement following Gd-DTPA occurs in most cerebral metastases.
D   Hyperintense lesions on T1-WSE scans are a feature of melanoma metastases.
E   Areas of increased signal on T1- and T2-WSE scans are a feature of haemorrhagic metastases.

7.13. Concerning the MRI features of intracranial cystic lesions
A   arachnoid cysts are isointense to CSF on all pulse sequences.
B   epidermoid cysts are typically hypointense to CSF on all pulse sequences.
C   dermoid cysts are characteristically hypointense on T1-WSE scans.
D   colloid cysts are invariably of high signal intensity on T2-WSE scans.
E   porencephalic cysts cannot be differentiated from arachnoid cysts.

7.14. Which of the following are true of intracanalicular acoustic neuromas?
A   They arise from the vestibular division of the VIIIth cranial nerve.
B   CT air cisternography is more sensitive than T1-WSE MRI in their detection.
C   MRI cannot demonstrate enlargement of the internal auditory canal.
D   On MRI, T1-WSE scans produce the best contrast between tumour and CSF.
E   On MRI, enhancement is a typical feature following Gd-DPTA injection.

7.15. Concerning CT in the assessment of intracranial infections
A   increased density of the choroid plexus is a recognized finding in untreated meningitis.
B   there is an association between subdural empyema and opacification of the middle ear.
C   gyral cortical enhancement adjacent to a subdural empyema is a recognized finding on postcontrast scans.
D   hypodense areas in the lenticular nucleus are a typical feature of adult herpes simplex encephalitis.
E   in herpes simplex encephalitis hyperdense areas in the temporal lobe on precontrast scans are a typical finding.

7.16. Concerning the CT features of cerebral infections
A   the wall of a pyogenic bacterial abscess is typically thick and irregular.
B   abscesses due to haematogenous spread are typically situated at the grey–white matter junction.
C   hyperdensity of the basal cisterns is a feature of tuberculous meningitis.
D   central calcification is a feature of tuberculous granuloma.
E   cryptococcosis is a cause of lesions showing 'ring' enhancement on postcontrast scans.

7.17. Concerning MRI in the assessment of intracranial inflammatory processes
A the capsule of a bacterial abscess can only be seen following Gd-DTPA.
B meningeal enhancement following Gd-DTPA is a feature of coccidioidomycosis.
C increased signal intensity in the basal cisterns on T1-WSE scans is a recognized feature of meningitis.
D a hypointense lesion on T2-WSE scans is a recognized feature of tuberculoma.
E meningeal sarcoid is a cause of hypointense thickening of the subarachnoid space on T2-WSE scans.

7.18. Concerning the neuroimaging manifestations of AIDS
A diffuse cerebral atrophy is a feature.
B primary CNS lymphoma is the commonest cause of a mass lesion.
C 'ring' enhancement on postcontrast CT is feature of primary CNS lymphoma.
D areas of increased signal intensity in the white matter on T2-WSE scans are a recognized finding.
E increased CSF signal intensity on T1-WSE scans is a feature of ependymitis.

7.19. Concerning the CT features of cerebral infarction
A precontrast scans are usually normal within 24 hours of a middle cerebral territory infarction.
B an initially hypodense area that becomes isodense after 2–3 weeks was not due to infarction.
C on postcontrast scans, enhancement in cortical infarcts typically has a gyral pattern.
D lacunar infarcts are typically less than 3 mm in size.
E lacunar infarcts are most commonly seen in the cerebellum.

7.20. Concerning MRI in the assessment of cerebral infarction
A acute infarcts are typically hypointense to grey matter on T1-WSE scans.
B chronic infarcts typically have high signal intensity on PDW scans.
C MRI is more sensitive than CT in the demonstration of lacunar infarcts.
D central CSF intensity is a recognized finding in lacunar infarcts.
E enhancement following Gd-DTPA is a typical feature of acute infarct.

7.21. Concerning the CT features of primary intracerebral haemorrhage
A the putamen and adjacent internal capsule is an atypical site.
B intraventricular extension is not a feature.
C it cannot be distinguished from haemorrhagic infarction.
D 'ring' enhancement on postcontrast scans is a recognized feature of resolving haematoma.
E old haematomas cannot be distinguished from old infarcts.

7.22. Concerning CT in the assessment of cranial extradural and subdural haematomas

A   an associated skull fracture is present in the majority of extradural haematomas.
B   a crescentic extra-axial collection is a recognized appearance of extradural haematoma.
C   associated cerebral contusion is a feature of extradural haematoma.
D   a hypodense subdural collection is a recognized finding in acute subdural haematoma.
E   administration of intravenous contrast medium is contraindicated in patients with chronic extradural haematoma.

7.23. Concerning MRI of intracranial haematomas imaged at 1.5 Tesla

A   a haematoma less than 3 hours old is hypointense on T2-WSE scans.
B   a haematoma at 24 hours typically appears uniformly hyperintense on T2-WSE scans.
C   at 2 weeks, a haematoma typically appears hyperintense on T1-WSE scans.
D   by 2 months, a hypointense rim is seen on T2-WSE scans.
E   GE sequences are more sensitive than SE sequences in the detection of haematomas.

7.24. Recognized CT features of multiple sclerosis (MS) include which of the following?

A   Foci of decreased attenuation in the periventricular white matter.
B   Spontaneous resolution of low-density lesions.
C   Enhancement of plaques on postcontrast scans indicating acute demyelination.
D   Demonstration of some plaques only on postcontrast scans.
E   Ventricular dilatation.

7.25. Concerning MRI in the assessment of MS

A   plaques are typically hyperintense to white matter on T2-WSE scans.
B   plaques are not seen in the grey matter of the cerebral cortex.
C   atrophy of the corpus callosum is a recognized finding.
D   enhancement of inactive plaques following Gd-DTPA is a feature.
E   optic neuritis cannot be demonstrated.

7.26. Which of the following are true of otosclerosis?

A   It is usually bilateral.
B   It results in only conductive hearing loss.
C   A hypodense band around the cochlea is a recognized CT finding.
D   Sclerotic foci in the margins of the oval window are a feature on CT.
E   Identical CT features are seen with osteogenesis imperfecta.

7.27. Concerning MRI in the assessment of acute cerebral trauma and post-traumatic sequelae

A   a subdural hygroma displays CSF signal characteristics.
B   chronic subdural haematomas appear hyperintense on T2-WSE scans.
C   signal void is seen in the majority of chronic subdural haematomas.
D   non-haemorrhagic contusions appear hypointense on T1-WSE scans.
E   diffuse axonal injury is typically seen at the grey–white matter junction.

7.28. Which of the following CT and MRI features suggest hydrocephalus rather than cerebral atrophy?
A   Early dilatation of the temporal horns.
B   CSF signal void in the cerebral aqueduct.
C   Rounded enlargement of the third ventricle.
D   Loss of grey–white matter differentiation.
E   Effacement of the cerebral sulci.

7.29. Concerning intracerebral calcifications as demonstrated by CT
A   calcified subependymal lesions are a finding in tuberous sclerosis.
B   serpiginous cortical calcifications are characteristic of Sturge–Weber syndrome.
C   choroid plexus calcification in the anterior part of the third ventricles is a feature of neurofibromatosis.
D   caudate nucleus calcification is typically physiological.
E   a calcified pineal gland should measure less than 1 cm.

7.30. Concerning CT in the assessment of the spinal cord
A   at the cervical enlargement the maximal AP diameter is 15 mm.
B   syringohydromyelia is differentiated from hydromyelia by its accumulation of contrast medium following intrathecal injection.
C   irregular expansion of the cord is typically due to primary intramedullary neoplasm.
D   astrocytoma is the commonest primary tumour in the thoracic cord.
E   haematomyelia can be demonstrated.

7.31. Concerning the CT assessment of intradural–extramedullary lesions of the spine
A   meningiomas are typically associated with hyperostosis of adjacent vertebral bodies.
B   meningioma and neurofibroma can be distinguished by attenuation values on precontrast scans in the majority of cases.
C   a lesion extending into and expanding an intervertebral foramen is diagnostic of neurofibroma.
D   there is an association between dermoid tumours and thickening of nerve roots.
E   lipomas are associated with a thickened filum terminale.

7.32. Concerning the MRI features of spinal cord tumours
A   astrocytomas and ependymomas are distinguishable in most cases.
B   T2-WSE scans typically demonstrate an area of high signal within the cord.
C   high signal on T1-WSE scans is a recognized finding.
D   enhancement of the solid part of a tumour following Gd-DTPA is diagnostic of haemangioblastoma.
E   intramedullary metastases do not cause cord expansion.

7.33. Concerning MRI in the assessment of cavitating cord lesions
A   in primary syringomyelia, the cyst fluid is isointense to CSF.
B   increased parenchymal signal on T2-WSE scans adjacent to the cavity indicates a neoplastic aetiology.
C   irregular parenchymal thickening adjacent to the cavity is a feature of neoplastic syrinx.
D   enhancement of the cyst wall following Gd-DTPA is a typical feature of primary syrinx.
E   absence of signal within the cyst on flow-sensitive pulse sequences is characteristic of a neoplastic syrinx.

7.34. Concerning MRI of the spinal cord
A   MS plaques are typically hypointense on T1-WSE scans.
B   high signal foci on T2-WSE scans are a recognized feature of sarcoid.
C   spinal cord infarction is a recognized cause of increased signal on T2-WSE scans.
D   focal areas of low signal on T2-WSE scans are a feature of spinal arteriovenous malformation (AVM).
E   acute transverse myelitis is typically associated with diffuse low signal in the cord on T1-WSE scans.

7.35. Concerning MRI in the assessment of intradural–extramedullary lesions of the spinal cord
A   presence of a layer of fat intensity between a lesion and the cord is a typical finding.
B   neurofibromas are typically hyperintense to cord on T2-WSE scans.
C   meningiomas and nerve sheath tumours are typically distinguished on the basis of their signal characteristics.
D   intradural drop metastases usually appear as focal nodular masses.
E   intradural drop metastases are a recognized cause of homogeneous signal increase in the CSF.

7.36. Concerning the CT features of Graves' ophthalmoplegia
A   extraocular muscle enlargement is usually asymmetrical.
B   an increase in the degree of enhancement of extraocular muscles on postcontrast scans is a feature.
C   the muscles are typically most enlarged at their insertions into the globe.
D   uveoscleral thickening is seen in the majority of cases.
E   obliteration of the retrobulbar fat is typical.

7.37. Concerning CT of the orbits
A   cavernous haemangiomas are usually intraconal masses.
B   a soft-tissue density mass adjacent to the lacrimal gland is a recognized appearance of a dermoid tumour.
C   optic nerve glioma is a cause of uniform thickening of the optic nerve.
D   diffuse uveoscleral thickening is a feature of malignant melanoma.
E   calcification of the globe is seen in retinoblastoma.

7.38. Concerning CT in the assessment of the orbits

A enlargement of the superior ophthalmic vein is a feature of carotid–cavernous fistula.

B bone destruction is seen with carcinoma of the lacrimal gland.

C metastases to the orbit are invariably extraconal masses.

D orbital lymphoma is a recognized cause of enlargement of the optic nerve.

E rhabdomyosarcoma is a recognized cause of an intraorbital soft-tissue mass in childhood.

7.39. Concerning MRI of the orbits

A the choroid and sclera cannot be differentiated.

B the optic nerve is typically hyperintense to brain on T2-WSE scans.

C retinoblastoma is usually hyperintense to vitreous on T1-WSE scans.

D choroidal melanoma and metastasis cannot be differentiated.

E meningiomas of the optic nerve sheath enhance following Gd-DTPA.

7.40. Concerning MRI in the investigation of the orbit

A high signal intensity on T2-WSE scans is a feature of cavernous haemangioma.

B orbital pseudotumour typically has high signal intensity on all pulse sequences.

C increased signal intensity of the extraocular muscles on T1-WSE scans is a typical feature of endocrine ophthalmopathy.

D carcinoma results in decreased signal in the lacrimal glands on T1-WSE scans.

E high signal intensity within a mass on T2-WSE scans excludes the diagnosis of orbital varix.

Central nervous system: Answers

7.1. A False
Intracranial lipomas characteristically occur in a midline or paramedian location, most commonly between the frontal lobes (in which case they may be associated with partial agenesis of the corpus callosum). Other locations include the quadrigeminal plate and the hypothalamus. CT demonstrates well-defined, lobulated masses of fat density (-50 to -150 HU). Curvilinear calcification is often identified at the lateral margins of corpus callosum lipomas.

7.1. B False
On precontrast CT, lymphomatous and leukaemic cerebral masses appear as isodense or slightly hyperdense lesions of varying size and shape. Multiple lesions are seen in approximately half the cases. On postcontrast scans, enhancement is typically homogeneous and marked but may be ring-like in larger lesions. Peritumoral oedema is usually not marked with lymphoma or leukaemia unless the lesions are very large. Although pathologically leptomeningeal involvement is common, it is often not demonstrated on CT.

7.1. C True
Medulloblastomas are commonest in the first decade of life and are virtually all midline. Less than 10% occur in the second and third decades, in which case they tend to be located posterolaterally in the cerebellum. Precontrast CT demonstrates an isodense or hyperdense mass that extends forward into the fourth ventricle, posteriorly into the cisterna magna and laterally into the cerebellar hemispheres. Cysts and calcification are seen in 10–20% of cases.

7.1. D False
PNETs (previously termed primary intracerebral neuroblastoma) are rare, highly malignant, primary tumours of childhood. They are commonly located in the peripheral cerebral hemispheres and pineal region. CT demonstrates predominantly hypodense masses which frequently show cysts and large calcifications. Contrast enhancement is typically heterogeneous.

7.1. E False
Chordomas of the clivus arise near the spheno-occipital synchondrosis. On precontrast CT, the tumour is commonly homogeneously isodense with surrounding brain. Destruction of the clivus and calcifications are also seen. Extension into the nasopharynx, pontine and suprasellar cisterns and parasellar region can all occur. Following contrast, moderate patchy enhancement is seen.

(Reference 9)

7.2. A False
Gliomas account for 40–45% of all intracranial tumours, and low-grade (I and II) astrocytomas account for only 15–20% of all gliomas. They are usually solid masses which are poorly circumscribed. Small and large cysts are occasionally found. They usually arise in the frontal, parietal and temporal lobes.

7.2.   B False
Astrocytomas (low-grade gliomas) are typically hypodense on precontrast CT, having attenuation values of 15–30 HU. Also, since enhancement is frequently not present on postcontrast CT, it may not be possible to distinguish the lesion from surrounding oedema. However, extensive peritumoral oedema is unusual with these lesions. Intratumoral calcifications may be seen.

7.2.   C True
Giant cell astrocytomas are associated with tuberous sclerosis and are typically located in a subependymal paraventricular location, commonly near the foramen of Munro. On precontrast CT, they appear as isodense or hyperdense masses which often calcify. Compression of the foramen of Munro may result in obstructive hydrocephalus. Intense enhancement on postcontrast scans is characteristic and indicates their neoplastic nature.

7.2.   D True
This appearance is seen with multicentric glioblastoma, which accounts for approximately 2% of cases of glioblastoma multiforme. A more typical appearance of glioblastoma on precontrast CT is of a poorly defined, heterogeneous mass with nodular areas of normal or slightly increased density intermixed within and/or surrounding larger areas of decreased attenuation. Peritumoral white matter oedema is almost invariable.

7.2.   E False
Contrast enhancement of the peripheral viable tumour rim and of the intratumoral isodense and hyperdense regions is nearly universal in glioblastoma. The central, necrotic areas of the tumour may also show increased attenuation on delayed scans. Glioblastoma multiforme is the commonest primary tumour of the CNS, comprising more than half of all intracranial gliomas. It characteristically involves the cerebral white matter.

(Reference 9)

7.3.   A True
Calcification is seen in approximately 90% of oligodendrogliomas. CT typically demonstrates large, irregular calcifications scattered throughout a poorly defined, hypodense tumour mass usually located in the frontal lobes. Isodense or slightly hyperdense nodules at the margins of the tumour may exhibit mild contrast enhancement. Larger tumours have areas of necrosis and cystic degeneration.

7.3.   B True
CT may show enlargement of both the intraorbital and intracranial portions of the optic nerve by a glioma. The tumour may spread into the optic chiasm and tract and also superiorly into the hypothalamus. The tumour is usually isodense to surrounding tissues but larger lesions may be hypodense. Calcifications may be seen in approximately 15% of cases. Contrast enhancement is variable and irregular.

7.3.   C True
Absence of the greater wing of sphenoid is a classic mesodermal defect in neurofibromatosis, which is associated with both gliomas and meningiomas of the optic nerve (in children). Approximately 10–20% of children with optic nerve gliomas will have neurofibromatosis.

7.3.   D False
Supratentorial ependymomas most commonly arise near the atrium of the lateral ventricle, and unlike infratentorial ependymomas are commonly extraventricular, causing either compression or obstruction of the ventricle. CT features include a well-defined, hypodense mass, cystic areas, calcification (in 33–50%) and peritumoral oedema. Moderate to intense contrast enhancement is characteristic.

7.3.   E False
Choroid plexus papillomas are intraventricular lesions and therefore cause localized expansion of the ventricle. The atria of the lateral ventricles (in children) and the fourth ventricle (in adults) are the commonest sites. CT features include hydrocephalus, a large isodense or hyperdense mass that engulfs the normal choroid plexus and marked, homogeneous contrast enhancement.

(Reference 9)

7.4.   A False
Colloid cysts may be either isodense or hyperdense or precontrast CT but are situated in the anterior aspect of the third ventricle next to the foramina of Munro. They are frequently associated with obstructive hydrocephalus of the lateral ventricles. Postcontrast CT may result in mild, peripheral enhancement, but usually the tumour does not enhance. However, enhancement has been reported after surgery.

7.4.   B True
Germinoma is the commonest tumour at this site but is also frequently situated in the suprasellar region (so-called ectopic pinealoma). Both sites may be affected simultaneously. Precontrast CT demonstrates a spherical, hyperdense mass which exhibits dense contrast enhancement. Intratumoral calcifications are uncommon although the tumour may displace or incorporate existing pineal calcification.

7.4.   C False
Although the posterior aspect of the third ventricle is the commonest site of intracranial teratomas, they appear on CT as isodense or mixed isodense/hypodense masses. Fat and bone density (due to ossification or tooth formation) are also features. These tumours may also occur in the suprasellar region. Contrast enhancement is typically patchy.

7.4.   D True
Both pineoblastomas and pineocytomas may appear as mildly hyperdense or isodense lesions precontrast and demonstrate marked enhancement on postcontrast scans. This diagnosis should particularly be considered in females. Seeding through the CSF is a recognized feature of pineoblastoma.

7.4. E True
A vein of Galen aneurysm appears on precontrast CT as a well-defined, round, hyperdense mass connected to the torcular Herophili by a dilated straight sinus. Marked enhancement is seen on postcontrast scans. With time, calcification may occur.

(Reference 9)

7.5. A True
Approximately 50% of intracranial meningiomas occur adjacent to the frontal and parietal convexities and parasagittal regions, while 35% are related to the sphenoid wing, olfactory groove and suprasellar region. Multiple tumours are found in about 6–9% of cases. Large parasagittal meningiomas may compress or invade the superior sagittal sinus.

7.5. B False
The majority of meningiomas are slow growing and well-encapsulated tumours but a small percentage (less than 4%) may be locally invasive, being referred to as malignant meningiomas. Oedema of the adjacent white matter is variable and there is no correlation between tumour size and extent of oedema. Mild to moderate oedema is usual.

7.5. C True
On precontrast CT, meningiomas typically have attenuation values of 40–50 HU, while some reach 70 HU. Their hyperdensity is related to a high concentration of calcified psammoma bodies. Approximately 10% are isodense. Marked homogeneous enhancement is characteristically seen on postcontrast scans. However, in 5–15% of cases focal areas of reduced attenuation are present, correlating with areas of necrosis.

7.5. D True
Meningioma-en-plaque describes a flat form of tumour that tends to invade the adjacent bone, resulting in marked hyperostosis. Meningiomas arising adjacent to the cribriform plate, planum sphenoidale and high cerebral convexity also tend to provoke bony thickening and sclerosis but are more typically rounded in shape.

7.5. E True
Intraventricular meningiomas cause expansion of the lateral ventricle and otherwise have similar features to other intracranial lesions. The normally compact glomus calcification may appear fragmented and spread by the expanding tumour mass.

(Reference 9)

7.6. A True
Fat may be present in the cyst fluid of a craniopharyngioma, resulting in the MRI appearances. CT typically demonstrates a heterogeneous mass with calcification and enhancement of the solid areas on postcontrast scans. The cyst fluid may have attenuation values of 0–25 HU on CT due to their mixed fat and keratin content or higher attenuation values due to a high protein content.

7.6. B False
Calcification is uncommon in pituitary adenomas (see question 7.9) and much more frequent in craniopharyngioma, being seen in up to 80% of childhood tumours. Also, pituitary adenomas appear much more homogeneous than craniopharyngioma, with a less marked cystic component.

7.6. C False
A lobulated, irregular border is more typical of craniopharyngioma than suprasellar meningioma, which is characteristically well-defined and homogeneous precontrast and shows uniform contrast enhancement. Adjacent hyperostosis may also be seen with meningioma.

7.6. D False
Giant internal carotid aneurysms are typically eccentrically located, spherical, homogeneous masses with rim calcification, which typically show intense central enhancement on postcontrast scans.

7.6. E True
Arachnoid cysts appear on precontrast CT as well-defined lesions that have CSF density and show no enhancement on postcontrast scans. Also, calcification is not a feature. Cysts in the suprasellar region are less common than in the temporal region or posterior fossa and may be associated with hydrocephalus and precocious puberty. Temporal fossa cysts may be associated with localized thinning and expansion of the cranial vault.

(Reference 9)

7.7. A True
Cystic astrocytomas account for 70–80% of cerebellar astrocytomas in children. CT demonstrates a well-defined cystic lesion surrounded by a peripheral rim of tumour or a tumour nodule in the wall. The solid parts of the tumour enhance on postcontrast scans. A small proportion of cerebellar astrocytomas are solid, poorly defined tumours that have a much worse prognosis than the cystic variety.

7.7. B True
Tumour nodules of metastatic medulloblastoma may be demonstrated in the ventricles and subarachnoid space intracranially and in the spinal canal. These nodules demonstrate contrast enhancement similar to the primary tumour.

7.7. C False
Fourth ventricle ependymomas are usually hypodense or isodense on CT. Cyst formation and calcification are common. The tumour may spread to involve the cerebellopontine angle, and seeding to the CSF occurs in approximately 30% of cases. Tumour growth eventually leads to occlusion of the fourth ventricle and obstructive hydrocephalus.

7.7. D True
Haemangioblastomas most commonly arise in the cerebellum but may also be found in the brain stem. The majority are small, solid lesions measuring 5–10mm in size. Larger lesions may have a cystic component with a peripherally located tumour nodule. Intense enhancement of the solid parts of

the tumour is characteristic. Haemangioblastomas are reported to be the commonest primary infratentorial tumours in adults and may be multiple.

7.7. E True
Pontine gliomas may appear on precontrast CT as diffuse enlargement of the brain stem and show patchy enhancement on postcontrast scans. CT may also be completely normal. They may spread inferiorly into the medulla and spinal cord, anteriorly into the pontine cistern and posteriorly to involve the cerebellar peduncles. On precontrast scans they may be hypodense or isodense, in which case their detection is based upon posterior displacement of the fourth ventricle or compression of the pontine cistern.

(Reference 9)

7.8. A True
Calcification is virtually unknown in acoustic or trigeminal neuromas but is seen in meningiomas. Meningioma of the cerebellopontine angle is also associated with hyperostosis of the petrous apex. Although less than 10% of meningiomas occur in the posterior fossa, they are the second commonest tumours in the cerebellopontine angle.

7.8. B True
Acoustic neuromas are the commonest tumours of the cerebellopontine angle. On precontrast CT they typically appear as homogeneously isodense masses which may obliterate the cerebellopontine angle cistern and cause contralateral displacement of the fourth ventricle. Oedema in the adjacent cerebellum may be extensive. On postcontrast CT, enhancement is usually homogeneous but rim enhancement is also seen.

7.8. C True
Trigeminal neuromas arise from the Gasserian ganglion in the middle cranial fossa. However, they typically extend posterosuperiorly under the tentorial incisura into the cerebellopontine angle, producing an isodense mass on precontrast CT that enhances homogeneously following contrast. The midbrain and pons are often displaced contralaterally, rotated, and compressed.

7.8. D False
Epidermoid tumours appear on precontrast CT as homogeneous, hypodense lesions that have CT numbers of 0–20 HU. Calcification is rare and no enhancement is seen following contrast. These tumours are also found in a parasellar region.

7.8. E True
Glomus tumours are highly vascular lesions and characteristically demonstrate intense rapid enhancement with an early high contrast concentration peak followed by rapid washout. Precontrast CT reveals a dome-like hyperdense mass which may be associated with destruction of the jugular canal. The tumours arise from the chemoreceptor bodies related to the IXth and Xth cranial nerves. The majority are benign, with less than 10% being malignant.

(Reference 9)

7.9. A False
Pituitary adenomas account for approximately 7% of intracranial tumours and usually present in the fourth and fifth decades of life. They typically arise from the anterior lobe of the gland. Pituitary macroadenomas are usually hypodense

or isodense on precontrast CT and typically show homogeneous enhancement on postcontrast scans. Occasionally ring enhancement may be seen and some show no contrast enhancement.

7.9.   B False
On precontrast CT, pituitary microadenomas may cause erosion of the floor of the pituitary fossa or result in enlargement of the gland which is manifest as a convex upper border to the gland and/or deviation of the pituitary infundibulum away from the side of the tumour. A microadenoma has been defined as a purely intrasellar tumour with a diameter of 10mm or less.

7.9.   C False
Punctate intratumoral calcification is uncommon in pituitary adenomas, occurring in less than 10%. It is likely related to areas of necrosis. Areas of decreased attenuation are seen in approximately 10–20% and represent regions of necrosis or cystic degeneration. These regions show no enhancement on postcontrast CT.

7.9.   D True
Microprolactinomas may become hyperdense or isodense to the rest of the gland on delayed scans. Other functioning microadenomas show a varying degree of contrast enhancement.

7.9.   E True
Some pituitary adenomas, particularly chromophobe adenomas, can become very large and grow cranially to obstruct the foramen of Munro, causing obstructive hydrocephalus. Extension may also be anterior beneath the optic chiasm, laterally resulting in displacement of the cavernous sinus and internal carotid artery and posteriorly into the interpeduncular cistern.

(Reference 9)

7.10.   A True
The extent of vasogenic oedema surrounding cerebral metastases is quite commonly disproportionate to the size of the lesion. Metastases from breast, bronchus and renal primaries may be hypodense precontrast and indistinguishable from peritumoral oedema. However, enhancement on postcontrast CT is almost invariable, with one study demonstrating lack of enhancement in less than 4% of cases.

7.10.   B False
Small metastases appear as homogeneous masses. However, virtually all large metastases undergo a degree of central cystic degeneration and appear on postcontrast CT as multiple ring lesions, typically with an irregular, thick wall.

7.10.   C True
Metastases from choriocarcinoma and some thyroid and renal secondaries are also hyperdense on precontrast CT. The increased density is likely related to haemorrhage within the tumour (or colloid in the case of thyroid secondaries). Subependymal tumour spread is also a feature of metastatic melanoma but is

more commonly seen with glioblastoma, lymphoma, ependymoma, and malignant forms of germinoma and teratoma.

7.10. D False
Leptomeningeal carcinomatosis may occur due to parenchymal tumour breaching the ependymal lining of the ventricles or the pia mater over the cerebral cortex, with subsequent seeding of tumour cells into the CSF. Haematogenous metastases to the meninges is a less common cause. In several series, CT has been able to demonstrate meningeal disease in less than one-third of cases. On postcontrast scans, thick, homogeneous meningeal enhancement obliterating the subarachnoid spaces may be seen.

7.10. E False
Cranial vault metastases are associated with intracerebral lesions in only 5–10% of cases. However, extradural spread of tumour from cranial vault lesions is commoner, appearing on CT as a band of soft-tissue density with a well-defined inner margin.

(Reference 9)

7.11. A False
Gliomas characteristically infiltrate along white matter tracts and therefore have poorly defined margins on unenhanced MRI scans. They are typically hypointense on T1-WSE scans and hyperintense on T2-WSE scans. Typically there is ring-like or nodular enhancement following Gd-DTPA, with non-enhancing necrotic foci within the tumour.

7.11. B True
The classic MRI appearance of haemangioblastoma is a cystic mass with a tumour nodule that enhances following Gd-DTPA. Surrounding oedema is typically minimal. The tumour nodules are hypervascular and a characteristic flow void in the vascular pedicle is often seen.

7.11. C True
Primary CNS lymphoma typically appears as homogeneous, slightly hyperintense to isointense masses on T2-WSE MRI. The mild prolongation of T2 relaxation times may be related to dense cell packing within the tumour, with relatively little space for the accumulation of interstitial water. Homogeneous enhancement is common following Gd-DTPA.

7.11. D False
Meningiomas are typically hypointense to white matter on T1-WSE scans and rarely isointense or hyperintense. On T2-WSE and PDW scans, they appear uniformly hyperintense to white matter in those portions that are not heavily calcified. Heterogeneity of signal intensity on T2-WSE scans may be related to vascularity, cystic changes and calcification.

7.11. E True
On unenhanced MRI, pituitary microadenomas are slightly hypointense to normal gland on T1-WSE scans and slightly hyperintense on T2-WSE scans. However, on T1-WSE scans immediately after Gd-DTPA, the normal gland enhances brightly, demonstrating the lesion more clearly as a hypointense

focus. On delayed postcontrast scans, the microadenoma may become isointense.

(Reference 3)

7.12.   A False
Bony metastases to the skull vault are clearly demonstrated by MRI. On T2-WSE and PDW scans they appear as slightly hyperintense masses that have replaced the normal low signal of the diploic space and cortical bone. On T1-WSE scans, they appear as low-signal-intensity lesions in the hyperintense marrow fat.

7.12.   B True
Most cerebral metastases have high signal intensity on T2-WSE scans and may therefore be obscured by the hyperintense tumour-associated vasogenic oedema. Oedema usually follows white matter boundaries and typically does not cross the corpus callosum or extend into the cortex, features that help in the differentiation from infiltrative primary brain tumours. Oedema is not a significant feature of cortical metastases.

7.12.   C True
On precontrast T1-WSE MRI, metastases usually appear as relatively hypointense lesions due to their increased relaxation times. Further areas of central low signal intensity may be seen due to cystic changes. On Gd-DTPA enhanced T1-WSE scans small metastases typically show nodular enhancement, while large metastases show ring-like enhancement with a necrotic non-enhancing centre. The enhancing wall is typically thick and irregular, unlike the thin, smooth wall usually seen with benign conditions such as abscess.

7.12.   D True
Most non-haemorrhagic, melanoma metastases are hyperintense on T1-WSE scans and of low signal intensity on T2-WSE scans, possibly due to the presence of paramagnetic stable free radicals within melanin. The appearance will be altered by the presence of haemorrhage within the lesions.

7.12.   E True
In the presence of subacute haemorrhage in the metastases, high signal will be seen on both T1- and T2-WSE scans due to methaemoglobin. The high signal intensity is typically seen in the centre of the lesion. The commonest lesions to bleed are metastases from choriocarcinoma, melanoma and renal cell carcinoma.

(References A, E, 3; B, C, D, 27)

7.13.   A True
Since arachnoid cysts contain CSF, they will demonstrate CSF signal characteristics on all pulse sequences (hypointense on T1-WSE scans and hyperintense on T2-WSE scans). They are well-defined, compress adjacent brain and cause localized thinning and expansion of the skull vault. Common sites include the anterior aspect of the middle cranial fossa (approximately 66% arise here), and also the frontal convexity region, the suprasellar and quadrigeminal cisterns, and the foramen magnum region.

7.13. B False
The signal intensity of epidermoid cysts on MRI varies according to the amount of keratin, cholesterol and water they contain. On T1-WSE scans, they usually demonstrate mild hypointensity, between that of CSF and the brain parenchyma. On T2-WSE scans, they are markedly hyperintense, similar to or greater than CSF. They usually exhibit a degree of heterogeneity and typically have a lobulated shape.

7.13. C False
Dermoid cysts typically have high signal intensity on T1-WSE scans due to their fat content, and consequently appear relatively hypointense on T2-WSE scans. They may rupture into the subarachnoid space or into the ventricular system, resulting in fat–fluid levels and chemical shift artefacts. They most commonly occur in a midline location in the posterior fossa, but also in the suprasellar cistern and within the ventricular system.

7.13. D False
Colloid cysts may demonstrate variable intensity patterns. They may be hypo- or hyperintense on either T1- or T2-WSE scans, presumably due to varying concentrations of contained paramagnetic substances, free water and mucoid material. A thin wall can commonly be identified. Ring-like peripheral enhancement may occur following Gd-DTPA.

7.13. E False
Although both arachnoid cysts and porencephalic cysts have the same signal characteristics (that of CSF), the latter communicate with the ventricular system whereas the former do not.

(References A, B, C, D, 27; E, 3)

7.14. A True
Intracanalicular acoustic neuromas originate within the internal auditory canal. Symptoms at presentation include sensorineural hearing loss, vertigo, tinnitus, unsteadiness and facial weakness.

7.14. B False
Unenhanced T1-WSE MRI is better than both contrast enhanced CT and CT air cisternography in detecting small intracanalicular acoustic neuromas. CT air cisternography involves intrathecal injection of air via a lumbar puncture and, with appropriate positioning of the patient's head, air fills the internal auditory canal outlining the enlarged nerve.

7.14. C False
MRI can demonstrate an enlarged canal filled either with tumour or CSF. Thin sections and a small field of view are helpful. Canal enlargement and bony erosion are best demonstrated on axial sections.

7.14. D True
Intracanalicular acoustic neuromas are typically hyperintense to CSF on T1-WSE scans. On T2-WSE scans, the tumours are also hyperintense and may therefore be indistinguishable from the high signal intensity CSF. Heterogeneity

of signal intensity may be due to cystic changes or haemorrhage, while those tumours that are very vascular or contain nodular calcifications may appear relatively hypointense.

7.14. E True
Following Gd-DTPA, there is an approximately 50% reduction in the T1 relaxation time of the tumour, resulting in marked enhancement. Lateral extension of the tumour can also be better defined on postcontrast MRI.

(References A, C, E, 3; B, 3 and 9; D, 27)

7.15. A True
In the early phase of bacterial meningitis or with prompt treatment, the CT scan is usually normal. In advanced cases, CT may demonstrate hyperdensity of the basal cisterns, interhemispheric fissure and choroid plexus (indicating ventriculitis). On postcontrast scans, enhancement of the cisterns and a gyral pattern of cortical enhancement may be seen.

7.15. B True
Opacification of the middle ear cavity is a CT feature of acute otitis media which predisposes to subdural empyema. Other predisposing causes include paranasal sinusitis, osteomyelitis, penetrating skull trauma and infection of a subdural effusion secondary to meningitis.

7.15. C True
Gyral cortical enhancement on postcontrast scans in patients with a subdural empyema can be due either to associated infarction or cerebritis. Its presence may indicate that there is a dural sinus or cortical venous thrombophlebitis. This sign may also occur with an acute or chronic subdural haematoma, due to ischaemia from the mass effect of the collection.

7.15. D False
Herpes simplex encephalitis is a necrotizing meningoencephalitis that typically involves the temporal lobes and inferior portions of the frontal lobes. Type 1 affects adults, and precontrast CT demonstrates hypodense areas with mass effect in the temporal and frontal lobes, which may be bilateral. Sparing of the lenticular nucleus is said to be characteristic.

7.15. E False
With herpes simplex encephalitis, hyperdense areas on precontrast CT may be due to acute haemorrhage. Although petechial haemorrhages are commonly seen pathologically in herpes simplex encephalitis, they are rarely evident on CT. On postcontrast CT, streaky linear areas of enhancement in the region of the Sylvian fissure may be seen.

(Reference 9)

7.16. A False
On precontrast CT a bacterial abscess typically appears as a thin, well-defined, uniform ring structure with surrounding oedema. The wall enhances following contrast injection. With successful medical therapy there may be a reduction in the degree of enhancement and oedema. The medial wall of the abscess is

thinner than the lateral and therefore abscess rupture is typically medial into the ventricular system.

7.16. B True
Abscesses from systemic sites spread via an arterial route and have a distribution similar to that of cerebral metastases. Abscess from local sites spread via a venous route and are typically situated in the subcortical white matter.

7.16. C True
Precontrast CT findings in tuberculous meningitis also include hydrocephalus. On postcontrast scans, there may be intense enhancement of the involved basal cisterns. Several years later, the basal cisterns may show calcification.

7.16. D True
Tuberculomas are associated with extracranial tuberculosis in 50% of cases. CT typically show hypodense, isodense or hyperdense lesions precontrast which typically show ring-like or homogeneous enhancement on postcontrast scans. The non-caseating granulomas of sarcoid also demonstrate uniform contrast enhancement.

7.16. E True
Multiple ring-like enhancing lesions in immunocompromised patients have a wide differential diagnosis, including bacterial abscesses, tuberculomas, aspergillosis, coccidioidomycosis, cryptococcosis, candidiasis and nocardiosis. The finding of opacification of the paranasal sinuses and ring-like lesions in the frontal lobes is very suggestive of mucormycosis, especially in a diabetic.

(Reference 9)

7.17. A False
On both T1- and T2-WSE scans, the capsule of a bacterial abscess can be seen as a relatively isointense (to white matter) structure. On T1-WSE scans, the abscess contents is hypointense, as is the surrounding oedema. On T2-WSE scans the abscess contents and oedema are both of high signal intensity. The capsule does enhance following Gd-DTPA.

7.17. B True
Coccidioidomycosis usually causes a meningitis but abscess and granuloma formation also occur. Regions of meningeal and cerebral inflammation are hyperintense on T2-WSE scans. These areas enhance following intravenous Gd-DTPA.

7.17. C True
Intermediate CSF signal intensity on T1-WSE scans and hyperintensity of the basal cisterns on PDW scans may be seen in severe cases of meningitis. In the majority of cases, however, MRI is either normal or demonstrates minimal hydrocephalus.

7.17. D True
Intracranial tuberculomas are rare and may appear hypointense on T2-WSE scans when mature and inactive. Active lesions may undergo central necrosis, resulting in a hyperintense centre with a hypointense rim on T2-WSE scans.

Tuberculous meningitis is commoner and may result in striking meningeal enhancement on MRI.

7.17. E True
Intracranial disease occurs in approximately 15% of patients with sarcoidosis, usually manifest as granulomatous leptomeningitis, but also as a single or multiple intracerebral masses. On T1-WSE MRI, leptomeningeal disease commonly appears as isointense thickening of the subarachnoid space, either focally or diffusely. On T2-WSE scans, the meningeal thickening is commonly hypointense. Following Gd-DTPA, diffuse homogeneous enhancement is seen.

(References A, B, C, D, 3; E, 27)

7.18. A True
HIV is neurotropic, causing a subacute encephalitis manifest as impaired memory and concentration, and psychomotor slowing progressing to dementia, the so-called AIDS dementia complex. Generalized cerebral atrophy is felt to be a direct effect of the virus.

7.18. B False
In AIDS patients, toxoplasmosis is the commonest cause of a mass and affects approximately 10% of patients. CT typically reveals multiple lesions with surrounding oedema which show ring-like or nodular enhancement on postcontrast scans.

7.18. C True
Primary CNS lymphoma in AIDS patients has different features from that in non-AIDS patients. In AIDS-related lymphoma, 50% of lesions are multiple and may show either homogeneous or ring-like enhancement on postcontrast scans.

7.18. D True
White matter abnormalities in AIDS patients are usually due to viruses and are very sensitively demonstrated by T2-WSE MRI. HIV, CMV and the papovavirus (the cause of progressive multifocal leucoencephalopathy or PML) can all result in white matter changes. In PML, these changes are characteristically in the subcortical white matter.

7.18. E True
Increased CSF signal intensity on both T1- and T2-WSE scans can result from increased protein content that may accompany ependymitis. HIV, CMV, toxoplasmosis, fungi (especially cryptococcus), mycobacteria and syphilis can all cause leptomeningeal and ependymal disease in AIDS.

(Reference 7)

7.19. A False
With acute cortical cerebral infarction, areas of decreased attenuation are the initial CT features usually seen between 12 and 48 hours but occasionally as early as 3 hours after the event. The larger the infarct, the earlier it is likely to

be identified. Transient ischaemic attacks are usually associated with normal pre- and postcontrast scans.

7.19. B False
An area of infarction may become isodense on precontrast CT at this stage due to increased cellularity within the infarcted region, as necrotic tissue is being phagocytosed and oedema is subsiding. The infarct may be detected on postcontrast scans.

7.19. C True
Approximately 60–70% of infarcts show contrast enhancement at some stage, typically between 1 and 4 weeks after the event. Focal dense contrast enhancement within the nuclear grey matter is a typical feature of basal ganglia (lacunar) infarcts.

7.19. D False
Small lacunar infarcts are typically 6–10 mm in size, with larger lesions being greater than 1 cm in size. CT can demonstrate infarcts as small as 5 mm in size. There appears to be a high degree of correlation between the CT examination and symptoms when the area of infarction identified on CT is at least 7 mm in size.

7.19. E False
Lacunar infarcts describe the small infarcts in the region of the basal ganglia and internal capsule (typically in the posterior limb). They are thought to be due to obstruction of the lenticulostriate arteries. Approximately 50% of patients are hypertensive.

(Reference 9)

7.20. A True
MRI is more sensitive than CT in demonstrating acute cerebral infarcts, with approximately 80% of scans being abnormal in the first 24 hours. On T1-WSE scans, gyral swelling is also seen. T1-WSE scans are the least sensitive for detection of non-haemorrhagic strokes. T2-WSE scans and PDW scans both demonstrate regions of increased signal intensity confined to a vascular territory. In the acute phase (within 24 hours), abnormal signal intensity is usually confined to the cortex, extending into the adjacent white matter thereafter.

7.20. B True
The stage of chronic cerebral infarction is usually considered to have begun when the integrity of the blood–brain barrier has been restored. This is usually accomplished by 3–6 weeks. MRI typically demonstrates a smaller, better-defined zone of altered signal intensity than was seen on previous scans. Also, the T2 and PDW signal intensity is greater, due to increased water content. Focal atrophy is also seen.

7.20. C True
In one study, CT could only demonstrate 28% of the lacunar infarcts identified by MRI. Lacunar infarcts appear as slit-like or oval lesions that are hyperintense to parenchyma on T2-WSE scans. They may be differentiated

from dilated Virchow–Robin spaces (VRS) by their size (usually greater than 5 mm), their signal intensity (usually not isointense to CSF on all pulse sequences), and their site (typically located in the upper two-thirds of the putamina).

7.20. D True
Central CSF intensity may be seen in lacunar infarcts due to the development of cavitation. This feature is best identified on PDW scans in which the liquid centre is isointense or hypointense to brain and is surrounded by a thin rim of high signal intensity, representing either oedema or gliosis.

7.20. E False
On postcontrast MRI, subacute infarcts typically demonstrate serpiginous cortical enhancement similar to the gyral pattern seen on postcontrast CT. Gd-DTPA enhancement may help distinguish infarction from neoplasm or infection and is seen between 2 and 6 weeks, at a time when there is a breakdown of the blood–brain barrier.

(References A, B, 27; C, 3 and 27; D, E, 3)

7.21. A False
As many as 50% of patients may have haemorrhages affecting this site. Other common sites include the central white matter, thalamus, pons and the cerebellar hemispheres. Most commonly haemorrhage occurs from rupture of the branches of the lenticulostriate or thalamoperforating arteries.

7.21. B False
Intraventricular extension of primary intracerebral haemorrhage is a common feature and the CSF may be blood-stained in 90% of cases. Rupture through the cerebral cortex is not seen. With a small bleed distant from the ventricular system, the CSF may remain clear.

7.21. C False
CT features that characterize a haemorrhagic infarct are a peripheral typically vascular distribution, extensive areas of surrounding low attenuation, less mass effect and a more heterogeneous density within the centre of the lesion. Intracerebral haematoma appears as an area of increased attenuation measuring 40–90 HU and surrounded by a zone of decreased attenuation.

7.21. D True
Rim enhancement may occur with resolving haematomas about 20–80 days after the event. Initially the enhancement is due to a breakdown of the blood–brain barrier.

7.21. E False
Calcification is a rare finding in some old haematomas, distinguishing them from old infarcts. Also, the site and shape of the residual CT lesion helps to differentiate the two conditions. However, some old haematomas appear as hypodense lesions with associated focal cerebral atrophy, similar to old infarction.

(Reference 9)

7.22.   A True
The commonest site for an extradural haematoma is the temporoparietal region and is due to a tear of the middle meningeal artery. Extradural haematomas can also result from tears of major veins such as the meningeal veins, dural sinuses and diploic veins.

7.22.   B True
Although the classical appearance of an extradural haematoma is of a well-defined, hyperdense, biconvex collection, it may also appear as a crescentic collection with an irregular inner margin, mimicking a subdural haematoma.

7.22.   C True
Underlying cerebral contusion may be seen with both extradural and subdural haematomas and may account for marked mass effect and midline shift which is out of proportion to the size of the extra-axial collection.

7.22.   D True
Although chronic subdural haematomas are typically hypodense, acute haematomas may have this appearance in anaemic patients or if CSF has mixed with the blood to reduce its density. Dependent areas of higher attenuation in a hypodense collection may indicate fresh haemorrhage in a chronic collection. Also, hypodense areas within a hyperdense (acute) collection may indicate fresh bleeding.

7.22.   E False
In isodense collections, contrast enhancement may aid in the diagnosis of extradural haematoma by demonstrating inward displacement of the enhanced cortical vessels and dura. In patients with chronic subdural haematomas, an enhancing inner membrane may be seen on postcontrast CT.

(Reference 9)

7.23.   A True
A 3 hour old haematoma has the magnetic resonance characteristics of any proteinaceous fluid collection, being dark to slightly hyperintense on T1-WSE scans and of intermediate to high signal intensity on T2-WSE scans. At this stage it contains only oxyhaemoglobin and no paramagnetic substances.

7.23.   B False
At 24 hours, reduction of oxygen tension within the haematoma results in the formation of intracellular deoxyhaemoglobin and methaemoglobin within intact RBCs. These paramagnetic substances produce T2 shortening, resulting in a hypointense appearance on T2-WSE scans. The haematoma may be surrounded by an area of bright signal due to oedema.

7.23.   C True
At 2 weeks, there is RBC lysis and extracellular methaemoglobin which results in T1 shortening, causing the haematoma to appear bright on T1-WSE scans and to a lesser extent on T2-WSE scans. The increased signal intensity initially appears as a bright rim which gradually extends inwards.

7.23. D True
At 2 months, phagocytosis of haemoglobin breakdown products and conversion into ferritin and haemosiderin has occurred. These products are superparamagnetic and result in T2 shortening, manifest as a dark rim to the haematoma, most marked on T2-WSE scans but also on T1-WSE scans.

7.23. E True
GE sequences are more sensitive than SE sequences in the detection of both acute and chronic haematomas. Acute haematoma appears hypointense on GE images, even at low field strength.

(Reference 3)

7.24. A True
Periventricular hypodensities are the commonest CT abnormality in MS. They represent the plaques of demyelination and are typically multiple, ranging in size from 5 mm to 70 mm. Mass effect is unusual. The density values of the plaques may be almost as low as CSF. Many of the smaller plaques may be missed due to partial volume averaging.

7.24. B True
The low density in the plaques is due to demyelination and oedema. Some plaques may resolve completely on subsequent CT scans if significant regeneration of myelin has occurred or if the lesion gets smaller. Other plaques are permanent sclerotic lesions which show no change with time.

7.24. C True
Acute demyelination is associated with breakdown of the blood–brain barrier and consequently with enhancement on postcontrast scans. Enhancing plaques are typically small, round and homogeneous. Their enhancement may be abolished by steroids. A less common appearance is of a doughnut-shaped lesion.

7.24. D True
The enhancing plaques of MS are usually hypodense precontrast. Some MS plaques are isodense precontrast and can only be seen after contrast administration. Other plaques may only be demonstrated by delayed high-dose contrast CT.

7.24. E True
Generalized ventricular and sulcal dilatation may be seen on CT and usually indicates chronic disease. Less commonly, a single sulcus may be enlarged.

(Reference 9)

7.25. A True
MS plaques also appear hyperintense on PDW scans and hypointense on T1-WSE scans. They are usually discrete, well-defined, homogeneous foci without evidence of haemorrhage, cystic change or necrosis. Oedema and mass effect are also uncommon.

7.25. B False
MS plaques may occur in the grey matter. However, they are more usual in a periventricular location, particularly along the lateral aspects of the atria and occipital horns. Other common sites include the corpus callosum, corona radiata, internal capsule and centrum semiovale.

7.25. C True
Atrophy of the corpus callosum has been demonstrated in 40% of cases with chronic MS and is best seen on sagittal T1-WSE scans.

7.25. D False
Enhancement following Gd-DTPA will be seen with active plaques. It may be either nodular or ring-like early after contrast adminstration, but on delayed scans the central areas tend to fill in. Enhancement can be present for up to 8 weeks after acute demyelination.

7.25. E False
The lesions of optic neuritis are not reliably demonstrated by MRI. However, T2-WSE scans can show optic neuritis as a high signal intensity plaque on the nerve without enlargement of the nerve sheath.

(Reference 3)

7.26. A True
Otosclerosis is bilateral in up to 90% of cases. It usually begins before puberty, has a familial tendency and affects females twice as commonly as males. The primary bone disease is fairly common and may occur in up to 10% of the population.

7.26. B False
Conductive deafness occurs if otosclerosis affects the oval window and stapes footplate. Involvement of the cochlea results in sensorineural hearing loss. Occasionally there is a combination of both types of deafness.

7.26. C True
This CT feature represents the active phase of cochlear (vestibular) otosclerosis and may affect the basal turn only, the entire cochlea or the whole otic capsule. The hypodensity seen on CT represents areas of demineralization involving the enchondral layer of the cochlea or the otic capsule.

7.26. D True
In fenestral otosclerosis, CT demonstrates sclerotic foci around the margin of the oval window and stapes footplate. The active changes are less well demonstrated at this site.

7.26. E True
Osteogenesis imperfecta produces CT changes in the oval window and otic capsule identical to otosclerosis, but the features may be much more extensive, and in advanced cases the cochlea may be impossible to distinguish from the surrounding petrous bone.

(Reference 9)

7.27. A True
A subdural hygroma consists of collections of CSF in the subdural space, presumably due to a traumatic arachnoid tear or following ventricular shunting. They are a particular feature of skull trauma in children and show CSF signal characteristics on all pulse sequences.

7.27. B True
A subdural haematoma also appears hyperintense on PDW images. The signal intensity on T1-WSE scans depends on the age of the haematoma. They are often isointense to grey matter, and any high T1 signal suggests rebleeding.

7.27. C False
Signal void is not a typical feature of subdural haematoma either due to low macrophage activity or removal of the haemosiderin that has been formed. It may be seen if there are many episodes of bleeding. Membranous strands coursing through the extradural collection are another feature of chronic haematomas.

7.27. D True
On T2-WSE scans, non-haemorrhagic contusions appear hyperintense due to associated oedema. Haemorrhagic contusions will have variable signal intensities depending on the age of the injury.

7.27. E True
Diffuse axonal injuries are also seen in the corpus callosum, brain stem (mostly the midbrain and rostral pons), centrum semiovale and cerebellum. The appearance on MRI is similar to that of cerebral contusions. They are usually multiple, ovoid and parallel to white matter fibre bundles.

(Reference 3)

7.28. A True
Temporal horn enlargement may be the first sign of hydrocephalus, and is differentiated from temporal horn dilatation due to atrophy by the absence of associated medial and lateral temporal lobe atrophy, or enlargement of the Sylvian, choroidal and hippocampal fissures. In atrophy, temporal horn dilatation is a late feature.

7.28. B False
CSF signal void in the cerebral aqueduct is a normal feature that may, however, be accentuated in normal pressure hydrocephalus (NPH). While the presence of moderate or marked CSF signal void does not help to distinguish NPH from atrophic ventriculomegaly, the absence of either is evidence against a diagnosis of NPH.

7.28. C True
In hydrocephalus, the third ventricle tends to become ballooned but in cerebral atrophy the enlargement is typically by uniform widening with parallel walls to the ventricle. The degree of enlargement of the lateral ventricles tends to be greater than that occurring in cerebral atrophy. The demonstration of fourth ventricle dilatation may be seen with both hydrocephalus and cerebral atrophy.

7.28. D ~~False~~ *True*
Normal or accentuated grey matter–white matter differentiation is (typical) of hydrocephalus. CT may also demonstrate periventricular oedema, especially around the frontal horns, in cases of acute and subacute hydrocephalus.

7.28. E True
However, the presence of sulcal enlargement does not exclude associated hydrocephalus, since the block to CSF flow may be at the level of the high convexity or the Pacchionian granulations, resulting in sulcal and fissural dilatation due to damming of fluid proximal to the block.

(References A, C, D, E, 9; B, 27)

7.29. A True
This finding represents the tubers (hamartomas) of tuberous sclerosis. They are located particularly along the basal ganglionic margins of the lateral ventricles and the anterior third ventricle adjacent to the foramen of Munro. Approximately 75% are multiple and they are usually bilateral. Cortical and periventricular calcifications also occur.

7.29. B True
In the Sturge–Weber syndrome, cortical calcification usually involves the parietal, temporal or occipital cortex. It is unilateral in 85% of cases (ipsilateral to the facial naevus) and may also be associated with ipsilateral cortical atrophy and abnormal calcification in the choroid plexus.

7.29. C True
Neurofibromatosis is the commonest pathological cause of extensive choroid plexus calcifications. Calcified intraventricular meningiomas may also be present in this disease. Conversely, physiological choroid plexus calcification is reported to be confined to the posterior half of the choroid between the glomus and foramina of Munro. The glomus is the most commonly calcified portion.

7.29. D False
Physiological basal ganglia calcification only involves the globus pallidus and is considered normal in adults over 40–50 years old. Calcification in the globus pallidus under the age of 40, or elsewhere in the basal ganglia or cortex, should be considered pathological.

7.29. E True
In a large CT series, no normal pineal calcification was found in children under 6.5 years of age but was seen in 8–11% between the ages of 8 and 14 years, 30% at age 15 years and 40% between ages 17 and 20 years. The calcification should be compact and granular in appearance.

(Reference 9)

7.30. A False
The maximal AP diameter of the cord at the cervical enlargement is approximately 8 mm and the maximum lateral diameter is 13 mm. The narrowest portion of the cord is in the midthoracic region, where the upper

limits of normal AP and lateral diameters are approximately 6 mm and 8 mm respectively.

7.30. B False
Hydromyelia refers to cystic expansion of the central canal of the cord and is associated with the Arnold–Chiari malformation. Syringomyelia refers to a cyst lined by glial cells resulting from cavitation within the cord, and is associated with trauma, neoplasm or degeneration within the cord. The two frequently coexist. CT prior to contrast shows the cysts as low-density areas within the cord. With intrathecal injection, contrast medium can reach the cyst either via the fourth ventricle or by diffusion through the cord.

7.30. C True
Irregular cord expansion is almost invariably due to a primary cord tumour. CT may also show reduced density of the cord due to a combination of oedema, tumour infiltration and expansion. Cystic areas and calcifications are other features.

7.30. D True
Overall, ependymomas are the commonest primary cord tumours, commonly affecting the conus medullaris and filum terminale. Spontaneous SAH is a feature of ependymomas due to their highly vascular nature. Other primary intramedullary tumours include haemangioblastoma, lipoma, dermoid, epidermoid, neurinoma and metastasis.

7.30. E True
Haematomyelia typically appears on CT as a poorly defined area of increased attenuation within the cord, which may be associated with symmetrical or asymmetrical distortion of the contours of the cord. Causes include trauma, or haemorrhage from an AVM or tumour such as haemangioblastoma.

(Reference 9)

7.31. A False
Hyperostosis is a rare feature of spinal meningiomas but when seen is relatively specific. Calcification is also uncommon. Meningiomas are commonest in the thoracic or cervical region, typically in middle-aged females.

7.31. B False
Both meningiomas and neurofibromas are slightly hyperdense to the cord on precontrast scans and may enhance following intravenous contrast medium. Spinal neurofibromas are commonest in the lumbar region but may involve any part of the spinal canal.

7.31. C False
So-called dumbbell neurofibromas are detected due to enlargement of the neural foramen into which they extend. However, meningiomas can also extend through the dura into the intervertebral foramen. Both lateral meningoceles and conjoint nerve roots may produce mass effect that erodes bone. These can be identified following intrathecal administration of contrast medium.

7.31. D True
Dermoids and epidermoids are rare in the intradural compartment. They may demonstrate typical attenuation values of −20 to −50 HU due to fat. Leakage of lipid into the subarachnoid space is a recognized feature and the resulting arachnoiditis appears on CT as thickening of nerve roots which may be adherent to the dura. Intrathecal contrast medium is usually necessary to demonstrate these changes.

7.31. E True
Most spinal lipomas occur in the cervical or thoracic region but are the commonest tumours of the filum terminale. CT typically reveals a lobulated mass with attenuation values of −50 to −100 HU. Associated abnormalities are seen in 15% of patients with spinal lipomas and include spinal dysraphism, syringomyelia, meningocele and tethering of the cord. In the latter condition, CT may show thickening of the filum terminale.

(Reference 9)

7.32. A False
The MRI appearances of primary cord tumours is generally non-specific, the site of the tumour being more helpful. Ependymomas (accounting for 60%) are commoner in the conus region, whereas astrocytomas (accounting for 25%) are slightly commoner in the cervical region. The two occur with about equal frequency in the thoracic region.

7.32. B True
The hallmark of an intramedullary tumour on T1-WSE scans is expansion of the cord. T2-WSE scans show high signal intensity within the expanded region and may also show high signal intensity in adjacent non-expanded areas of the cord. The latter represents either tumour-associated oedema or microinfiltration of tumour.

7.32. C True
Focal areas of either increased or decreased signal intensity may be present within the expanded parenchyma on T1-WSE scans due to the presence of cysts, necrotic cavities or haemorrhage. Cavities with a high protein content or areas of subacute haemorrhage cause reduced T1 relaxation times and high signal intensity.

7.32. D False
Enhancement following Gd-DTPA will be seen in the solid portions of astrocytomas, ependymomas, haemangioblastomas and intramedullary metastases. Evidence of hypervascularity on non-contrast scans suggests haemangioblastoma.

7.32. E False
Haematogenous metastases from breast, bronchial or melanoma primaries may produce similar MRI appearance to those of primary cord tumours, including irregular focal cord expansion. However, metastases tend to involve several segments of the cord and associated subarachnoid spread is common.

(Reference 3)

7.33.   A True
Other MRI features of a primary syrinx include thinning of the cord parenchyma adjacent to the cyst, smooth internal margins to the cyst and the presence of septations resulting in a haustrated appearance. However, cavities with these characteristics may be associated with an intra- or extramedullary neoplasm distant from the cavity.

7.33.   B False
In one large series, at least one-third of patients with non-neoplastic cavities had associated high signal intensity in the adjacent cord parenchyma or in an area proximal or distal to the cavity. The high signal intensity has been attributed to gliosis, oedema, microcystic malformation or demyelination.

7.33.   C True
Other MRI features that are found with tumour-associated syrinxes include increased signal intensity of the cyst fluid on T1- and T2-WSE scans due to high protein content and the presence of necrotic debris, irregular margins to the cyst wall and the presence of high signal intensity in the adjacent cord parenchyma on T2-WSE scans due to oedema, tumour cells or both.

7.33.   D False
Gd-DTPA enhancement helps to distinguish simple from neoplastic cysts. The failure of T1-WSE scans in the sagittal and axial planes following Gd-DTPA to demonstrate an enhancing solid nodule makes a cord neoplasm unlikely.

7.33.   E True
A CSF signal flow void on flow-sensitive pulse sequences has been stated to be a reliable indicator of a non-neoplastic cyst. However, the sign is unreliable in cysts less than 3 mm in diameter since such small channels may normally not demonstrate signal void.

(Reference 3)

7.34   A False
Unlike intracerebral MS plaques which may show low signal intensity on T1-WSE scans, plaques in the cord are not typically hypointense. On T2-WSE scans, plaques appear as hyperintense lesions which may be unifocal or multifocal and may involve long or short segments of the cord. Occasionally, focal mild cord expansion may be seen.

7.34.   B True
High signal intensity foci in the cord on T2-WSE scans are non-specific and may be seen in other causes of granulomatous myelitis such as Wegener's, in post-traumatic myelomalacia and in association with AVMs. High T2 signal may also be seen in areas of spondylotic spinal cord compression.

7.34.   C True
Spinal cord infarction typically produces areas of high signal in the cord on T2-WSE scans due to oedema, and focal cord swelling. The abnormality involves a recognizable vascular territory. Areas of high signal on T1-WSE scans may be seen in haemorrhagic infarction. With time the infarct evolves to leave a region of myelomalacia and cord atrophy.

7.34. D True
Spinal AVM has a variable appearance on MRI. T1- and T2-WSE scans may show areas of low or high signal intensity due to flow-related signal void or enhancement. Focal areas of low signal on T2-WSE scans may be present due to haemosiderin deposition following haemorrhage. Large vessels within the cord or on the pial surface may give the cord a scalloped appearance.

7.34. E False
Acute transverse myelitis has been reported to cause cord swelling with increased intramedullary signal intensity on T2-WSE scans. The abnormal signal may extend above the level of clinical deficit and may show mild enhancement after contrast administration. MRI following recovery may show the cord to have returned to a normal appearance.

(References A, B, C, D, 3; E, 27)

7.35. A False
This finding indicates that the lesion is extradural. MRI also differentiates extradural from intradural–extramedullary lesions by the fact that the former compress both the thecal sac and the cord, resulting in narrowing of the subarachnoid space on both sides of the cord, whereas the latter widens the ipsilateral subarachnoid space, leaving a cap of CSF both above and below the lesion.

7.35. B True
Nerve sheath tumours (neurofibromas and neurilemmomas) are typically slightly hypointense or isointense to cord on T1-WSE scans and hyperintense on T2-WSE scans. They enhance uniformly and brightly following Gd-DTPA. MRI also demonstrates bone erosion as scalloping of vertebral bodies and enlargement of neural foramina.

7.35. C False
Meningiomas often have similar signal characteristics to nerve sheath tumours on T1-WSE scans and typically become slightly hyperintense to the cord on T2-WSE scans. Signal intensities are therefore unreliable in differentiating the two lesions. Features that suggest nerve sheath tumour rather than meningioma include location (anterior rather than posterolateral), number (commonly multiple), and the presence of a low-intensity centre on T2-WSE scans, which is not seen with meningioma.

7.35. D True
Intradural drop metastases usually occur from subarachnoid seeding of primary brain tumours (medulloblastoma, ependymoma, germinoma, choroid plexus papilloma, teratoma, glioblastoma and pineoblastoma) or less commonly from brain metastases (lymphoma, melanoma, breast, lung and renal primaries). Their signal intensities are similar to those of neurofibromas and they can range in size from a few millimetres to 1 cm. They typically enhance following Gd-DTPA.

7.35. E True
Increased CSF signal intensity due to metastases is seen predominantly in the lumbar region and is probably due to a combination of increased CSF protein content, malignant cells and damping of CSF pulsation. Intradural metastases

may also cause pseudoexpansion of the cord due to diffuse coating of the spinal cord with tumour. Scanning following Gd-DTPA readily differentiates this from an intramedullary tumour.

(References A, B, D, E, 3; C, 27)

7.36. A False
In one series of 116 cases, extraocular muscle enlargement was symmetrical in 70% of patients. Unilateral enlargement was seen in 6%. The inferior and medial rectus muscles were involved in 75% of cases and the superior and lateral rectus muscles in 50%. In 30% of cases, all four muscles were enlarged and in 9% only the inferior rectus was involved. Isolated enlargement of the lateral rectus was not seen.

7.36. B True
On precontrast CT, the affected muscles may have lower attenuation than normal muscles. An increased volume of retro-orbital fat has been demonstrated in some cases and was associated with proptosis.

7.36. C False
In Graves' ophthalmoplegia, the muscle enlargement is typically more posterior near the apex of the orbit and the muscle attachment to the globe is normal. In contrast, the orbital myositis or pseudotumour typically causes enlargement at the insertion of the muscle into the globe. Orbital pseudotumour is typically unilateral and enlargement of a single extraocular muscle is a well-recognized finding.

7.36. D False
Uveoscleral thickening is a feature of orbital pseudotumour and was seen in 53% of cases in one series after contrast enhancement. This finding can also be seen with orbital trauma, after surgery, with melanoma and with lymphoma.

7.36. E False
Graves' disease is not associated with masses in the retro-orbital fat but pseudotumour may result in partial or complete obliteration of retro-orbital fat by either well-defined or irregular soft-tissue masses. These may be indistinguishable from orbital tumours but the demonstration of either partial or complete remission following steroids confirms the diagnosis. Proliferation of retrobulbar fat may be seen in dysthyroid eye disease.

(Reference 9)

7.37. A True
Cavernous haemangiomas are the commonest primary tumours of the orbit. They are usually intraconal but may be extraconal. On CT, small tumours are smooth and well-defined but larger lesions may obscure other orbital structures. Homogeneous enhancement is seen on postcontrast scans.

7.37. B True
The upper, outer quadrant of the orbit is the commonest location for dermoid tumours, which may occasionally have a dumbbell configuration with an intracranial component attached via a bony defect in the wall of the orbit.

Both dermoid and epidermoid tumours of the orbit are smoothly demarcated lesions which may display density values ranging from those of fat to soft tissue.

7.37. C True
On precontrast CT, optic nerve gliomas may also cause a solitary, fusiform enlargement or a solid, irregular thickening of the nerve. Calcification may occasionally occur. Some tumours may show marked enhancement on postcontrast scans. Meningiomas of the optic nerve are associated with hyperostosis of the optic canal and usually show a more fusiform, smooth enlargement of the nerve with intense peripheral enhancement around a relatively less enhanced nerve.

7.37. D True
On CT, the normal scleral–uveal rim has a uniform thickness throughout its entirety, and any eccentric thickening is diagnostic of a lesion in the retina, choroid or sclera (which cannot be separated on CT), or in the adjacent retrobulbar space. Ocular melanoma affects the choroid in 93% of cases, the ciliary body in 4% and the iris in 3%. The majority of tumours are unilateral. CT usually demonstrates a fairly well-defined lesion along the uveal tract which may be associated with diffuse uveoscleral thickening.

7.37. E True
Demonstration of calcification in retinoblastoma is considered a good prognostic sign since it appears to be commoner in tumours confined to the globe, whereas contrast enhancement has been associated with a less favourable outcome since it is far commoner with extraocular spread of tumour. Extension along the optic nerve and widespread metastases are commonly seen.

(Reference 9)

7.38. A True
Carotid–cavernous fistulas may be either spontaneous or traumatic in origin and when they involve the eye may result in pulsatile exophthalmos. CT demonstrates enlargement of the inferor ophthalmic vein in some cases also. Enlargement of the superior ophthalmic vein also occurs in the Wyburn–Mason syndrome, which consists of multiple AVMs that may involve the retina.

7.38. B True
Primary epithelial tumours account for 50% of masses involving the lacrimal gland. Of these, 50% are benign mixed tumours, 40% represent adenocystic carcinoma and 10% are malignant mixed or other carcinomas. The other 50% of lacrimal gland masses are due to various types of lymphoid infiltration, including the Mikulicz syndrome and lymphoma.

7.38. C False
Metastases can occur in the choroid, soft tissues or bony walls of the orbit. Retrobulbar metastases can be either intraconal or extraconal and usually appear as irregular, hyperdense masses which show slight contrast enhancement. The commonest site for bony metastases is the greater wing of sphenoid,

and these lesions frequently have a soft-tissue component that extends into the lateral extraconal compartment of the orbit and the middle cranial fossa.

7.38.   D True
Approximately 11% of all orbital tumours are lymphomas. They have variable CT appearances ranging from localized, circumscribed masses to more diffuse lesions. Most lymphomas occur in the anterior portion of the orbit. Uveoscleral thickening has also been described.

7.38.   E True
Rhabdomyosarcomas are the commonest primary orbital tumours in childhood, when they are typically of the embryonal type. Pleomorphic rhabdomyosarcomas are rare intraorbital tumours seen in adults which arise in striated muscle. Embryonal rhabdomyosarcomas appear on CT as soft-tissue masses that are not typically related to the muscles. They may be destructive. Orbital neuroblastoma may have identical CT appearances.

(Reference 9)

7.39.   A False
On T1-WSE and PDW scans, the combined choroid and retina have higher signal intensity than the inner vitreous and outer sclera. The sclera is hypointense on all pulse sequences. Both the vitreous and aqueous humor have a high water content and therefore display low signal on T1-WSE scans, moderately low signal on PDW scans and high signal on T2-WSE scans.

7.39.   B False
The optic nerve has similar MRI signal characteristics to brain. High-resolution, thin-section MRI can demonstrate the intraocular portion of the nerve, unlike CT. On T2-WSE scans, a small amount of fluid can normally be seen as high signal within the dural sheath of the nerve contrasting with the relatively hypointense nerve and retro-orbital fat.

7.39.   C True
Retinoblastoma is typically hyperintense to vitreous on PDW scans also and hypointense to vitreous on T2-WSE scans. T1- and T2-WSE scans can also demonstrate any associated hyperintense subretinal effusion. Tumour extension into the optic nerve is manifest as a mass, focal thickening or altered signal within the nerve.

7.39.   D False
Choroidal melanoma typically has high signal intensity on T1- and low signal intensity on T2-WSE scans. Conversely, choroidal metastases (most commonly from breast, bronchus, gastrointestinal or genitourinary primaries) usually have moderate signal intensity on T1-WSE scans, becoming hyperintense on T2-WSE scans. Melanoma is also associated with hyperintense subretinal effusions.

7.39.   E True
On precontrast MRI, meningiomas characteristically have similar signal intensities to the optic nerve on all pulse sequences. This helps to distinguish them from other diseases associated with enlargement of the nerve sheath

(such as optic neuritis, pseudotumour, papilloedema) which typically have high signal intensity on T2-WSE and PDW scans.

(Reference 3)

7.40. A True
On T1-WSE scans, cavernous haemangioma has low signal intensity which contrasts it with the hyperintense retro-orbital fat. It typically appears as a well-defined, ovoid mass lying lateral to the optic nerve and is the commonest benign intraorbital tumour.

7.40. B False
Orbital pseudotumour is typically hypointense on both T1- and T2-WSE scans, which helps to differentiate it from lymphoma, which has high signal intensity on T2-WSE scans. Orbital pseudotumour, like lymphoma, can affect the intra- or extraconal compartments.

7.40. C False
The MRI signal characteristics of involved muscles in Graves' disease are indistinguishable from those of pseudotumour. However, the two conditions can be differentiated by the site of muscle involvement (see question 7.35 and answer).

7.40. D True
The lacrimal glands normally have intermediate signal intensity on all pulse sequences, and carcinoma results in increased signal on T2-WSE scans. Dermoid tumours can also appear hyperintense on T2-WSE scans but also have high signal on T1-WSE scans. MRI may demonstrate fat–fluid levels. The presence of fluid excludes the diagnosis of lipoma, and epidermoid tumours are differentiated from dermoids since they are usually hypointense on T1-WSE scans.

7.40. E False
High flow in vascular lesions such as orbital varix, AVM and carotid–cavernous fistula usually results in signal void on all pulse sequences, but in the presence of slow flow increased signal intensity on T2-WSE scans can be seen due to flow-related enhancement.

(Reference 3)

8 Paediatrics

8.1. Concerning ultrasound in the assessment of congenital anomalies of the brain
A separation of the lateral ventricles is a feature of agenesis of the corpus callosum.
B dilatation of the third ventricle is a feature of agenesis of the corpus callosum.
C anterolateral displacement of the occipital horns is a feature of Dandy–Walker syndrome.
D dilatation of the third ventricle is a feature of Dandy–Walker syndrome.
E fusion of the frontal horns is invariable in semilobar holoprosencephaly.

8.2. Concerning cranial ultrasound findings in congenital anomalies of the brain
A absence of the septum pellucidum is a feature of septo-optic dysplasia
B an echogenic, midline posterior fossa mass is a feature of vein of Galen aneurysm.
C hydranencephaly is associated with absence of the falx.
D enlargement of the massa intermedia is a recognized feature of the Chiari type II malformation.
E hydrocephalus is invariably present at birth in the Chiari type II malformation.

8.3. Concerning cranial ultrasound in the assessment of neonatal intracranial haemorrhage
A acute haemorrhage (less than 1 week old) typically appears hyperechoic relative to the choroid plexus.
B subependymal haemorrhage (SEH) characteristically occurs in the caudothalamic groove.
C intraventricular extension of SEH is an atypical finding.
D intraparenchymal haemorrhage (IPH) is most commonly seen in the occipital lobe.
E posthaemorrhagic hydrocephalus is seen at some stage in a minority of babies.

8.4. Which of the following are true of neonatal cranial ultrasound?
A Periventricular leucomalacia (PVL) initially results in hyperechoic foci.
B Cystic changes within PVL typically occur within 5 days.
C Cerebral atrophy is a recognized consequence of PVL.
D Diffuse parenchymal hyperechogenicity is a recognized finding in cerebral oedema.
E Slit-like ventricles are a specific finding in cerebral oedema.

8.5. Concerning ultrasound in the assessment of paediatric chest masses
A the normal thymus is typically hyperechoic relative to the thyroid.
B an echogenic mass adjacent to the diaphragm is a recognized finding with bronchopulmonary sequestration.
C bronchopulmonary sequestration cannot be distinguished from cystic adenomatoid malformation (CAM).
D absence of cystic spaces in a mass excludes CAM.
E bronchial atresia is a recognized cause of a solid/cystic mass.

8.6. Concerning CT in the assessment of the paediatric mediastinum
A under the age of 10 years, the thymus is typically quadrilateral in shape.
B the normal thymus does not abut the sternum.
C under the age of 5 years, the azygo-oesophageal recess is typically concave laterally.
D thymoma is the commonest cause of a calcified anterior mediastinal mass.
E under the age of 1 year, ganglioneuroblastoma is the commonest cause of a posterior mediastinal mass.

8.7. Concerning imaging of the paediatric airway and chest
A in choanal atresia, CT usually demonstrates a bony plate as the cause of obstruction.
B opacification of the maxillary antrum is a feature of juvenile nasal angiofibroma.
C destruction of the antral walls is typical of antrochoanal polyp.
D subglottic haemangioma typically results in symmetrical narrowing of the subglottic space on the AP radiograph.
E pulmonary blastoma is a cause of complete opacification of one hemithorax.

8.8. Which of the following are true concerning neonatal hepatitis and biliary atresia?
A On ultrasound, the liver parenchyma typically appears hypoechoic.
B Demonstration of the gallbladder excludes biliary atresia.
C Choledochal cyst is an associated finding in biliary atresia.
D In neonatal hepatitis, hepatic uptake of ^{99m}Tc-HIDA is typically normal.
E In biliary atresia, absence of bowel activity is a feature of ^{99m}Tc-HIDA scintigraphy.

8.9. Concerning imaging of the paediatric biliary tract
A choledochal cyst is associated with intrahepatic bile duct dilatation.
B in cases of choledochal cyst, ^{99m}Tc-HIDA scintigraphy shows no bowel activity.
C on ultrasound, Caroli's disease is characterized by central biliary dilatation.
D spontaneous perforation of the CBD is a cause of a cystic mass in the porta hepatis.
E rhabdomyosarcoma typically appears as an intrahepatic mass.

8.10. Concerning the ultrasound features of paediatric liver masses

A   hepatoblastoma is differentiated from hepatocellular carcinoma (HCC) by the presence of calcification.
B   undifferentiated embryonal sarcoma usually appears as a multiseptate cystic mass.
C   mesenchymal hamartoma typically appears as a solid, hyperechoic mass.
D   dilatation of the proximal abdominal aorta is a recognized finding in haemangioendothelioma.
E   a complex solid/cystic mass is a recognized association in children with glycogen storage disease.

8.11. Concerning CT of the liver in children

A   metastatic stage IV S neuroblastoma typically produces multifocal masses.
B   early central enhancement is characteristic of haemangioendothelioma on postcontrast CT.
C   enhancement is not a feature of mesenchymal hamartoma on postcontrast CT.
D   hepatoblastomas typically become hyperdense to liver on postcontrast CT.
E   multicentric tumour is a recognized feature of HCC.

8.12. Concerning ultrasound in the diagnosis of hypertrophic pyloric stenosis in term infants

A   the hypertrophied pyloric muscle is usually hypoechoic relative to liver.
B   a pyloric muscle thickness of greater than 4 mm is abnormal.
C   a pyloric canal length of 20 mm is within normal limits.
D   the stomach is typically atonic.
E   following pyloromyotomy, the pyloric wall thickness usually returns to normal within 1 week.

8.13. Concerning ultrasound in the assessment of the paediatric gastrointestinal tract

A   antral dyskinesia syndrome and hypertrophic pyloric stenosis can be distinguished.
B   gastric teratomas typically appear as completely intraluminal masses.
C   in duodenal atresia, dilatation of the whole duodenal loop is typical.
D   the superior mesenteric vein lying to the left of the superior mesenteric artery is a feature of small bowel malrotation.
E   hyperperistalsis of the duodenum is a feature of midgut volvulus.

8.14. Concerning ultrasound in the assessment of paediatric gastrointestinal tract masses

A   an echogenic mass is a recognized appearance of a duplication cyst.
B   an echogenic rim is a recognized finding in a duplication cyst.
C   a hyperechoic mass in the right anterior pararenal space is a recognized finding following blunt abdominal trauma.
D   a 'target' lesion with double hypoechoic rings is a recognized finding in intussusception.
E   Ultrasound cannot demonstrate meconium ileus.

8.15. Concerning CT in the assessment of blunt abdominal trauma in children

A subcapsular splenic haematomas are usually associated with parenchymal injury.

B a congenital cleft in the spleen cannot be differentiated from a splenic laceration.

C following renal trauma, enhancement of perirenal fluid collections on delayed postcontrast CT is a recognized finding.

D renal trauma is associated with pre-existig renal anomalies in less than 1% of cases.

E a pancreatic pseudocyst is a recognized finding.

8.16. Concerning ultrasound of the adrenal gland in neonates and children

A in the neonate, the adrenal cortex and medulla can be differentiated.

B normal adrenal size excludes congenital adrenal hyperplasia (CAH).

C adrenal calcification is a feature of Wolman disease.

D adrenal haemorrhage is the commonest cause of an adrenal mass in the neonate.

E free intraperitoneal fluid is not a feature of adrenal haemorrhage.

8.17. Concerning imaging of the adrenal glands in infants and children

A on CT, the neonatal adrenal has convex outer borders.

B a solid mass is most likely to be a non-functioning adenoma.

C on ultrasound, neuroblastoma typically appears as a homogeneous, hypoechoic mass.

D IVC invasion is a typical feature of adrenal neuroblastoma.

E on CT, calcification is seen in less than 50% of adrenal neuroblastomas.

8.18. Concerning the imaging features of primary neuroblastoma

A the adrenal gland is the commonest site for the primary tumour.

B a single cyst is a recognized ultrasound finding.

C T2-WSE MRI typically demonstrates a mass of high signal intensity.

D uptake of ^{99m}Tc-MDP by the primary tumour is a recognized feature.

E uptake of ^{131}I-MIBG by the tumour is a specific finding.

8.19. Concerning ultrasound of the kidneys in infants and children

A in the full-term neonate, the renal cortex is typically hyperechoic relative to the liver.

B in the neonate, the medullary pyramids are normally hypoechoic relative to the cortex.

C by 6 months of age, the renal cortex should be hypoechoic relative to the liver.

D demonstration of a triangular hyperechoic focus on the superior anterior surface of the renal cortex indicates renal scarring.

E the interrenicular septum is more commonly seen in the right kidney.

8.20. Concerning ultrasound in the assessment of the paediatric renal tract
A   in pelviureteric junction (PUJ) obstruction a dilated ureter is seen in 50% of cases.
B   uniform ureteric dilatation is typical of primary megaureter.
C   in multicystic dysplastic kidney (MCDK), normal renal parenchyma is seen between the cysts.
D   bilateral nephromegaly is a feature of Beckwith–Wiedemann syndrome.
E   demonstration of cysts differentiates adult from infantile polycystic kidney disease (PCKD).

8.21. Concerning the ultrasound features of paediatric renal masses
A   mesoblastic nephroma typically appears as a multiloculated cyst.
B   renal vein thrombosis is a recognized cause of a unilateral enlarged kidney.
C   diffuse renal leukaemia typically results in enlarged hypoechoic kidneys.
D   multiple hypoechoic parenchymal masses are a recognized finding in nephroblastomatosis.
E   in multilocular cystic nephroma, the cysts are non-communicating.

8.22. Concerning CT in the assessment of renal masses in children
A   septal enhancement is a feature of multilocular cystic nephroma on postcontrast CT.
B   calcification is a feature of paediatric renal cell carcinoma.
C   extrarenal extension is not a feature of mesoblastic nephroma.
D   clear-cell sarcoma typically results in bilateral renal masses.
E   nephroblastomatosis is a recognized cause of multiple hypodense renal masses on postcontrast CT.

8.23. Concerning the imaging features of Wilms' tumour
A   bilateral tumours typically occur at a younger age than unilateral tumours.
B   a hyperechoic rim is a recognized ultrasound finding.
C   on CT, areas of fat density within a mass exclude the diagnosis.
D   encasement of the IVC is a characteristic feature.
E   a mediastinal mass is a recognized finding.

8.24. Concerning imaging in the assessment of paediatric urinary tract infection
A   on fluoroscopic voiding cystourethrography (VCUG), reflux into a non-dilated renal pelvis indicates Grade 4 reflux (International System).
B   on fluoroscopic VCUG, intrarenal reflux indicates Grade 5 reflux (International System).
C   radionuclide VCUG has a lower radiation dose than fluoroscopic VCUG.
D   a wedge-shaped hypodense area in the kidney on postcontrast CT is a recognized finding.
E   ultrasound is as sensitive as ^{99m}Tc-DMSA scintigraphy in the detection of renal scars.

8.25. Concerning ultrasound in the assessment of the paediatric female genital tract

A in the newborn, demonstration of an endometrial cavity echo is always abnormal.

B prior to the menarche, the uterine cervix should be equal in length to the uterine body.

C in the newborn, the ovary typically lies anterior to the internal iliac vessels.

D hydrometrocolpos typically appears as an anechoic, spherical, midline pelvic mass.

E follicular cysts are the commonest cause of ovarian cysts in the neonate.

8.26. Which of the following are true of paediatric germ cell tumours?

A They are the commonest cause of a malignant ovarian mass.

B Non-hereditary sacrococcygeal teratomas are typically associated with sacral anomalies.

C Familial sacrococcygeal teratomas appear as entirely presacral masses.

D Intraspinal extension is a feature of sacrococcygeal teratoma.

E Teratoma is a recognized cause of a fat-containing retroperitoneal mass.

8.27. Which of the following are true of the imaging features of rhabdomyosarcoma?

A The orbit is the commonest primary site in the head.

B On CT, tumour calcification is an atypical feature.

C Extremity tumours are typically associated with adjacent bone destruction.

D A solid scrotal mass is a recognized ultrasound finding.

E Irregularity of the bladder base is a recognized ultrasound finding.

8.28. Concerning neonatal hip ultrasound in the assessment of hip instability

A on the transverse view, the centre of the femoral head lies directly lateral to the junction of the ischium and triradiate cartilage.

B posterior subluxation of 5 mm during stress may be a normal finding in the newborn hip.

C on coronal scanning, the acetabular labrum appears uniformly echogenic.

D on the coronal view, the bony acetabulum should accommodate over half the diameter of the femoral head.

E the femoral head ossification centre is seen earlier with ultrasound than with radiography.

8.29. Concerning imaging in the assessment of Ewing sarcoma

A long-bone lesions usually occur in the metaphysis.

B primary tumours in flat bones are characteristically associated with adjacent soft-tissue masses.

C CT demonstrates calcification in the tumour mass in 50% of cases.

D presence of a soft-tissue mass after cessation of therapy indicates residual tumour.

E metastases are most commonly to the lungs.

8.30. Which of the following are true concerning ultrasound of the paediatric spine?
A   On longitudinal scans, the posterior dura appears as an echogenic line.
B   The substance of the cord is isoechoic to CSF.
C   With the child prone, the anterior and posterior subarachnoid spaces are usually of equal depth.
D   The nerve roots cannot be identified.
E   Longitudinal movements of the cord are seen with flexion–extension of the neck.

8.31. Which of the following are true concerning spinal dysraphism?
A   Coronal cleft vertebra is a recognized association.
B   Absence of skin over a back mass is typical of lipomyelomeningocele.
C   If the conus lies at the level of L3 in the newborn, the cord is tethered.
D   A tethered cord is a recognized finding in myelomeningocele.
E   Hydromyelia is a recognized association of myelomeningocele.

8.32. Concerning spinal dysraphism
A   on T1-WSE MRI, a hyperintense mass is a recognized finding with lipomyelomeningocele.
B   a myelocystocele is most commonly demonstrated in the cervical region.
C   there is an association between diastematomyelia and lipoma of the filum terminale.
D   there is an association between dorsal dermal sinus and intraspinal dermoid.
E   a filum terminale thickness of 4 mm is a recognized finding.

8.33. Which of the following are true concerning the skeletal manifestations of non-accidental injury (NAI)?
A   Metaphyseal corner fractures are typically associated with periosteal reactions.
B   Diaphyseal periosteal reactions are a recognized finding.
C   Rib fractures typically occur at the costochondral junctions.
D   Parietal skull fractures are characteristic of NAI.
E   Fractures involving the acromion process are usually due to NAI.

8.34. Concerning the differential diagnosis of NAI to the paediatric skeleton
A   physiological diaphyseal periosteal reactions do not extend to involve the metaphysis.
B   metaphyseal fractures are a feature of congenital insensitivity to pain.
C   fractures in full-term infants under 6 months of age are typical of copper deficiency.
D   osteoporosis is a recognized finding in copper deficiency.
E   the presence of Wormian bones is diagnostic of osteogenesis imperfecta.

8.35. Which of the following are true concerning paediatric skeletal trauma?
A   The majority of physeal injuries are Salter–Harris type II injuries.
B   Osteoporosis of the femoral neck is a recognized feature of slipped capital femoral epiphysis.
C   A torus fracture is most commonly seen in the proximal tibial shaft.
D   The triplane fracture of the distal tibia typically occurs in adolescence.
E   A diaphyseal fracture in a baby that cannot walk is a feature of NAI.

8.36. Concerning the cranial CT findings in NAI
A subdural haematomas are invariably unilateral.
B interfalcial haematomas are a recognized finding.
C a depressed occipital skull fracture is a recognized feature.
D haemorrhages at the grey–white matter junction are a recognized feature.
E generalized decreased attenuation of cerebrum on precontrast scans is a recognized finding following anoxia.

8.37. Concerning CT in the assessment of paediatric neck masses
A cystic hygroma usually appears as a mass in the posterior triangle.
B dermoid cysts are typically demonstrated in the midline.
C malignant teratomas are typically located in relation to the thyroid.
D primary cervical neuroblastomas usually occur in the paraspinal location.
E squamous carcinoma is the commonest cause of a soft-tissue density mass.

8.38. Which of the following are true of extracranial peripheral neuroectodermal tumours (PNET)?
A They arise exclusively in bones.
B Bone lesions are typically associated with cortical destruction.
C A chest wall mass is a recognized finding.
D Calcification in the primary tumour is a typical feature.
E Metastases are most commonly to the brain.

8.39. Which of the following are true concerning paediatric lymphoma?
A Under the age of 10 years, Hodgkin's disease is commoner than non-Hodgkin's lymphoma (NHL).
B Tracheobronchial compression is a recognized feature of mediastinal Hodgkin's disease.
C Thymic enlargement after cessation of chemotherapy is a recognized finding.
D Omental disease is a feature of Burkitt's lymphoma.
E Primary bone lymphoma usually affects a single bone.

8.40. Concerning imaging in the assessment of AIDS in children
A air bronchograms are not a feature of *Pneumocystis carinii* pneumonia.
B miliary nodules on chest radiography are most commonly due to TB.
C oesophageal hypoperistalsis is a recognized manifestation of candidal oesophagitis.
D Kaposi's sarcoma is the commonest cause of an abdominal mass.
E basal ganglia calcification is a recognized finding on cranial CT.

Paediatrics: Answers

8.1. A True
Agenesis of the corpus callosum may be partial or complete. Also the frontal horns may be narrowed (unless hydrocephalus is present), and on coronal scans the lateral peaks of the frontal horns and bodies of the lateral ventricles appear sharply angled. There is also relatively greater dilatation of the occipital horns (colpocephaly). The medial border of the lateral ventricles appears concave.

8.1. B True
Agenesis of the corpus callosum is also associated with varying degrees of dorsal extension of the third ventricle and interposition between the lateral ventricles. There is elongation of the foramina of Munro and a radial arrangement of the medial cerebral sulci around the roof of the third ventricle. These sulci extend through the zone normally occupied by the corpus callosum.

8.1. C True
The Dandy–Walker syndrome is a congenital cystic dilatation of the fourth ventricle associated with dysgenesis of the cerebellum and vermis. Ultrasound demonstrates a posterior fossa cyst that is continuous with the fourth ventricle (allowing differentiation from a posterior fossa arachnoid cyst). The posterior fossa is enlarged and the tentorium elevated, resulting in displacement of the occipital horns.

8.1. D True
In the Dandy–Walker syndrome, there is variable dilatation of the third and lateral ventricles. Ultrasound also demonstrates hypoplasia of the cerebellar hemispheres, which are displaced anterolaterally. The condition is associated with encephalocele, agenesis of the corpus callosum, and holoprosencephaly.

8.1. E True
Holoprosencephaly is classified as alobar, lobar and semilobar. Ultrasound in alobar holoprosencephaly shows a small amount of midline cerebral tissue with no separation into hemispheres, a large horseshoe-shaped single ventricular cavity, and fusion of the thalami. In lobar and semilobar forms, there is partial separation of the hemispheres, but the frontal horns are always fused and the falx cerebri partially developed. Facial anomalies such as cleft lip and cyclopia are seen.

(Reference 29)

8.2. A True
Septo-optic dysplasia (DeMorsier syndrome) is characterized by absence of the septum pellucidum, an enlarged anterior recess of the third ventricle and hypoplasia of the optic nerve, chiasm and infundibulum. Ultrasound may also demonstrate dilatation of the lateral ventricles, and on coronal views the

anteromedial aspect of the frontal horns has an angular appearance. The falx is present.

8.2. B True
On ultrasound, vein of Galen aneurysm appears as a midline mass posterior to the third ventricle. The mass appears anechoic when filled with liquid blood, or echogenic if filled with thrombus. Pulsation of the mass or dilated feeding vessels are seen with real-time ultrasound, while Doppler will identify flow. Other findings include hydrocephalus, cerebral atrophy and parenchymal calcifications due to cortical ischaemia.

8.2. C False
Hydranencephaly results from massive destruction of the cerebral hemispheres due to intrauterine bilateral occlusion of the supraclinoid internal carotid arteries. The falx is intact. Only the brain stem and the portion of the occipital lobes receiving blood from the basilar artery are normally developed.

8.2. D True
Enlargement of the massa intermedia has been described in 82–90% of babies with the Chiari II malformation. It may almost fill the third ventricle, which is usually mildly dilated. Posterior fossa anomalies include downward displacement of the fourth ventricle, medulla and cerebellar tonsils, with obliteration of the cisterna magna.

8.2. E False
In the Chiari II malformation, the lateral ventricles may be of normal size or variably enlarged at birth. Serial scans usually demonstrate gradually increasing ventricular size after birth until a shunting procedure is performed. The lateral ventricles are frequently asymmetric, and the occipital horns and atria tend to be more dilated than the frontal and temporal horns. The choroid plexus is unusually prominent.

(Reference 29)

8.3. A True
On ultrasound, acute intracranial haemorrhage (ICH) appears homogeneous, hyperechoic to the choroid and without associated acoustic shadowing. With time, the haematoma becomes less echogenic, inhomogeneous, smaller, and its central region becomes anechoic. The normal echogenic periventricular halo may be mistaken for ICH. However, this region is less echogenic than choroid, has poorly defined lateral borders and its echogenicity may disappear if scanned in an axial or true perpendicular plane.

8.3. B True
SEH is the commonest type of ICH in the premature infant, usually arising in the germinal matrix of the caudothalamic groove. On ultrasound, SEH appears as an echogenic mass that elevates the choroid at its attachment in the caudothalamic groove. It should be viewed in two planes for confident diagnosis. SEH may result in the development of a subependymal cyst in a minority of cases.

8.3.   C False
SEH may rupture into the lateral ventricle in up to 80% of cases, since the haematoma is separated from the ventricle only by the thin ependymal lining. Acute intraventricular haemorrhage (IVH) is highly echogenic and may appear as a bulky choroid or as a CSF–clot fluid level in the occipital horn. Also, choroid plexus is not normally seen in the frontal or occipital horns, and therefore echogenic material in these sites is suggestive of IVH.

8.3.   D False
IPH usually results from extension of an SEH and is usually seen in the frontal and parietal lobes, since the most common site of SEH is the caudothalamic groove. IPH initially appears as a uniformly hyperechoic parenchymal region with irregular margins, then develops a sonolucent centre and eventually results in a porencephalic cyst. Such an appearance is considered by some to be due to venous infarction rather than haemorrhage.

8.3.   E False
Obstructive posthaemorrhagic hydrocephalus develops in approximately 70% of babies with IVH, typically within the first 2 weeks after the haemorrhage. Hydrocephalus will resolve in approximately one-third of babies, remain stable in one-third, and be progressive in the rest, requiring ventricular shunting.

(Reference 29)

8.4.   A True
PVL represents necrosis of the periventricular white matter and is a result of cerebral ischaemia. The initial ultrasound may be normal but within 2 weeks of the insult areas of increased echogenicity develop in the affected white matter, typically lateral to the frontal horns and extending posteriorly to the trigone area above the lateral ventricles.

8.4.   B False
Cystic changes within PVL appear at approximately 2–3 weeks. The cysts are characteristically multiseptate. They may resolve, but usually enlarge or remain unchanged, and rarely they develop into actual porencephaly. It should be noted that a normal cranial ultrasound or CT scan several months after birth does not exclude the absence of earlier PVL.

8.4.   C True
Cerebral atrophy is manifest on cranial ultrasound as widening of the interhemispheric fissure (the most reliable sign), widened cerebral sulci, or mild enlargement of the lateral ventricles. The distinction between PVL-associated ventriculomegaly and obstructive hydrocephalus is an important one since the latter may respond to shunting whereas the former will not.

8.4.   D True
Cerebral oedema may result as a consequence of hypoxic–ischaemic encephalopathy, which is associated with a high incidence of major neurological sequelae. Diffuse brain infarction may also occur. Ultrasound can demonstrate these abnormalities as diffuse increased parenchymal echogenicity, but is less sensitive than CT or MRI. Focal areas of hyperechogenicity and decreased vascular pulsations may also be seen.

8.4. E False
Slit-like ventricles are a non-specific feature and may be a normal finding on neonatal cranial ultrasound. Other abnormalities that can be seen with cranial ultrasound include extradural and subdural collections, and subarachnoid haemorrhage, which although difficult to detect may present as widening and increased echogenicity of the Sylvian fissure.

(Reference 29)

8.5. A False
The normal thymus gland has a homogeneous echotexture and is normally slightly hypoechoic relative to the normal thyroid gland. It is best scanned via a suprasternal or trans-sternal approach. On longitudinal ultrasound scans, it appears as a triangular or oval mass, while on transverse scans it has either a trapezoidal or bilobate shape.

8.5. B True
When a bronchopulmonary sequestration does not connect with the bronchial tree, it typically appears as an echogenic mass in the posterior aspect of a lower lobe. Occasionally, dilated fluid-filled bronchi may be seen within the mass. If the mass contains air (usually following infection), it appears echogenic and will demonstrate reverberation artefacts.

8.5. C False
The demonstration of anomalous vessels (arterial supply from the aorta and venous drainage into either the azygos or pulmonary circulation) is highly suggestive of bronchopulmonary sequestration since such vessels are very rarely seen with cystic adenomatoid malformation.

8.5. D False
Approximately 10% of CAMs appear on gross pathology as solid masses of tissue composed of multiple microscopic cysts. These are type III lesions and appear on ultrasound as solid, echogenic masses. Type I CAM consists of a single or multiple cysts greater than 2 cm in diameter and accounts for about 50% of lesions, while type II lesions consist of multiple cysts under 1 cm in size and make up the remaining 40% of cases. Ultrasound demonstrates type I and II lesions as complex solid/cystic masses, or a single cyst.

8.5. E True
Bronchial atresia appears on ultrasound as an echogenic mass containing multiple hypoechoic tubular structures representing dilated fluid-filled bronchi. The sonographic features can mimic those of either bronchopulmonary sequestration or CAM.

(Reference 28)

8.6. A True
Under the age of 10 years, the thymus gland appears quadrilateral in shape with gently convex lateral borders. It is isodense or slightly hyperdense to muscle on precontrast CT and has homogeneous density before puberty. After

puberty it may contain foci of low attenuation due to fat. It also assumes a triangular shape. Multilobularity is not a feature and is a fairly sensitive sign of an abnormal gland.

8.6.   B False
Under the age of 20 years, the thymus may lie directly against the sternum and the paucity of mediastinal fat in children makes it difficult to separate the gland from the intercostal muscles. The mean thickness of the gland for children between 0 and 10 years is 1.5 cm and 1.05 cm for the 10–20-year age group.

8.6.   C False
In children under 5 years, the heart is relatively large compared to the AP diameter of the chest, causing the oesophagus and azygos vein to bulge laterally into the azygo-oesophageal recess. The recess is therefore either straight or convex laterally, assuming the adult configuration after the age of 10 years.

8.6.   D False
A prominent thymus gland is the commonest cause of a pseudomass in the anterior mediastinum. However, tumours of the thymus (and thyroid) are unusual in infants and children, making teratoma and lymphoma the major differential diagnoses of a paediatric anterior mediastinal mass. Teratomas are characterized on CT by the demonstration of areas of soft-tissue density, fat density and calcification. The latter is rarely seen in treated or untreated lymphoma.

8.6.   E False
Up to 95% of paediatric posterior mediastinal masses are neurogenic tumours (neuroblastoma, ganglioneuroblastoma and ganglioneuroma). Neuroblastoma is much commoner than ganglioneuroblastoma and ganglioneuroma and usually occurs in children less than 2 years old. CT typically demonstrates a soft-tissue paraspinal mass which may be calcified (40% of neuroblastomas; 20% of ganglioneuromas).

(References A, B, C, 32; D, E, 31)

8.7.   A True
Choanal atresia is usually bilateral, and in over 90% of cases the obstruction is due to a bony plate. In these cases, CT shows medial bowing and thickening of the lateral wall of the nasal cavity, enlargement of the vomer and fusion of these bony elements. In membranous atresia, accumulation of nasal secretions may result in a falsely thickened appearance of the narrowed segment.

8.7.   B True
Juvenile nasal angiofibroma appears on CT as a densely enhancing naso-pharyngeal mass which can extend into the pterygopalatine fossa, sinuses and the cranial cavity. Bony changes include anterior bowing of the posterior wall of the maxillary antrum, erosion of the sphenoid bone and skull base, and deviation of the nasal septum. Opacification of the antra is due to tumour or

obstruction. MRI may distinguish the two since the latter will show higher signal on T2-WSE scans.

8.7. C False
Antrochoanal polyps are benign inflammatory masses that may expand into the nasopharynx and obstruct the contralateral choana. The mass may also expand into the oral cavity. Characteristically, there is mild expansion of the maxillary antrum without destruction, and a soft-tissue mass in the antrum which can be seen extending into the oropharynx on the lateral radiograph.

8.7. D False
Subglottic haemangioma typically occurs in the first 3 months of life. The mass is usually eccentric and causes asymmetrical distortion of the subglottic portion of the trachea. However, some cases of viral croup may also cause asymmetrical narrowing. Some haemangiomas may involute spontaneously, while others can expand rapidly, resulting in further airway compromise.

8.7. E True
Pulmonary blastomas are peripherally located, solitary masses, usually occurring in children under the age of 4 years. They cause little in the way of bronchial obstruction, and are commonly large at presentation. Imaging demonstrates a large tumour that may opacify the hemithorax and result in contralateral mediastinal shift. CT reveals a soft-tissue mass with hypodense areas due to haemorrhage and necrosis.

(Reference 31)

8.8. A False
On ultrasound, the hepatic parenchymal echogenicity is either increased or normal and the intrahepatic bile ducts may be normal in calibre. In biliary atresia, the obstruction is at or above the porta hepatis in approximately 85% of cases. These babies may benefit from a Kasai procedure (hepatoportoenterostomy). The success of the operation is dependent upon the timing of surgery, with bile flow being established in approximately 90% of infants under 2 months of age. The success rate drops to about 20% in those babies older than 90 days at the time of operation.

8.8. B False
A normal gallbladder can be seen in approximately 10% of babies with biliary atresia. In the remaining cases, it is either absent or small. The presence of a normally functioning gallbladder suggests patency of the biliary tract, making the diagnosis of biliary atresia unlikely.

8.8. C True
Biliary atresia may coexist with choledochal cyst. Other recognized associations with biliary atresia include the polysplenia syndrome, preduodenal portal vein, azygos continuation of the IVC, situs inversus and hydronephrosis.

8.8. D False
The HIDA scan in babies with neonatal hepatitis usually demonstrates reduced uptake of isotope but there is usually evidence of passage of isotope into the bowel, excluding obstruction.

8.8.   E True
Absence of bowel activity on the HIDA scan indicates biliary tract obstruction, as occurs in biliary atresia. Uptake of isotope by the liver is normal. The level of biliary obstruction is variable, being in the distal common bile duct in 15–25% of cases.

(References A, 28 and 29, B, C, D, E, 28)

8.9.   A True
Choledochal cyst is characterized by cystic dilatation of the extrahepatic bile duct. Ultrasound is diagnostic when it demonstrates a well-defined cystic structure in the porta hepatis, separate from the gallbladder, with a dilated bile duct entering the cyst. Intrahepatic bile duct dilatation has been reported to occur in 46% of cases and is usually most prominent in the central bile ducts. Diverticulum of the CBD and choledochocele are other types of choledochal cyst.

8.9.   B False
^{99m}Tc-HIDA scintigraphy in patients with choledochal cyst demonstrates early uptake of isotope by the liver, with a photopenic area in the region of the porta hepatis. Subsequently there is accumulation of activity in the dilated bile duct and cyst. Excretion into the bowel is present but delayed.

8.9.   C False
Caroli's disease is characterized by segmental saccular dilatation of the intrahepatic bile ducts. There is an association with renal tubular ectasia and congenital hepatic fibrosis. Ultrasound typically reveals multiple cysts of various sizes distributed throughout the liver. The commonest complications are recurrent cholangitis and stone formation.

8.9.   D True
Spontaneous perforation of the CBD usually occurs at the junction of the cystic and common bile ducts, and is frequently associated with distal CBD obstruction. It characteristically occurs in neonates. The bile leak results in an inflammatory reaction with the development of a pseudocyst in the porta hepatis. Ultrasound may also demonstrate free fluid in the abdominal cavity.

8.9.   E False
Hepatobiliary rhabdomyosarcoma most commonly arises from the CBD, and imaging typically reveals a polypoid or infiltrating mass in the porta hepatis. On ultrasound, the mass is usually hypoechoic to liver parenchyma. Rarely, the tumour may arise from the intrahepatic bile ducts and appears as a solid liver lesion with no distinguishing features. The gallbladder is also a rare primary site.

(References A, B, C, D, 29; E, 30)

8.10.   A False
Plain films may show calcifications in approximately 40% of cases of hepatoblastoma and HCC. The ultrasound features are very variable and non-specific. Intravascular tumour thrombus may be identified and local lymphadenopathy is commonly seen at presentation. Hepatoblastoma usually

occurs in children under 3 years of age who have a previously normal liver, whereas HCC is more commonly seen over the age of 5 years and is often associated with a prior hepatotoxic process such as glycogen storage disease and tyrosinaemia.

8.10. B True
Undifferentiated embryonal sarcoma usually occurs in the 6–10 year age group. It is thought to be the malignant counterpart of mesenchymal hamartoma. It is a rapidly growing malignant tumour that frequently contains necrosis and cystic degeneration, appearing on ultrasound as a multicystic, septate mass.

8.10. C False
The mesenchymal hamartoma is a benign tumour that presents in the 4-month to 2-year age group with painless abdominal enlargement or an upper abdominal mass. Ultrasound demonstrates a spectrum of cystic changes ranging from small cysts with thick septa to a multilocular cystic mass with thin septa.

8.10. D True
More than 85% of haemangioendotheliomas present before 6 months of age. Hepatomegaly is a common presentation. High-output cardiac failure is an uncommon complication. Ultrasound reveals a diffuse lesion of variable echogenicity, but solitary or multifocal masses are also seen. If significant arteriovenous shunting is present, a dilated proximal abdominal aorta and distension of the hepatic veins may be seen.

8.10. E True
There is an increased incidence of hepatic adenomas in children with glycogen storage disease. The liver appears diffusely hyperechoic due to fatty infiltration. Adenomas may appear as homogeneous hypoechoic lesions, which because of their tendency to bleed, may also have a complex solid-cystic nature. Malignant degeneration into hepatoma is manifest as increasing size and loss of the well-defined contour of the adenoma.

(Reference 29)

8.11. A False
Stage IV S neuroblastoma is almost exclusively seen in children under 1 year of age. Metastases are characteristically to the liver (83–94%), skin (18–26%) and bone marrow (33–50%). The liver is typically diffusely infiltrated by tumour and CT reveals hepatomegaly. In older children, with other stages of disease, liver metastases tend to be well-defined, single or multifocal lesions.

8.11. B False
The CT features of haemangioendothelioma following a bolus of intravenous contrast medium are similar to those of haemangioma, with early peripheral enhancement and delayed central opacification over a period of several minutes to half an hour. If the tumour contains central thrombosis or fibrosis,

these areas will remain hypodense on postcontrast scans. Fine calcification is also a feature.

8.11. C False
Mesenchymal hamartoma is commoner in boys. CT typically demonstrates a well-defined, multiloculated mass with low-density areas separated by areas of solid density due to septa and stroma. The solid regions of tumour may enhance on postcontrast CT. Occasionally, the mass may be completely solid, especially in the very young. The tumour is commoner in the right lobe of the liver.

8.11. D False
Both hepatoblastoma and HCC are usually hypodense on precontrast CT, containing areas of further reduced attenuation due to haemorrhage and necrosis. On postcontrast scans, they enhance to a lesser degree than normal liver parenchyma. Scanning in the arterial phase following a rapid bolus injection of contrast medium may result in peripheral rim enhancement, but this is a non-specific finding.

8.11. E True
Hepatoblastoma and paediatric HCC may be solitary lesions that are found more commonly in the right lobe than the left. However, both tumours may be multicentric or diffuse. HCC commonly involves both lobes of the liver (in 33–72% of cases), either due to direct spread, or due to multicentric tumour. Multicentric tumour is less common with hepatoblastoma.

(Reference 30)

8.12. A True
Ultrasound in hypertrophic pyloric stenosis typically shows the thickened pyloric muscle to be hypoechoic relative to the liver, although it may occasionally appear isoechoic. The pyloric mucosa is identified as a central echogenic area which contain linear echolucencies, giving rise to the 'double track' sign.

8.12. B True
The normal pyloric muscle thickness is approximately 2.0–2.2 mm, with a thickness of 4 mm or more being characteristic of pyloric stenosis. A cross-sectional diameter of the entire pylorus of 15 mm or greater is also taken as evidence of muscle hypertrophy.

8.12. C False
A pyloric canal length of 18 mm or greater is considered abnormal. It is now felt that out of all three measurements muscle thickness is the most specific criteria for diagnosing pyloric stenosis.

8.12. D False
In pyloric stenosis, the stomach is typically dilated, fluid filled and hyperperistaltic. Extension of the hypertrophied, elongated pyloric canal into the gastric antrum, delayed gastric emptying and failure to visualize the descending duodenum are other sonographic findings.

8.12. E False
Following pyloromyotomy, 2–6 weeks are usually required for the muscle hypertrophy to resolve. Therefore, in babies with persistent vomiting, ultrasound is not a reliable investigation of incomplete pyloromyotomy in the early postoperative period.

(Reference 28)

8.13. A True
Infants with antral dyskinesia syndrome present with recurrent non-bilious vomiting. Barium meal may demonstrate a funnel-shaped antrum with absent peristalsis and delayed gastric emptying. On ultrasound, the pyloric canal is elongated but the pyloric muscle thickness is normal. Sluggish peristalsis also helps to distinguish the condition from pyloric stenosis.

8.13. B False
Sonography in infants with gastric teratomas typically demonstrates a predominantly hypoechoic mass containing calcifications. Both intraluminal and large extraluminal components are usually present. Teratomas are the commonest gastric tumours in the first year of life.

8.13. C False
Duodenal atresia is associated with dilatation of the stomach and duodenal cap, which are fluid filled, resulting in the ultrasound equivalent of the 'double-bubble' sign. The level of obstruction is near the ampulla of Vater. A complete duodenal diaphragm will produce similar ultrasound findings.

8.13. D True
In the normal situation, the superior mesenteric vein lies anterior and to the right of the superior mesenteric artery. In small bowel malrotation, this arrangement may be reversed or the vein may lie directly anterior to the artery. A normal relationship of the vessels does not exclude malrotation.

8.13. E True
In infants with small bowel malrotation, obstruction can be due either to midgut volvulus or associated peritoneal (Ladd's) bands. The level of obstruction is typically at the descending or transverse parts of the duodenum. Sonography demonstrates dilatation of the proximal duodenum, which is hyperperistaltic. Both antegrade and retrograde peristalsis may be seen.

(Reference 28)

8.14. A True
Bowel duplications (enteric cysts) are the most common gastrointestinal mass in the neonate, with 85% occurring under the age of 1 year. The mesenteric border of the terminal ileum is the commonest location. Stomach duplications are usually located along the greater curve. Ultrasound may demonstrate an anechoic mass in contiguity with the gastrointestinal tract. The cyst may alternatively be echogenic due to mucoid content.

8.14.   B True
Duplication cysts may have a highly echogenic rim corresponding to the mucosal lining, but this may be destroyed by extensive ulceration. The presence of gastric mucosa can be confirmed by ^{99m}Tc-pertechnetate scintigraphy. Symptoms in such cases may be due to ulceration, haemorrhage or perforation. Bowel obstruction may also occur. The imaging features of distal ileal duplications may be mimicked by Meckel's diverticulum.

8.14.   C True
A duodenal haematoma initially appears on ultrasound as an echogenic mass in the right anterior pararenal space, which may be associated with elevation of the superior mesenteric artery. The mass may become less echogenic on follow-up scans. Duodenal and small bowel haematomas are usually due to blunt abdominal trauma, but other causes include haemophilia, Henoch–Schönlein purpura and anticoagulation therapy.

8.14.   D True
On ultrasound, an intussusception may produce a characteristic 'bull's-eye' or 'target' lesion. The thickened hypoechoic rim is thought to represent the oedematous wall of the intussusceptum surrounding a hyperechoic centre due to compressed mucosa. A double hypoechoic ring may be due to either the entering and exiting walls of the intussusceptum, or to the walls of the intussusceptum and intussusipiens. Such an appearance has been correlated with irreducible intussusception.

8.14.   E False
Occasionally, meconium ileus may present as a right iliac fossa (RIF) mass lacking the classic plain radiographic findings (soap bubble appearance in the RIF with relative paucity of air–fluid levels). The plain film may demonstrate small bowel obstruction and an RIF mass displacing the bowel into the left upper quadrant. In such cases, ultrasound can provide the diagnosis by demonstrating echogenic, thick meconium filling dilated distal small bowel loops.

(Reference 28)

8.15.   A True
Subcapsular splenic haematomas are rarely isolated and are usually associated with either a splenic parenchymal haematoma or a splenic laceration. On postcontrast CT, haematomas appear as well-defined, hypodense, intra-parenchymal masses, whereas lacerations appear as low-density, linear or stellate lesions.

8.15.   B False
Congenital splenic clefts are commonly seen on CT in children and can be distinguished from lacerations by their horizontal orientation, often involving the anterior and inferior surface of the spleen, their smooth contours, and by the absence of perisplenic clot.

8.15.   C True
Renal trauma may be associated with rupture of the collecting system, and on postcontrast CT extravasation of contrast enhanced urine into the perirenal collection will result in increased density.

8.15. D False
Several large series indicate a 9–12% incidence of predisposing renal anomalies in children with renal trauma. Such anomalies should be considered if CT demonstrates an unusual appearance to the kidney or if the CT findings are out of proportion to the degree of trauma. Demonstrated anomalies include PUJ obstruction, hydronephrosis of other aetiology, horseshoe kidney, and solid lesions such as Wilms' tumour and angiomyolipoma.

8.15. E True
CT in children with pancreatic trauma may demonstrate irregular clefts of decreased attenuation, with or without peripancreatic fluid collections. Complete transection of the pancreas is uncommon.

(Reference 32)

8.16. A True
In the neonate, ultrasound demonstrates the adrenal as a thin hyperechoic core surrounded by a thick hypoechoic cortex. With increasing age, the cortex becomes thinner, and by 5–6 months the gland appears uniformly hyperechoic with poor corticomedullary differentiation. After 1 year, the gland is uniformly hypoechoic.

8.16. B False
In infants with CAH, the adrenals are usually bilaterally enlarged or at the upper limit of normal size. The enlarged glands appear triangular rather than flattened and corticomedullary differentiation is less prominent than normally.

8.16. C True
In Wolman disease, ultrasound typically demonstrates bilaterally enlarged, hyperechoic adrenal glands with distal acoustic shadowing due to diffuse punctate calcification throughout the cortex. Ultrasound may also show hepatosplenomegaly.

8.16. D True
Conditions associated with neonatal adrenal haemorrhage include birth trauma, perinatal stress, hypoxia, coagulopathies, septicaemia and shock. Sonography may demonstrate an enlarged round or triangular adrenal gland, which may be echogenic, hypoechoic or complex depending on the age of the haemorrhage. Residual calcification is also a feature.

8.16. E False
Adrenal haemorrhage is usually intracapsular, but if there is capsular rupture blood may extend into the retroperitoneal or intraperitoneal spaces. Adrenal haemorrhage may also cause focal enlargement of the gland.

(Reference 28)

8.17. A True
In neonates and infants, the adrenal glands are also relatively large. By the age of 3 years, the limbs of the gland have either straight or concave outer borders. In older children, as in adults, the presence of convex outer borders should be considered abnormal.

8.17.   B False
Non-functioning adrenal adenomas are extremely rare in children. Neuroblastoma and ganglioneuroblastoma are the commonest adrenal tumours in children, with other lesions accounting for less than 5% of adrenal masses. Of these, adrenal carcinoma is the commonest, with phaeochromocytoma and adenoma being rarer.

8.17.   C False
On ultrasound, neuroblastoma usually appears as a large mass with mixed echotexture. Hyperechoic areas may be seen due to calcification and fresh haemorrhage, whereas anechoic regions may be present due to old haemorrhage and necrosis, which are common features of the tumour. The heterogeneity of the mass may help to distinguish it from Wilms' tumour. Stages I and IV S lesions are localized to the gland and appear as well-defined, round or oval masses.

8.17.   D False
Primary adrenal neuroblastoma spreads by extension through the perirenal space to displace and encase the aorta and IVC. Spread may then occur along the major branches of the aorta. The tumour can cross the midline and invade the kidneys via the renal hilum. Spread across the midline indicates stage III disease. Actual vascular invasion, however, is uncommon.

8.17.   E False
CT may identify calcification in neuroblastoma in up to 90% of cases. Precontrast scans reveal a mass of predominantly soft-tissue density, with hypodense areas due to haemorrhage and necrosis. On postcontrast CT, solid areas of tumour enhance. Features of a retroperitoneal mass include obliteration of the perinephric fat between the kidney and the psoas and anterior displacement of the IVC, aorta, renal vessels, duodenum, or adjacent colon.

(References A, B, 32; C, D, E, 30)

8.18.   A True
Neuroblastoma can be found at any site along the sympathetic chain or in the adrenal medulla. The sites of primary tumour are the adrenal gland (40%), other intra-abdominal sites (25%), posterior mediastinum (15%), the neck and pelvis (5% each). In approximately 10% of cases, no primary site can be identified. At diagnosis, metastatic spread is present in up to 70% of cases.

8.18.   B True
Cystic neuroblastoma is a rare occurrence, having been described in the abdomen, chest and neck, usually in the neonate. Ultrasound may demonstrate a single or multiple cysts, which are usually well-defined and echo free. Adjacent abnormal soft tissue may or may not be identified, with tumour tissue seen in the cyst wall only at histological examination.

8.18.   C True
In general, neuroblastoma results in prolongation of T1 and T2 relaxation times, appearing relatively hypointense on T1-WSE scans and hyperintense on T2-WSE scans. Areas of calcification will not be seen unless very large.

Heterogeneous signal intensity within the mass is common. Ability to image in the coronal plane is particularly useful for identification of extradural extension of tumour. MRI is also very sensitive in the detection of bone marrow metastases.

8.18. D True
The major indication for ^{99m}Tc-diphosphonate scintigraphy is to detect skeletal metastases. However, the primary tumour may also show uptake in 66–74% of cases, and occasionally may identify the site of a previously undetected tumour. The uptake of isotope is proportional to the size of the tumour but independent of the stage and amount of calcium seen in the tumour.

8.18. E False
MIBG (*meta*-iodobenzylguanidine) is metabolized by neuroblastoma and therefore taken up by the tumour. Scans are obtained at 48 and 72 hours after injection of isotope. Other tumours that will result in a positive scan include carcinoid, phaeochromocytoma and some thyroid tumours. However, all of these are very rare in the paediatric age group, and therefore uptake in a young infant is highly suggestive of neuroblastoma.

(Reference 30)

8.19. A False
In the full-term neonate and infant, the cortex of the right kidney is isoechoic to the liver in approximately 65% of cases, and that of the left kidney is isoechoic to the spleen in about 50% of cases. In the remainder, the renal cortex is slightly hypoechoic to liver or spleen. In premature neonates, the renal parenchyma may appear hyperechoic relative to the liver or spleen.

8.19. B True
In the neonate, the medullary pyramids appear hypoechoic and prominent due to the larger medullary and smaller cortical size relative to that of older children and adults. The renal sinus is also less echogenic than that of adults due to the relative paucity of renal sinus fat.

8.19. C True
The cortex of the right and left kidneys becomes less echogenic than the liver and spleen, respectively, by 2–3 months of age. By this time, the renal pyramids have also become less prominent in the majority of babies.

8.19. D False
A triangular echogenic focus on the superior anterior aspect of the kidney is a normal anatomical finding and represents the junctional parenchymal defect. A similar appearance may be seen at the posterior inferior aspect of the renal cortex.

8.19. E True
The interrenicular septum appears as an echogenic linear shadow which extends from the junctional parenchymal defect to the renal hilum. It is seen three times more commonly in the right kidney compared to the left (as is the

junctional parenchymal defect). These two structures are thought to represent the site of fusion of the two metanephric elements that form the kidney.

(Reference 28)

8.20.   A False
The majority of cases of PUJ obstruction occur on the left side and 10–25% are bilateral. There is an association with contralateral multicystic dysplastic kidney. Ultrasound demonstrates pyelectasis and caliectasis, with lack of visualization of the ureters. The latter feature helps to distinguish it from conditions associated with more distal obstruction, and from reflux where a dilated ureter is usually identified.

8.20.   B False
The classical ultrasound appearance of primary megaureter is a dilated renal collecting system and proximal ureter, with a 'rat-tail' narrowing of the distal end. The ureter is tortuous and relatively small for its last few centimetres before it inserts into the bladder.

8.20.   C False
The commonest non-obstructing mass in the neonate is the MCDK. The classic (pelvoinfundibular) form appears as a mass composed of multiple, non-communicating cysts of varying size arranged in a disorganized manner. No normal renal parenchymal tissue can be seen. The majority of such kidneys do not appear to change in size with time, although a small percentage either enlarge or regress.

8.20.   D True
In the Beckwith–Wiedemann syndrome, the kidneys are typically enlarged with an otherwise normal sonographic architecture. The condition is also associated with an increased incidence of nephroblastomatosis and Wilms' tumour until the age of 6, when the incidence is no greater than the general population. These children should be screened with ultrasound regularly to identify the development of tumours.

8.20.   E False
The neonate who presents with bilateral renal masses most likely has polycystic kidney disease, either the recessive (infantile) or dominant (adult) type. Neither of these lesions is necessarily cystic, although small or large cysts can be found. Ultrasound in children with recessive PCKD has revealed a 'salt and pepper' pattern in the renal parenchyma, and in older children a cortical rim of normal echogenicity as opposed to the rest of the kidney, which is echogenic.

(Reference 29)

8.21.   A False
Mesoblastic nephroma is the commonest solid renal mass in the neonate. The mass is unilateral, involving 60–90% of the kidney and appearing on ultrasound as an echogenic lesion. Necrotic regions may be identified within the tumour, and the ultrasound features may be indistinguishable from Wilms' tumour. Surgery is associated with an excellent prognosis.

8.21. B True
Renal vein thrombosis typically occurs in the neonate in the setting of dehydration and haemoconcentration. The affected kidney is large and there may be haematuria. Ultrasound shows a large kidney with loss of corticomedullary differentiation. DTPA scintigraphy may show reduced perfusion and loss of function.

8.21. C False
In children with renal lymphoma or leukaemia, ultrasound usually demonstrates the renal cortex to be relatively hyperechoic, with accentuation of the corticomedullary junction. The hyperechogenicity may be due either to cellular infiltration or associated uric acid nephropathy. Multiple hypoechoic masses may also be seen, especially in Burkitt's lymphoma.

8.21. D True
Nephroblastomatosis refers to the persistence of immature renal tissue beyond the time of cessation of nephrogenesis. Renal ultrasound may show multiple anechoic, hypoechoic or hyperechoic parenchymal masses. There is an increased association with Wilms' tumour.

8.21. E True
Multilocular cystic nephroma is an uncommon intrarenal mass occurring in children less than 2 years of age. The tumour is well encapsulated and usually benign, although some contain partially differentiated Wilms' tumour. Ultrasound characteristically demonstrates a mass consisting of multiple non-communicating cysts with brightly echogenic septations, and areas of solid tissue interspersed between them.

(References A, C, D, E, 29; B, 31)

8.22. A True
Multilocular cystic nephroma appears on precontrast CT as a well-defined, multicystic mass. The cysts have varying size and attenuation due to the presence of myxoid material and haemorrhage. Curvilinear calcification is also reported. On postcontrast scans, enhancement of the solid portions of the tumour may be seen. Treatment is by surgical resection, and the prognosis is usually excellent.

8.22. B True
Renal carcinoma occurs in an older age group than Wilms' tumour (mean age 10–12 years). It is associated with tuberous sclerosis and von Hippel–Lindau syndrome. Calcification may be seen in 25% of primary tumours, and tends to be denser and more homogeneous than in Wilms' tumour. The mass is relatively small at presentation and there is a higher incidence of bilateral tumours than occurs with Wilms' tumour. The presence of bone metastases is another differentiating feature. Calcification may also be identified in lymph node metastases.

8.22. C False
Although considered a benign neoplasm, mesoblastic nephroma does not have a capsule and is locally invasive, commonly penetrating through the renal capsule and perirenal fascia. Invasion of adjacent organs such as pancreas and

colon has been reported. Also, since the mass may contain functioning nephrons, enhancement may be seen on postcontrast CT scans. CT usually underestimates the extent of the tumour.

8.22.   D False
Clear-cell sarcoma is considered a variant of Wilms' tumour but has some differentiating features (no capsule, never bilateral, worse prognosis). It represents about 4% of paediatric renal tumours. On CT, a large, unilateral, mixed attenuation mass is seen, but there are no specific imaging features to distinguish it from classical Wilms' tumour. Another differentiating feature is the high incidence of bone metastases (42–76% of cases).

8.22.   E True
Nephroblastomatosis may be either focal or diffuse. Focal disease will result in a fairly well-defined mass which has no features to distinguish it from a small Wilms' tumour. Diffuse nephroblastomatosis results in bilateral nephromegaly and caliceal distortion. The masses enhance poorly on postcontrast scans and are typically subcapsular.

(Reference 30)

8.23.   A True
Wilms' tumours are bilateral in approximately 13% of cases, and involvement is usually synchronous. Children with increased risk of bilateral tumours include those in a younger age group, those with associated congenital anomalies (particularly bilateral sporadic aniridia) and those with a family history of Wilms' tumour. Bilateral tumours almost always arise in kidneys that are affected with nephroblastomatosis.

8.23.   B True
On ultrasound, Wilms' tumour appears as a large mass with mixed echotexture, although the degree of heterogeneity is characteristically less than that of neuroblastoma. Anechoic regions are due to old haemorrhage or necrosis, while echogenic areas may be due to fresh haemorrhage, fat or calcification. A hyperechoic rim represents a pseudocapsule and may be surrounded by a hypoechoic zone due to compressed renal tissue.

8.23.   C False
Fat density may be seen in Wilms' tumour in up to 10% of cases, and calcification in 7–13% of patients. Precontrast CT demonstrates a large, well-defined, mixed-attenuation mass that is hypodense relative to the normal renal parenchyma. On postcontrast scans, solid areas of tumour show less enhancement than adjacent renal tissue, which may appear particularly dense if the tumour is causing obstruction.

8.23.   D False
Wilms' tumour may displace the IVC but encasement is not a feature. IVC invasion occurs in approximately 6% of cases and tumour may extend as far cranially as the right side of the heart and also caudally below the level of the renal veins. Tumour can also spread into the tributaries of the IVC. Budd–Chiari syndrome is a recognized complication.

8.23. **E** True
Metastatic spread from Wilms' tumour is characteristically to the lungs and less commonly to the liver. However, spread to the mediastinum, either blood-borne or by extension through the diaphragm, is a rare feature and frequently indicates a very poor prognosis. Imaging may also demonstrate regional lymphadenopathy, but cannot distinguish between tumour involvement and reactive hyperplasia.

(Reference 30)

8.24. **A** False
Reflux detected by fluoroscopic VCUG can be graded according to the standards of the International Commission on Reflux. Reflux into a non-dilated renal pelvis indicates grade II reflux, whereas grade IV reflux indicates reflux into a moderately dilated tortuous ureter with moderate dilatation of the pelvis and loss of the sharp angles of the caliceal fornices.

8.24. **B** True
Grade V reflux refers to reflux into a very dilated tortuous ureter with gross dilatation of the pelvis and gross distortion of the calices. Intrarenal reflux may be seen. Grade IV and grade V reflux may be associated with renal parenchymal damage even in the absence of infection (reflux nephropathy). They are associated with delayed emptying of the collecting systems on radionuclide examinations.

8.24. **C** True
Radionuclide VCUG has the advantage of reduced radiation dose (the dose to the ovaries is 1% of that from fluoroscopic VCUG) and since it is a continuous examination it is more sensitive in detecting reflux. It lacks anatomical specificity and cannot assess the urethra or determine the presence of intravesical pathology such as ureterocele or stone. In the event of reflux, it cannot detect a duplex system.

8.24. **D** True
Focal renal infection such as acute lobar nephronia appears on precontrast CT as an area of reduced parenchymal attenuation. After contrast, the infected area may show no enhancement or appear as a wedge-shaped area with a striated pattern. CT will also demonstrate renal and perirenal abscess.

8.24. **E** False
^{99m}Tc-DMSA scintigraphy is the most sensitive imaging technique for the demonstration of renal scarring. A photopenic area may be due either to acute infection or old scarring. Defects present during an acute infection may resolve following treatment. DMSA is also of value in assessing split renal function, whereas isotopes such as DTPA and MAG-3 are better for assessing obstruction.

(Reference 33)

8.25. **A** False
During the first weeks of life, the uterine cavity is a little more bulbous and prominent than throughout the remaining premenarchal years, and an echogenic endometrial cavity echo may be seen. These features are a response

to the relatively high levels of maternal or placental oestrogens. At birth the uterine length is approximately 3.5 cm.

8.25. B False
Before the menarche, the uterine shape may be either tubular (with the AP diameter of the corpus being equal to that of the cervix) or spade-like (with the AP diameter of the cervix greater than that of the corpus). After the age of 1 month, the uterus is 2.6–2.8 cm in length, with a maximal width of 0.5–1.0 cm. The cervix accounts for two-thirds of the total uterine length. During childhood, the uterus is typically high in the pelvis and in a neutral position.

8.25. C False
The ovaries are small and relatively high in the child's pelvis, making ultrasound evaluation difficult, especially in children under 2 years of age. At birth the ovaries are typically located at the upper margin of the broad ligament, although they may be anywhere from the lower pole of the kidney to the broad ligament. The neonatal ovaries are typically 0.5–1.5 cm long, 0.3–0.4 cm thick and 0.25 cm wide.

8.25. D False
Hydrometrocolpos results from distension of the vagina and uterine cavity by accumulated secretions. Ultrasound typically shows a predominantly cystic, tubular, midline mass lying between the bladder and rectum. It usually contains scattered echoes due to contained cellular debris, mucoid material, or blood. The causes include imperforate hymen (67% of cases), complete vaginal membrane and vaginal stenosis or atresia.

8.25. E True
Ovarian cysts smaller than 9 mm in diameter may be seen in 5% of newborns. The percentage increases with increasing age. Aberrant follicular development in the presence of maternal hormone stimulation is the accepted aetiology for the large ovarian cysts that may present as abdominal masses.

(Reference 29)

8.26. A True
Malignant gynaecological neoplasms account for 2–3% of all paediatric cancers. Germ cell tumours account for approximately 85% of ovarian malignancies and include dysgerminoma (commonest), endodermal sinus tumour, immature teratoma and embryonal carcinoma. CT demonstrates a predominantly cystic or solid mass which may contain calcifications. Extension occurs into the bladder, bowel or adjacent muscles.

8.26. B False
Sacrococcygeal teratoma is the commonest presacral mass in children (others include neuroblastoma, anterior meningocele and rarely chordoma, rhab-domyosarcoma or lymphoma). The majority are non-familial, are commoner in girls and are usually benign under the age of 2 years. Non-hereditary sacrococcygeal teratomas have only an external component, only a presacral component or a combination of the two. Sacral anomalies are infrequent.

8.26. C True
Familial sacrococcygeal teratomas have a low incidence of malignancy and are entirely presacral. Calcifications are seen in only 5% but sacral anomalies are present in 95% of cases. They may present with symptoms of anorectal stenosis, constipation or perirectal abscess.

8.26. D True
Intraspinal extension may occur with sacrococcygeal teratoma or with any other malignant presacral neoplasm. CT features of sacrococcygeal teratoma include a solid-cystic mass with calcification, fat, bone or teeth. Predominantly cystic masses are generally benign whereas more solid tumours are likely to be malignant.

8.26. E True
Teratoma is the third commonest retroperitoneal tumour in children. With increasing age, the incidence of malignancy increases. The CT features are similar to those previously described. If calcification is absent and there is a large lipid content, the tumour may be indistinguishable from a lipoma. Lymphangiomas are also seen in the retroperitoneal space and may have a high fat content.

(Reference 32)

8.27. A False
The commonest sites of rhabdomyosarcoma are the head and neck, the genitourinary system and the extremities. In the head, parameningeal tumours are the commonest. This term refers to those tumours that are close enough to the meninges to permit intracranial spread. The sites included are the nasopharynx, paranasal sinuses, ear, infratemporal fossa and mastoids.

8.27. B True
On precontrast CT, rhabdomyosarcomas appear as non-homogeneous soft-tissue masses that show patchy enhancement on postcontrast scans. The demonstration of calcification is an unusual finding and suggests an alternative diagnosis. Tumours in the head are associated with bone destruction, and consequently bone may be identified in the tumour mass.

8.27. C False
Extremity rhabdomyosarcomas typically present as large soft-tissue masses with no bone involvement. They may have variable appearances on CT, being isodense or hypodense to muscle on precontrast scans, and becoming hyperdense, or remaining hypodense or isodense on postcontrast scans. MRI similarly shows non-specific prolongation of T1 and T2 relaxation times. The signal intensity is heterogeneous due to the presence of haemorrhage and necrosis, and irregular enhancement is seen following Gd-DTPA.

8.27. D True
In the genitourinary system, rhabdomyosarcoma usually arises in the bladder base, prostate or high in the vaginal wall. Paratesticular rhabdomyosarcomas describe those tumours that arise in the scrotum and account for approximately 4% of tumours. It is frequently impossible to determine the site of origin of the

tumour (testis, epididymis, tunica or spermatic cord). Scrotal ultrasound reveals a solid mass.

8.27. E True
Tumours arising in the bladder base are usually botryoid (like a bunch of grapes) in shape and cause irregular lobularity of the bladder base that can be seen on ultrasound or urography. They may extend inferiorly into the prostate, displace the rectum and spread through the sciatic and obturator foramina. Tumours arising in the prostate gland may spread into the bladder base and be indistinguishable from a primary bladder tumour.

(Reference 30)

8.28. A True
On the transverse view, the three components of the acetabulum (the bony ischium posteriorly, the bony pubis anteriorly and the triradiate cartilage, which transmits sound, in the centre) can be identified by hip ultrasound. The femoral head appears as a hypoechoic structure in the centre of the acetabulum. The echogenic ligamentum teres may be seen between the femoral head and the triradiate cartilage. Transverse scanning is not usually performed in the UK.

8.28. B True
However, 6–7 mm of posterior subluxation during stress indicates significant subluxation or dislocation. In most cases of infant hip instability, the hips are normally positioned except when stressed. More severe hip dysplasia is associated with posterosuperior hip displacement without stress. Ultrasound may reveal a thickened ligamentum teres and echogenic material (pulvinar) at the base of the acetabulum.

8.28. C False
The correct coronal plane for neonatal hip ultrasound is identified when the iliac bone cephalad to the centre of the acetabulum appears straight in a craniocaudal direction. The upper femoral shaft appears straight and the femoral head will exhibit its maximal diameter. The acetabular labrum has a characteristic appearance. Its juxta-osseous portion is echogenic, as is its fibrocartilaginous tip. The midportion is composed of hyaline cartilage, and appears hypoechoic.

8.28. D True
If the bony acetabulum accommodates one-third or less of the femoral head, it should be considered dysplastic and shallow. Coronal views optimally demonstrate superolateral subluxation/dislocation, which is associated with cephalad displacement of the cartilaginous acetabular labrum. An inverted, interposed labrum can also be identified, and may act as a barrier to prevent relocation of the femoral head.

8.28. E True
The femoral head ossification centre is seen several months earlier by ultrasound than by radiography. It initially appears as an echogenic structure in the centre of the femoral head, without acoustic shadowing. As ossification continues, the centre increases in size and begins to cast a shadow, which will

obscure the acetabulum. Consequently, by about 1 year of age, hip ultrasound usually becomes impossible.

(Reference 29)

8.29. A False
In long bones, Ewing sarcoma is most commonly located in the metadiaphyseal region and then in the diaphysis. However, metaphyseal tumours are reported to occur in approximately 5% of cases. The classic radiographic appearance is of a symmetric lesion (indicating a medullary origin), with permeative bone destruction and a soft-tissue mass. However, there may be many variations to this pattern.

8.29. B True
Ewing sarcoma arising in flat bones such as the scapula and pelvis characteristically produces a large area of bone abnormality which may be lytic, blastic or mixed. Areas of sclerosis are more common than in long bones, and a large soft-tissue mass is commonly seen. Rib lesions commonly cause bone expansion and extensive lysis. Soft-tissue masses are also frequent and pleural implants are reported in 25% of cases.

8.29. C False
CT is better than plain radiography at identifying the extent of the soft-tissue component of the tumour. Calcification has rarely been reported and has been attributed to dystrophic calcification in dead tissue. MRI also demonstrates the medullary and soft-tissue component of the lesion. Marrow disease appears on T1-WSE scans as areas of reduced signal intensity. The soft-tissue mass is best delineated on T2-WSE scans, when it appears hyperintense to adjacent muscle.

8.29. D False
Radiographic features that indicate healing of Ewing tumour include solidification of the periosteal reaction, which becomes confluent and incorporated into the adjacent cortex. The cortex reconstitutes and the permeative bone destruction disappears. CT and MRI will show a reduction in the size of the associated mass, although there is frequently incomplete resolution. It is impossible in this situation to exclude viable tumour by either technique, although an increase in the size of the mass on follow-up scans is evidence of active tumour. Persistent marrow abnormality is also common despite a good response.

8.29. E True
Metastases are identified in 15–50% of patients with Ewing sarcoma at presentation, although micrometastases are present in the majority of cases. The commonest locations are the lung, bone or a combination of the two. Less common sites include lymph nodes, liver and the CNS. Metastases appear to be commoner in patients with axial primaries rather than extremity primaries.

(Reference 30)

8.30. A True
The posterior dura appears as an echogenic line that parallels the anechoic posterior subarachnoid space. The anterior dura is usually difficult to identify. On transverse scanning, the posterior dura appears as a curved echogenic

linear shadow. Inward displacement of the dura allows the detection of extradural masses by ultrasound.

8.30. B False
The CSF is normally anechoic. The cord is visualized as a relatively hypoechoic tubular structure with echogenic anterior and posterior walls, and a central echogenic complex representing the anterior median fissure. On transverse scanning, the cord is oval in outline.

8.30. C False
Sonography of the neonatal spinal cord is usually performed with the baby prone and the back slightly flexed. The cord normally occupies the ventral aspect of the spinal canal, and the posterior subarachnoid space should therefore be deeper than the anterior. When the cord is tethered, the conus medullaris has an abnormally low position and abuts the posterior surface of the lumbosacral spinal canal.

8.30. D False
The nerve roots can be visualized by ultrasound (optimally on transverse scans), and appear as paired echogenic anterior and posterior roots bordering the conus medullaris in the subarachnoid space. Further caudally, they are seen to surround a centrally located echogenic structure that represents the filum terminale.

8.30. E True
With real-time ultrasound, vascular pulsation can be identified within the cord and cauda equina, and anterior and posterior movement of the cord may occur with episodes of crying. Absence of arterial pulsations within nerve roots and absent or decreased motion of the distal cord are features of a tethered cord.

(Reference 28)

8.31. A True
Bone abnormalities associated with spinal dysraphism include defects of the posterior arches (spina bifida), typically involving the spinous processes and laminae. In myelomeningocele, the pedicles may assume a coronal orientation. Abnormalities of the vertebral bodies include hemivertebra, coronal or sagittal cleft vertebra and block vertebra.

8.31. B False
Spinal dysraphism may be associated with a non-skin-covered back mass (myelomeningocele, myelocele), a skin-covered back mass (posterior meningocele, lipomyelomeningocele, myelocystocele) or no back mass (diastematomyelia, dorsal dermal sinus, spinal lipoma, tight filum terminale syndrome, anterior sacral meningocele, lateral thoracic meningocele, hydromyelia, split notochord syndrome and the syndrome of caudal regression). The latter are causes of occult spinal dysraphism.

8.31. C False
A tethered cord refers to the situation where the cord is abnormally fixed by a mass, spur (bony, cartilaginous or fibrous) or a fibrous band. In the normal situation, the conus medullaris lies at the level of the body of L3 in the

newborn and up to the age of 3 months. After this, the conus should be at L1–2. If the conus or distal end of the cord lies below this level, it is low lying and usually tethered.

8.31. D True
In cases of myelomeningocele, there is a herniated sac containing neural tissue (placode) dorsally and CSF, pia, arachnoid and nerve roots ventral to this. The placode is cleft dorsally and protrudes beyond the plane of the back. The cord is tethered at the level of the exposed neural tissue. Myelocele differs in that the herniated sac is flush with the plane of the back. The bony defect is usually large and affects multiple levels.

8.31. E True
Recognized associations with myelomeningocele include the Chiari II malformation (99%), dysgenesis of the corpus callosum, dysplasia of the calvarium, meninges, cerebral hemispheres and cerebellum. Hydromyelia occurs in 40–60% of cases, spinal arachnoid cyst in 20%, and diastematomyelia in 30–40%. They can also develop kyphoscoliosis. The bony abnormalities are best assessed by plain radiography or CT and the neural changes by T1-WSE MRI in the sagittal and axial planes.

(Reference 46)

8.32. A True
Lipomyelomeningocele consists of a back mass that contains fat, neural tissue, CSF and meninges. A large bony defect is present. The lesion is characterized by the presence of a lipoma or lipomatous tissue that extends from the subcutaneous fat into a dorsal cleft in the spinal cord and into the canal. The cord is tethered by this mass. T1-WSE MRI identifies this mass due to the high signal intensity of the fat. Associations include Chiari I malformation (10%) and hydromyelia (10%). Sacral asymmetry and partial sacral agenesis also occur.

8.32. B False
A myelocystocele is a localized dilatation of the distal end of the central canal of the cord. This splays the distal end of the cord and is herniated through a posterior bony defect into a skin-covered back mass. The lumbosacral area is the commonest location. The spinal cord is low lying and tethered. The conus is not developed. Associations include hydromyelia, partial sacral agenesis and cloacal exstrophy.

8.32. C True
Diastematomyelia is characterized by partial or complete sagittal clefting of the spinal cord, usually in the lumbar or thoracic region. In 50% of cases, there is duplication of the dura and arachnoid at the level of the split. A cartilaginous or bony spur is always present and the cord is tethered. These children are symptomatic. In 50% of cases, there is no spur and no duplication of the dura or arachnoid. These children are asymptomatic. Bony anomalies are also present at the site of the cord cleft. Other associations include hydromyelia, intradural lipoma, and fibrous bands. These may cause tethering of the cord at a different level from the spur.

8.32.   D True
A dorsal dermal sinus consists of an epithelial lined tract that extends inward from the skin surface of the back. The locations in decreasing order of frequency are sacrococcygeal, lumbosacral, occipital, thoracic and cervical regions. Sinuses above the sacrococcygeal region commonly extend into the spinal canal or onto the dorsal surface of the cord. In 50% of cases, the sinus terminates in an epidermoid or dermoid cyst in the spinal canal or conus. The cord is usually tethered by these cysts.

8.32.   E True
The normal filum terminale should not measure more than 2 mm in thickness. The filum may be thickened by fatty or fibrous tissue and results in tethering of the cord (tight filum terminale syndrome). The cord is abnormally low in 50% of cases and symptoms are due to cord ischaemia. On sagittal T1-WSE MRI, the filum may have increased signal intensity (if thickened by fat) and the conus hugs the posterior wall of the thecal sac.

(Reference 46)

8.33.   A False
Commonly, with NAI, the periosteum of the adjacent metadiaphysis is not damaged and periosteal reaction does not result, making dating of these fractures difficult. Metaphyseal corner fractures are most commonly seen at the knees, wrists and ankles.

8.33.   B True
Diaphyseal periosteal reactions are the result of torsional forces applied at the site of injury and represent healing of subperiosteal haemorrhages. Repeated injuries may result in layered periosteal reaction.

8.33.   C False
Rib fractures most commonly occur through the posterior ends of adjacent ribs when caused by a side-to-side compressive force. If the compression is from front to back, fractures typically occur in the mid-axillary line. Acute rib fractures may not be evident on the chest radiograph. Occasionally, pleural reaction or pulmonary contusion are the initial radiographic findings and bone scintigraphy may then indicate the presence of fractures.

8.33.   D False
Skull fractures that are suggestive of NAI include those that are greater than 5 mm in width, those that are multiple, non-parietal, depressed, and especially those in the occipital region. It is usually not possible to date skull fractures.

8.33.   E True
In infants, certain types of fracture are so rarely due to accidental injury that they should raise the suspicion of NAI. These include fractures of the acromion process with associated avulsion of the distal end of the clavicle, fractures of the sternum, pelvic bones and fractures of the feet.

(Reference 36)

8.34. A True

Physiological diaphyseal periosteal reactions are not an uncommon finding in children under 4 months of age. Whereas the diaphyseal periosteal reactions due to NAI commonly extend to involve the adjacent metaphysis, physiological periosteal reactions do not. They are usually also bilateral and symmetrical, appearing as a single fine line along the diaphysis. They do not have the multilayered appearance that is seen following repeated trauma.

8.34. B True

Children with congenital insensitivity to pain can also have periosteal reactions and fractures of long-bone shafts. However, rib and skull fractures, subdural haematomas and retinal heamorrhages are not features. Metaphyseal fractures are also seen in copper deficiency and Menke's syndrome.

8.34. C False

Maternally derived copper stores make copper deficiency a very unlikely diagnosis in term infants under 6 months of age or premature infants under 2 months of age. Factors predisposing to copper deficiency include low birth weight, dietary deficiency, malnutrition and rarely peritoneal dialysis.

8.34. D True

Osteoporosis is the commonest radiographic abnormality in babies with copper deficiency. Other features include symmetrical cupping and irregularity of the metaphyses, especially around the wrists and knees. Sickle-shaped metaphyseal spurs are characteristic and may fracture. Fractures may be associated with excessive periosteal new bone formation.

8.34. E False

Wormian bones are seen in many conditions, including Menke's syndrome. Other features of this syndrome include osteoporosis, metaphyseal fractures and periosteal reactions. The changes are typically symmetrical.

(Reference 36)

8.35. A True

Salter–Harris type II fractures account for approximately 75% of all physeal injuries and typically occur in children over 10 years of age. Type I injuries account for about 6% of all fractures and are commonest in children under 5 years of age.

8.35. B True

Other radiographic features of slipped capital femoral epiphysis include indistinctness of the metaphyseal margin or 'widening' of the physis (equivalent to a Salter–Harris type I injury), and medial displacement of the epiphysis in relation to the metaphysis. In a typical case, the lateral view will demonstrate the epiphysis to have slipped posteriorly in relation to the metaphysis.

8.35. C False

Torus fractures most commonly affect the distal forearm bones and typically produce an angulation of the dorsal cortex of the distal radius. They are the result of a fall on the outstretched hand.

8.35.   D True
Triplane fractures occur due to incomplete closure of the physis. This injury results in fracture lines in the sagittal, coronal and axial planes, and may create either two or three fracture fragments. It may have the appearance of a Salter–Harris II injury on the lateral radiograph and a Salter–Harris III or IV injury on the AP radiograph.

8.35.   E True
Other types of fracture that are suggestive of NAI include metaphyseal corner fractures, and fractures of the ribs, sternum, pelvis, vertebrae or skull.

(Reference 34)

8.36.   A False
Subdural haematomas in children suffering NAI are commonly bilateral and may extend from the middle cranial fossa to the vertex. They are commonly frontoparietal in location. Extradural haematomas are a less common finding.

8.36.   B True
Interfalcial haematomas are typically small, biconvex and located in the posterior aspect of the falx cerebri. They are thought to result from a shearing force. Interhemispheric subdural haematomas are also characteristic of NAI and are due to tearing of cerebral surface veins as they enter the dural sinuses. Both injuries are associated with violent shaking of the child's head.

8.36.   C True
Depressed skull fractures at any site are more commonly due to NAI than accidental injury. In particular, a depressed occipital fracture unassociated with a road traffic accident is highly suggestive of child abuse.

8.36.   D True
Haemorrhages at the grey–white matter junction are most commonly seen in children under 2 years of age and are the result of severe rotational forces. They are associated with poor long-term prognosis since many of the white matter tracts are disrupted.

8.36.   E True
Diffuse hypodensity of the cerebral hemispheres is due to ischaemia following a severe anoxic episode. The cerebellum, brain stem and basal ganglia may appear to have increased attenuation on precontrast CT.

(Reference 31)

8.37.   A True
Approximately 80% of cystic hygromas involve the neck and 3–10% extend into the superior mediastinum. CT usually demonstrates a unilocular or multilocular, near water density mass, with thin or imperceptible walls, posterior to the sternocleidomastoid muscle. However, they can also occur in the floor of the mouth or anterior triangle. Extensive local soft-tissue infiltration is a frequent finding.

8.37. B True
Cervical dermoid cysts typically present as painless masses in children under 3 years of age. CT demonstrates a midline mass that is usually located above the hyoid bone, a feature which helps to distinguish them from thyroglossal duct cysts, which are usually below the level of the hyoid.

8.37. C True
The vast majority of primary cervical teratomas occur within or around the thyroid gland. They are commonly present at birth and almost always present in the first year of life. Maternal polyhydramnios is reported in 20% of cases.

8.37. D True
Cervical neuroblastomas appear on CT as fairly well-defined paraspinal soft-tissue masses that may be calcified. Extension into the spinal canal is a characteristic feature and can be demonstrated by MRI or CT myelography. Approximately 5% of all primary neuroblastomas occur in the neck. Children may present with a Horner's syndrome.

8.37. E False
Squamous carcinoma accounts for only 4% of head and neck malignancies in childhood, whereas lymphoma (both Hodgkin's and NHL) accounts for 50%. The second commonest soft tissue malignancy is rhabdomyosarcoma (10%) while fibrosarcoma, neurofibrosarcoma, thyroid carcinoma and primary neuroblastoma account for approximately 5% each.

(Reference 32)

8.38. A False
PNET is an undifferentiated small, round cell sarcoma that cannot be distinguished from Ewing sarcoma on imaging studies. PNET may arise in the CNS or at other sites, both in the bones or soft tissues. The commonest site for primary bone lesions is the leg, followed by the pelvis and humerus. Tumours have also been described in the thorax, abdomen and psoas. Children are usually in their early teens at presentation.

8.38. B True
The features on plain radiography of PNETs in long and flat bones are similar to those of Ewing sarcoma at the aggressive end of the spectrum. Lesions may be in the metaphysis or diaphysis and are poorly defined. Cortical destruction is seen in the majority of cases, as is periosteal reaction. Sclerotic bone lesions and pathological fractures are less common findings.

8.38. C True
PNET arising in the chest wall is referred to as the Askin tumour. It typically appears as a soft-tissue chest wall mass that may be large, and associated with rib destruction. Pleural effusions, pleural invasion and collapse or invasion of the underlying lung are all recognized features. Involvement of mediastinal and hilar lymph nodes is unusual.

8.38. D False
PNET at any site may be associated with a large soft-tissue component. Tumours may also arise in the soft tissues. They have no distinctive imaging features and calcification is an atypical occurrence.

8.38.   E False
Metastases from PNET are not uncommon. The most frequent sites are the lung and bones, followed by lymph nodes. Metastases have also been recorded in the liver, brain, bone marrow, adrenal and sympathetic chain. After initial therapy, recurrence is typically local.

(Reference 30)

8.39.   A False
Overall in the paediatric age group, Hodgkin's lymphoma is slightly more common than NHL, but under the age of 10 years NHL is much commoner than Hodgkin's disease. Hodgkin's disease shows its peak incidence in the late teens. NHL is disseminated at the time of presentation in over 75% of cases, and therefore imaging for staging is not as important as it is for Hodgkin's disease, which is typically localized at presentation.

8.39.   B True
In the chest, Hodgkin's disease is commoner than NHL and typically presents with mediastinal lymphadenopathy, with or without hilar node enlargement. Significant compression of the major airways occurs in up to 50% of children with Hodgkin's disease and may be life threatening. It is particularly important to detect prior to anaesthesia.

8.39.   C True
Thymic involvement by Hodgkin's disease only occurs in the presence of mediastinal nodal disease and may take the form of thymic enlargement or occasionally as thymic cysts. Thymic enlargement following cessation of chemotherapy is also a well-recognized phenomenon and is due to rebound hypertrophy. It must not be mistaken for recurrent disease. Gallium scintigraphy cannot distinguish between the two since it may be positive in cases of rebound hypertrophy.

8.39.   D True
Burkitt's lymphoma most commonly occurs in the abdomen (70%) but is also seen in the neck or pharynx (25%), chest (pleural, nodal disease), bone (10%) and kidneys (5%). Abdominal disease is characterized by a large abdominal mass, and omental disease appears as soft tissue lying between the bowel and anterior abdominal wall. Bowel lymphoma is almost always due to NHL and rarely to Hodgkin's disease, in which case it usually results from direct extension of tumour from involved lymph nodes. Mesenteric lymphadenopathy is also typically due to NHL.

8.39.   E True
Bone is occasionally the primary site of NHL and then usually involves a single bone. In the case of disseminated disease, bone involvement may be seen in up to 20% of cases and typically involves multiple bones. Bone involvement by Hodgkin's disease is reported in less than 1% of cases. The abnormality may be either lytic, sclerotic or mixed. Periosteal reaction is more likely to be seen in the presence of local soft-tissue involvement. Bone marrow disease is also common in NHL.

(Reference 30)

8.40. **A** False
The typical radiographic finding in children with *Pneumocystis carinii* pneumonia (PCP) consists of a mixture of alveolar and interstitial shadowing with air bronchograms. The shadowing may opacify the whole lung within hours or days.

8.40. **B** False
Lymphocytic interstitial pneumonitis (LIP) characteristically produces a diffuse reticulonodular or nodular pattern on the chest radiograph and is the commonest cause of such an appearance in a child with AIDS. The nodules are distributed symmetrically and are usually 2–3 mm in size. Hilar and mediastinal lymphadenopathy may also be seen. Such lymph node enlargement may also be seen with *Mycobacterium* tuberculosis, *Mycobacterium avium intracellulare* (MAI), CMV, Kaposi's sarcoma, lymphoma and fungal infections.

8.40. **C** True
Candidal oesophagitis is the commonest gastrointestinal infection in children with AIDS. Barium studies may show abnormal peristalsis (hypoactive to complete atony), thickened folds, pseudopolypoid mucosal irregularity and ulceration. The appearances usually return to normal 2–4 weeks after treatment.

8.40. **D** False
Tumour is a rare cause of an abdominal mass in children with HIV infection. Causes include Kaposi's sarcoma, lymphoma and hepatic tumours. Hepato-splenomegaly is a common finding in these children and may be due to infection (chronic active hepatitis, viral hepatitis, TB), tumour (fibrosarcoma of the liver) or cardiac failure. Abdominal lymphadenopathy is also common. Causes include MAI infection, benign hypertrophy, lymphoma and Kaposi's sarcoma.

8.40. **E** True
The commonest cranial CT abnormalities in children with AIDS include cortical atrophy and changes consistent with meningitis or viral infection. Basal ganglia calcification is a recognized finding, thought to be due to a vasculitis. Toxoplasmosis and CMV appear to be rare in children. CT may also demonstrate chronic parotid gland enlargement.

(Reference 33)

9 Obstetrics, gynaecology, breast

9.1. Concerning ultrasound in the first trimester of pregnancy
A an empty intrauterine gestational sac cannot be differentiated from the decidual cast of an ectopic pregnancy.
B the yolk sac is the first structure seen within a normal gestational sac.
C cardiac activity cannot be identified by transvaginal ultrasound (TVS) until the crown–rump length (CRL) measures 6 mm.
D by the end of the first trimester, the bi-parietal diameter (BPD) is more accurate than the CRL in the assessment of gestational age.
E a corpus luteum cyst is always unilocular.

9.2. Concerning ultrasound in the assessment of the first trimester of pregnancy
A on transabdominal ultrasound (TAS), a mean gestational sac diameter of 25 mm without a visible embryo is abnormal.
B a choriodecidual reaction of less than 2 mm in thickness is abnormal.
C between 5.5 and 9 weeks gestational age, the mean sac diameter should be at least 5 mm more than the CRL.
D the yolk sac should not measure more than 5 mm in diameter.
E demonstration of a subchronic haemorrhage is invariably associated with spontaneous abortion.

9.3. Concerning the ultrasound findings in the first trimester of pregnancy
A the pseudogestational sac (decidual cast) of an ectopic pregnancy is located in an intracavitary position.
B Doppler ultrasound can distinguish a pseudogestational sac from an intrauterine gestational sac.
C thickening of the central cavity complex without an intrauterine gestational sac indicates an ectopic pregnancy.
D demonstration of a viable intrauterine gestational sac excludes an ectopic pregnancy.
E on transvaginal scanning, an echogenic adnexal ring is a feature of ectopic pregnancy.

9.4. Concerning ultrasound of the first trimester of pregnancy
A at 8 weeks gestational age, a cyst is normally visible in the posterior fossa of the embryo.
B physiological herniation of the bowel is identified in the majority of fetuses at 12 weeks gestational age.
C at 6 weeks gestational age, the normal embryonic heart rate is 100 beats per minute.
D the fetal stomach cannot be identified at 12 weeks gestational age.
E monochorionic, diamniotic twins can be differentiated from dichorionic, diamniotic twins.

9.5. Concerning antenatal ultrasound in the assessment of the fetal CNS
A   if the glomus of the choroid plexus fills the ventricular atrium, ventricular dilatation is excluded.
B   a transverse ventricular atrial diameter of 12 mm is within normal limits.
C   ventricular dilatation is a feature of agenesis of the corpus callosum.
D   an AP cisterna magna measurement of 6 mm is invariably abnormal.
E   inability to demonstrate the cavum septum pallucidum is always abnormal.

9.6. Concerning antenatal ultrasound in the assessment of neural tube defects
A   anencephaly is associated with absence of the cranial vault cephalad to the orbits.
B   encephalocele is most commonly demonstrated in the frontal region.
C   spina bifida is associated with outward splaying of the dorsal ossification centres.
D   obliteration of the cisterna magna is a feature of the Arnold–Chiari malformation.
E   the 'lemon sign' is an associated finding in cases of spina bifida.

9.7. Which of the following are true concerning antenatal ultrasound?
A   Polyhydramnios is invariably associated with a fetal anomaly.
B   The greater the degree of polyhydramnios, the greater the likelihood of an associated fetal anomaly.
C   There is a recognized association between a cystic neck mass and fetal pleural effusion.
D   Cystic hygroma cannot be diagnosed until the second trimester.
E   Nuchal thickening greater than 5 mm is a feature of Down's syndrome.

9.8. Concerning the ultrasound assessment of the fetal thorax
A   Ebstein's anomaly can be demonstrated.
B   isolated atrial septal defect (ASD) is the easiest identifiable cardiac anomaly.
C   the demonstration of fetal hydrothorax is associated with a greater than 50% perinatal mortality rate.
D   a homogeneous echogenic mass is a recognized appearance of cystic adenomatoid malformation (CAM).
E   a cystic intrathoracic lesion that displaces the heart is characteristic of bronchogenic cyst.

9.9. Concerning antenatal ultrasound in the assessment of the fetal abdomen
A   ascites is most commonly associated with fetal hydrops.
B   demonstration of the fetal stomach excludes tracheo-oesophageal fistula.
C   duodenal atresia is typically diagnosed before 20 weeks gestation.
D   peritoneal calcification is the commonest ultrasound abnormality demonstrated in meconium peritonitis.
E   anorectal atresia may be associated with oligohydramnios.

9.10. Which of the following are true concerning antenatal ultrasound of the fetal renal tract

A   Normal kidneys are visualized by transabdominal scanning at 10 weeks gestation.

B   An AP renal pelvis diameter of 10 mm is within normal limits.

C   Unilateral hydronephrosis is a recognized feature of posterior urethral valves.

D   A normal appearance of the fetal kidneys excludes infantile polycystic kidney disease.

E   The commonest cause of an abnormally dilated renal pelvis is pelviureteric junction (PUJ) obstruction.

9.11. Which of the following are true of fetal antenatal ultrasound?

A   Choroid plexus cysts are invariably associated with chromosomal anomaly.

B   Polyhydramnios is commoner with distal than proximal bowel obstruction.

C   Gastroschisis typically occurs to the right of the umbilicus.

D   Omphalocele invariably contains small bowel.

E   Absence of diastolic flow in the umbilical artery is normal in the third trimester.

9.12. Concerning ultrasound assessment of the placenta

A   a placental thickness of 3 cm is always abnormal.

B   chronic intrauterine infection is a recognized cause of placental enlargement.

C   placenta membranacea is associated with antepartum haemorrhage.

D   bilobate placenta cannot be differentiated from succenturiate placenta.

E   velamentous insertion of the cord is a recognized cause of vasa praevia.

9.13. Concerning ultrasound of the placenta

A   intervillous thrombosis is a cause of hypoechoic areas in the placenta.

B   the majority of placental infarctions can be recognized.

C   chorioangioma usually appears as a hypoechoic mass deep within the placenta.

D   placental abruption is most commonly seen as elevation of the chorionic membranes.

E   placental abruption is a recognized cause of a hypoechoic retroplacental mass.

9.14. Which of the following are true concerning gestational trophoblastic disease?

A   Hydatidiform mole progresses to invasive mole in 50% of cases.

B   Ultrasound cannot differentiate hydatidiform mole from invasive mole.

C   Choriocarcinoma never follows a normal pregnancy.

D   Bilateral ovarian cysts are a recognized ultrasound finding.

E   The prognosis of choriocarcinoma is independent of the site of metastatic disease.

9.15. Concerning ultrasound of the ovaries and adnexa
A an ovarian volume of 10 ml is invariably abnormal.
B normal follicles may reach 5 cm in maximal diameter.
C the ovarian cortex and medulla can be distinguished by transvaginal ultrasound.
D enlarged multicystic ovaries are a recognized finding in cystic fibrosis.
E the ovaries are invariably enlarged in the Stein–Leventhal syndrome.

9.16. Concerning ultrasound in the assessment of ovarian masses
A internal septation excludes a diagnosis of uncomplicated follicular cyst.
B cystic ovaries are seen in the ovarian hyperstimulation syndrome.
C distal acoustic shadowing is a feature of benign cystic teratoma.
D ovarian metastasis usually produces a solid adnexal mass.
E ovarian fibroma typically appears as an echogenic adnexal mass.

9.17. Ultrasound demonstrates thickening of the endometrial echo complex. The differential diagnosis includes which of the following?
A Luteal phase endometrium.
B Adenomyosis.
C Endometrial hyperplasia.
D Endometrial carcinoma.
E Endometrial polyp.

9.18. Concerning the ultrasound features of pelvic infections
A a normal pelvic ultrasound excludes pelvic inflammatory disease (PID).
B a cystic adnexal mass is a recognized finding in PID.
C hydronephrosis is a recognized finding in PID.
D dilatation of the uterine cavity is a feature of endometritis.
E echogenic shadowing foci in the uterine cavity are a feature of pyometra.

9.19. Concerning ultrasound in the assessment of uterine tumours
A leiomyomas are typically associated with distal acoustic enhancement.
B diffuse increase in size of the uterus is a recognized feature of leiomyoma.
C enlargement during the third trimester of pregnancy is invariable with leiomyoma.
D leiomyosarcoma is differentiated from leiomyoma by the presence of central cystic areas.
E leiomyosarcoma usually appears as a solitary nodule in the uterus.

9.20. Which of the following are true concerning hysterosalpingography?
A A uterine body : cervix ratio of 2 : 1 is typical of the infantile uterus.
B The internal cervical os cannot be demonstrated.
C The internal cervical os is normally less than 6 mm in diameter.
D Mucosal folds are a normal finding in the cervical canal.
E An arcuate uterus is characterized by a concave shape to the fundus.

9.21. Concerning hysterosalpingography in the assessment of the uterine cavity
A stenosis of the internal os is a feature of intrauterine synechiae.
B mucous and fibrous adhesions of intrauterine synechiae can be distinguished.
C a normal appearance to the uterine cavity excludes adenomyosis.
D polypoid filling defects are a feature of endometrial hyperplasia.
E endometrial polyps are typically greater than 1 cm in size.

9.22. Concerning hysterosalpingography in the assessment of the Fallopian tubes
A   tubal diverticula are a feature of genital tuberculosis.
B   lymphatic intravasation is a specific feature of genital tuberculosis.
C   hydrosalpinx is a recognized finding in endometriosis.
D   salpingitis isthmica nodosa typically results in multiple filling defects in the tubes.
E   pooling of contrast medium around the ovary is a feature of peritubal adhesions.

9.23. Concerning CT in the assessment of the uterus
A   the myometrium and endometrial cavity cannot be distinguished.
B   diffuse enlargement of the uterine body is a feature of endometrial carcinoma.
C   the inguinal lymph nodes are the commonest demonstrated site of nodal metastasis from endometrial carcinoma.
D   extension of endometrial carcinoma into the cervix cannot be identified.
E   leiomyomas do not enhance on postcontrast CT.

9.24. Concerning CT in the assessment of carcinoma of the uterine cervix
A   carcinoma localized to the cervix can only be demonstrated on contrast enhanced CT.
B   fluid within the uterine cavity is a recognized finding.
C   obliteration of the periureteral fat plane indicates parametrial extension.
D   demonstration of hydronephrosis due to ureteric obstruction indicates stage IIIB disease.
E   the periaortic nodes are the initial site of metastatic node involvement.

9.25. Concerning CT in the assessment of the ovaries
A   normal sized ovaries cannot be demonstrated.
B   papillary projections within an ovarian cyst differentiate cystadenoma from cystadenocarcinoma.
C   nodular soft-tissue densities beneath the anterior abdominal wall are a characteristic feature of metastatic cystadenocarcinoma.
D   calcified peritoneal nodules are a recognized feature of metastatic cystadenocarcinoma.
E   lymphadenopathy is a typical feature of metastatic cystadenocarcinoma.

9.26. Concerning MRI of the uterus, cervix and vagina
A   the endometrium and myometrium are best distinguished on T2-WSE scans.
B   during the midsecretory phase of the menstrual cycle, the endometrium typically becomes hypointense on T2-WSE scans.
C   the endometrium is less clearly seen in women on the oral contraceptive pill.
D   the cervix appears isointense to the uterine body on all pulse sequences.
E   the vagina is usually isointense to muscle on all pulse sequences.

9.27. Concerning MRI of the uterus and cervix
A areas of high signal on T2-WSE scans are a feature of leiomyoma.
B endometrial carcinoma is a cause of relatively reduced signal intensity in the endometrium on T2-WSE scans.
C focal obliteration of the junctional zone is a feature of invasive endometrial carcinoma.
D cervical carcinoma is typically hyperintense on T2-WSE scans.
E extension of cervical carcinoma into the pelvic fat is seen more clearly on T2-WSE scans than T1-WSE scans.

9.28. Concerning MRI of the ovaries
A the ovaries normally show increased signal intensity on T2-WSE scans compared to T1-WSE scans.
B a simple ovarian cyst is hypointense on T1-WSE scans.
C chemical shift artefact is seen in the majority of benign cystic teratomas.
D pelvic inflammatory disease is a recognized cause of a hyperintense mass on T1- and T2-WSE scans.
E benign and malignant ovarian tumours are typically differentiated by their signal intensities.

9.29. Concerning MRI in the assessment of endometriosis and adenomyosis
A on T1-WSE scans, a cystic mass with a hypointense rim is a finding in endometriosis.
B loss of the outline of the uterine body is a recognized feature of endometriosis.
C signal void is a typical finding in adenomyosis.
D widening of the junctional zone is a feature of adenomyosis.
E focal adenomyosis appears as a hypointense mass on T2-WSE scans.

9.30. Which of the following statements are true in the assessment of mammograms?
A Increased vascularity indicates the presence of a carcinoma.
B A solitary dilated duct is associated with intraductal papilloma.
C Intramammary lymph nodes are typically seen in the upper inner quadrant of the breast.
D Radiolucent lesions are never malignant.
E Metastasis is a recognized cause of a radiolucent mass.

9.31. Which of the following are true of mammography in the assessment of breast carcinoma?
A Increasing size of the primary tumour is associated with increased incidence of axillary node involvement.
B A completely surrounding radiolucent 'halo' excludes the diagnosis.
C Architectural distortion is a specific feature.
D Spread to an intramammary lymph node is indicated by loss of its well-defined border.
E Carcinoma is the commonest cause of a spiculated mass.

9.32. Which of the following statements are true of breast calcifications?
A   Peripheral 'egg-shell' calcifications are a feature of fat necrosis.
B   Calcifications are seen in less than 5% of carcinomas.
C   The number of microcalcifications within a cluster is unrelated to the likelihood of malignancy.
D   Clustered microcalcifications without an associated mass are never associated with malignancy.
E   Malignant microcalcifications are usually less than 0.5 mm in size.

9.33. Which of the following types of breast calcifications are typically associated with malignancy?
A   Spherical calcifications with radiolucent centres.
B   Widespread, thick, rod-shaped calcifications.
C   Crescentic calcifications seen on the lateromedial view.
D   Irregular, branching, linear microcalcifications.
E   Irregular, branching microcalcifications limited to a segment of the breast.

9.34. Which of the following statements are true concerning mammography?
A   Thickening of Cooper's ligaments is an association of SVC obstruction.
B   Nipple retraction is most commonly due to carcinoma.
C   Radiotherapy is a recognized cause of diffuse skin thickening.
D   Paget's disease of the nipple is a cause of nipple calcifications.
E   Axillary lymph nodes are normally less than 2 cm in diameter.

9.35. Which of the following statements are true concerning breast ultrasound?
A   Microcalcifications can be demonstrated.
B   Skin thickening can be demonstrated.
C   Fat in the breast is hypoechoic relative to the breast parenchyma.
D   Cooper's ligaments cause acoustic shadowing only when abnormally thickened.
E   Infiltrating ductal carcinoma is a cause of a hypoechoic mass.

9.36. Which of the following are true concerning breast ultrasound?
A   Lipomas are typically hyperechoic relative to the breast parenchyma.
B   Duct ectasia may be diagnosed.
C   Galactocele is a recognized cause of an anechoic mass.
D   Phyllodes tumour typically appears as a hypoechoic mass.
E   Poorly defined, hyperechoic masses with distal acoustic shadowing are usually malignant.

9.37. Which of the following are true of fibroadenomas of the breast?
A   They are associated with an increased risk of breast carcinoma.
B   They usually calcify from the periphery.
C   A lobulated margin is a recognized finding on mammography.
D   They are the commonest cause of a hypoechoic mass on breast ultrasound.
E   Posterior acoustic shadowing on ultrasound excludes the diagnosis.

9.38. Concerning intraduct papilloma of the breast
A   it is associated with a bloody nipple discharge.
B   it is most commonly located in the subareolar region.
C   calcification is a typical finding on mammography.
D   it can be distinguished from papillary carcinoma by ductography in the majority of cases.
E   ultrasound typically demonstrates a hyperechoic mass.

9.39. Concerning mammography in the assessment of the breast
A   postsurgical scar typically does not progress 1 year after operation.
B   microcalcifications are a recognized feature of radial scar.
C   radiotherapy is a recognized cause of breast calcifications.
D   loss of the well-defined outline of a simple cyst usually indicates a complicating carcinoma.
E   asymmetric dense breast tissue is usually a benign finding.

9.40. Concerning mammography in the assessment of breast lesions
A   a capsule is a typical feature of fibroadenolipoma.
B   medullary carcinoma usually appears as a spiculated mass.
C   a subcutaneous mass is seen in the majority of cases of inflammatory carcinoma.
D   microcalcification is not a typical feature of phyllodes tumour.
E   lymphoma is a recognized cause of diffuse increase in breast density.

Obstetrics, gynaecology, breast: Answers

9.1. A False
The normal intrauterine gestational sac can be identified by the presence of a double decidual 'ring' sign produced by the sonographic visualization of the three layers of decidua of early pregnancy (decidua capsularis/chorion laeva, decidua vera and decidua basalis/chorion frondosum). However, this sign is not always present in normal intrauterine gestations.

9.1. B True
With transabdominal scanning, the yolk sac is often seen when the mean gestational sac diameter is 10–15 mm, and should always be seen with a mean gestational sac diameter of 20 mm. With transvaginal scanning, it should be identified with a mean gestational sac diameter of 8 mm. The demonstration of a yolk sac is an important feature in differentiating an early intrauterine gestation from a decidual cast.

9.1. C False
Cardiac activity should routinely be identified by transvaginal ultrasound at a CRL of 4 mm and can be seen with CRL values of 2–4 mm. However, with transvaginal scanning, an embryo with CRL of 1–2 mm can be identified immediately adjacent to the yolk sac, and inability to demonstrate cardiac activity at this stage is not abnormal.

9.1. D True
Gestational age can be assessed before demonstration of the embryo and is accurate to within 1 week. Once the embryo is identified (during the 5th menstrual week by transvaginal ultrasound), CRL is accurate to within 5–7 days. By the end of the first trimester, the BPD alone is more accurate than the CRL, since errors in CRL may occur due to fetal flexion and extension.

9.1. E False
The corpus luteum cyst is the commonest pelvic mass during the first trimester. It usually appears as a thin-walled, unilocular cyst, less than 5 cm in diameter. However, they may reach 10 cm in diameter, and transvaginal ultrasound frequently demonstrates septations and debris. The cyst wall and septations may be quite thick. Corpus luteum cysts usually resolve by 16 weeks gestation.

(Reference 11)

9.2. A True
With transabdominal ultrasound, a mean gestational sac diameter of 25 mm with no visible embryo or 20 mm without a yolk sac is considered abnormal. With transvaginal scanning, corresponding measurements of 16 mm with no embryo and 8 mm with no yolk sac are considered abnormal.

9.2. B True
The sonographic features of an abnormal gestational sac include an irregular shape of the sac, a thin (less than 2 mm thickness), weakly echogenic or irregular choriodecidual reaction, and absence of the double decidual 'ring' sign when the sac diameter exceeds 10 mm.

9.2. C True
In one study, when the mean sac diameter was less than 5 mm greater than the CRL, implying first trimester oligohydramnios, there was a 94% spontaneous abortion rate. In each of these cases, the embryonic heart rate was normal. The cause of first trimester oligohydramnios is not known.

9.2. D True
Embryonic death may be associated with an abnormal appearance to the yolk sac. Ultrasound may demonstrate a large size (greater than 5 mm), calcification or echogenic material within the sac, and a double ring appearance to the yolk sac.

9.2. E False
Up to 18% of women with vaginal bleeding in the first half of pregnancy have ultrasound evidence of a subchorionic haemorrhage as the cause of the bleeding. Ultrasound typically demonstrates an extrachorionic fluid collection located between the gestational sac and the uterine wall. The collection begins at the edge of the placenta and does not usually extend behind, within or in front of the placenta. The clinical significance of subchorionic haemorrhage is unclear, with some reporting an increased incidence of abortion, while others suggest that there is no adverse outcome.

(Reference 11)

9.3. A True
Approximately 20% of patients with an ectopic pregnancy have an intrauterine pseudogestational sac (an intrauterine fluid collection surrounded by high-level echoes). With transvaginal scanning, this sac is demonstrated to be within the uterine cavity, which distinguishes it from a normal or abnormal intrauterine gestational sac, even in the absence of a double 'ring' sign, since the latter is located in an intradecidual site.

9.3. B True
Doppler interrogation of the peritrophoblastic area (located between the external border of the gestational sac and the inner third of the myometrium) typically shows pulsatile flow with a prominent diastolic component in normal as well as abnormal intrauterine pregnancies. Such Doppler signals cannot be demonstrated from the region surrounding a pseudogestational sac.

9.3. C False
When the central uterine cavity complex is abnormally thickened (and often irregularly echogenic), the differential diagnosis includes intrauterine blood, retained products after an incomplete abortion, decidual changes secondary to

an early normal, but not yet visible intrauterine pregnancy, or a decidual reaction secondary to an ectopic pregnancy.

9.3.   D False
In patients without risk factors for an ectopic pregnancy (pelvic inflammatory disease, tubal scarring from other causes, chromosomal abnormalities, endocrine dysfunction, intrauterine contraceptive device (IUCD), and artificial insemination), the incidence of an associated ectopic pregnancy is approximately 1:30000. In those patients at risk, this incidence increases to about 1:2600 to 1:8000.

9.3.   E True
Transvaginal ultrasound may demonstrate echogenic adnexal rings in 65–76% of cases of ectopic pregnancy, and an extrauterine embryo or yolk sac in up to one-third of patients. Transvaginal scanning also identifies echogenic pelvic fluid, which correlates with the presence of haemoperitoneum. In one study, the association of echogenic pelvic fluid and absence of an intrauterine gestation sac had a 93% positive predictive value for an ectopic pregnancy.

(Reference 11)

9.4.   A True
From approximately 6–12 weeks gestational age, small cystic structures can be identified in the posterior aspect of the embryonic cranium. The first is seen at 6–8 weeks and represents the normal rhombencephalon, which eventually forms the fourth ventricle. By 9 weeks, the third and lateral ventricles can be demonstrated as three small cysts in the embryonic head.

9.4.   B False
The midgut normally herniates into the base of the umbilical cord at the beginning of the eighth week and appears on ultrasound as a small, 6–9mm, echogenic mass protruding into the cord. By 9 weeks, this mass reduces to approximately 5–6mm in size and usually disappears by 10–12 weeks. However, in up to 20% of normal embryos it may still be seen at 12 weeks.

9.4.   C True
The embryonic heart rate increases to approximately 140 beats per minute by 8–9 weeks. In one study, the demonstration of embryonic heart rates of less than 85 beats per minute at 5–8 weeks gestational age was associated with miscarriage, suggesting that embryonic bradycardia is a sign of impending abortion.

9.4.   D False
The fetal stomach can be seen in approximately 90% of cases at 12 weeks gestation. The fetal kidneys and adrenals may be visualized at 9 weeks, and by 12 weeks the bladder can be demonstrated in 50% of cases (on transvaginal scanning).

9.4.   E True
A dichorionic twin pregnancy is characterized by two separate gestational sacs that may initially be adjacent to, or separate from, one another. As the sacs enlarge, the chorion laeva/decidua capsularis complex thins and the space

between the sacs is obliterated, leaving a single membrane between them. During the first trimester this membrane is relatively thick, allowing easy differentiation between dichorionic diamniotic twinning and monochorionic diamniotic twinning.

(Reference 11)

9.5. A True
The position of the trigone (atrium) of the lateral ventricle is marked by the echogenic glomus of the choroid plexus. This normally fills, or nearly fills the whole of the atrium, and in this situation the lateral ventricle can be considered to be of normal size.

9.5. B False
The transverse atrial diameter is a useful measurement for assessing ventriculomegaly since this portion of the lateral ventricle remains constant in size throughout the second and third trimesters. The mean transverse atrial diameter is 6.5 mm and a measurement greater than 10 mm is considered abnormal.

9.5. C True
Three major pathological processes can result in ventricular dilatation: obstructive hydrocephalus (most commonly due to aqueduct stenosis), maldevelopment of the ventricle or surrounding tissues (e.g. agenesis of the corpus callosum), and destruction of the surrounding brain tissue (hydrocephalus ex vacuo).

9.5. D False
The depth of the cisterna magna is measured from the margin of the vermis to the inner margin of the occipital bone. The measurement is typically 5–6 mm, with a normal range of 3–11 mm. Prominence of the cisterna magna is seen in cerebellar hypoplasia (Dandy–Walker syndrome), while effacement is seen with myelomeningocele due to an associated Arnold–Chiari malformation.

9.5. E False
The cavum septum pallucidum is visualized in approximately 95% of cases and therefore inability to demonstrate it is not necessarily abnormal. Visualization of this structure ensures normal frontal midline development. Its absence is associated with complete agenesis of the corpus callosum, holoprosencephaly and septo-optic dysplasia.

(Reference 11)

9.6. A True
Anencephaly occurs with an incidence of approximately 1 : 1000 births and is four times commoner in females. Risk factors include a family history of neural tube defect (NTD) and twin pregnancy. Sonography is a very reliable method of diagnosis and demonstrates absence of the cranial vault and brain above the

orbits. The base of the skull is nearly always present and a variable amount of disorganized brain tissue is seen protruding from it.

9.6. B False
Encephalocele occurs with an incidence of approximately 1 : 4000 births. The occipital location is affected six times more frequently than other sites, followed by the frontal, ethmoid and parietal regions. Hydrocephalus has been reported in 80% of occipital meningoceles, 65% of occipital encephaloceles, and 15% of frontal cephaloceles. About 20% of cases have associated microcephaly. There is an association with the Meckel–Gruber syndrome (renal cystic dysplasia, occipital encephalocele and polydactyly).

9.6. C True
Open spina bifida occurs in about 1 : 1000 births and is seen most commonly in the lumbosacral region. Over 90% are myelomeningoceles and the rest are meningoceles. Approximately 80% are open defects, while 20% are closed. Sonography demonstrates the spinal dysraphic defect as outward splaying of the dorsal ossification centres. Other features include the (myelo)meningocele sac, and various associated cranial abnormalities, including the Arnold–Chiari malformation.

9.6. D True
Obliteration of the cisterna magna occurs in virtually all cases of the Arnold–Chiari malformation. The 'banana' sign refers to the shape of the cerebellar hemispheres, which are wrapped around the posterior midbrain.

9.6. E True
The 'lemon sign' refers to a pointed shape to the anterior calvarium and typically occurs when the spinal defect is small and difficult to visualize. It usually resolves by 34 weeks. It is only seen in those fetuses that have hydrocephalus associated with spina bifida. Also, 75% of fetuses with spina bifida will have ventricular dilatation by 24 weeks, and the combination of hydrocephalus and the 'lemon sign' is virtually pathognomonic of spina bifida.

(Reference 23)

9.7. A False
Polyhydramnios in the absence of fetal anomaly is usually idiopathic and most commonly reflects a large-for-dates baby, with or without associated maternal diabetes mellitus.

9.7. B True
Also, the association of polyhydramnios with intrauterine growth retardation (IUGR) is highly suggestive of an underlying congenital anomaly or chromosomal abnormality, even if ultrasound evaluation of the fetus is considered normal (apart from size). The commonest fetal anomalies associated with polyhydramnios are related to the fetal neuraxis, gastrointestinal tract and genitourinary tract. Multiple gestations and chromosomal abnormalities are also recognized causes.

9.7. C True
Cystic hygromas typically appear on ultrasound as fluid-filled structures in the back of the neck, commonly containing internal septations. They are occasionally large enough to fill the amniotic cavity. Severe cases are associated with pleural effusions and ascites due to fetal hydrops. They may also be localized or associated with diffuse lymphangiectasia of the whole body.

9.7. D False
Cystic hygromas can be detected as early as 12 weeks by transabdominal scanning. Chromosomal abnormalities are found in approximately 50% of cases in the late first trimester and 75% of cases in the second trimester. Turner's syndrome, trisomy 18 and trisomy 21 are the most common associations and these fetuses have a poor prognosis. However, isolated non-cervical cystic hygromas do not appear to carry a significant risk of chromosomal abnormality and have a favourable prognosis.

9.7. E True
Nuchal thickening of greater than 5 mm may be seen in less than 1% of normal fetuses but is the commonest single sonographic abnormality of Down's syndrome seen during the second trimester, being observed in 10–20% of fetuses with Down's.

(Reference 11)

9.8. A True
Congenital cardiac defects that can be readily demonstrated on the four-chamber view of the fetal heart include endocardial cushion (A–V canal) defects, single ventricle, hypoplastic left or right ventricle and Ebstein's anomaly.

9.8. B False
Cardiac defects that are most consistently missed in most antenatal ultrasound series include isolated ASD, small to moderate-sized ventricular septal defects, aortic and pulmonary stenoses, aortic coarctation, and total anomalous pulmonary venous drainage. Antenatal series suggest that 22–32% of fetuses with sonographically detectable cardiac defects will have a chromosomal abnormality.

9.8. C True
Fetal hydrothorax may be isolated or associated with trisomy 21, monosomy X, and with syndromes involving vascular and lymphatic abnormalities. Fetuses that have associated hydrops have an even greater mortality rate than those in which the hydrothorax is an isolated finding.

9.8. D True
CAM is subdivided into three types. Type 3 CAM may appear as a homogeneously echogenic intrathoracic mass. The survival of fetuses with CAM correlates with the presence or absence of pulmonary hypoplasia. Fetal

hydrops may occur and is associated with a particularly poor prognosis. Other causes of an intrathoracic echogenic mass include bowel from a congenital diaphragmatic hernia and bronchopulmonary sequestration.

9.8.   E False
Bronchogenic cysts appear on antenatal ultrasound scans as well-defined, unilocular, intrathoracic cystic lesions adjacent to the mediastinum, typically without associated mediastinal shift, polyhydramnios, or fetal hydrops. Congenital diaphragmatic hernia and CAM type 1 can both produce cystic intrathoracic masses.

(Reference 11)

9.9.   A True
Most cases of fetal hydrops now reflect non-immune hydrops, which is associated with various fetal anomalies or chromosomal abnormalities. Isolated ascites is not a feature of hydrops. Causes include urinary tract obstruction, meconium peritonitis, hydrometrocolpos, ruptured ovarian cyst and intrauterine infection.

9.9.   B False
In cases of tracheo-oesophageal fistula, antenatal ultrasound demonstrates polyhydramnios and absence of the fetal stomach in one-third of fetuses at 24 weeks gestation. Demonstration of polyhydramnios and a small stomach is a more sensitive but less specific finding. This combination of ultrasound features may be seen in other abnormalities that are associated with disorders of fetal swallowing.

9.9.   C False
As with tracheo-oesophageal fistula, the sonographic features of duodenal atresia are usually not present until after 24 weeks gestational age. Ultrasound demonstrates a 'double-bubble' sign representing the dilated fetal stomach and duodenum. Continuity between the two structures should be identified to distinguish the dilated duodenum from other cystic structures in the right upper quadrant (e.g. the gallbladder).

9.9.   D True
Peritoneal calcifications may be seen in up to 85% of cases of meconium peritonitis. Other ultrasound features include meconium pseudocysts, which result from contained perforations and appear as well-defined, hypoechoic masses with echogenic, calcified walls, polyhydramnios (in 64%), fetal ascites (in 54%), and bowel dilatation (in 24%).

9.9.   E True
Anorectal atresia has been demonstrated on antenatal ultrasound as a dilated colon in the lower abdomen or pelvis. The amniotic fluid volume may be normal, reduced (when associated with bilateral renal disorders or bladder outflow obstruction), or increased (when associated with tracheo-oesophageal fistula).

(Reference 11)

9.10. A False
The fetal kidneys can reliably be sonographically visualized at 14–16 weeks gestational age as hypoechoic areas each side of the spine. By 16–18 weeks the kidneys are the major source of amniotic fluid and therefore oligohydramnios can only suggest underlying renal anomaly as the cause after about 16 weeks. Also, reduced amniotic fluid requires the presence of a bilateral renal anomaly (or bladder outflow obstruction). Urine is usually seen in the fetal bladder at 14–16 weeks.

9.10. B False
The assessment of hydronephrosis is subjective but an AP renal pelvis diameter of 10 mm or a PD/KD ratio (AP diameter of the renal pelvis/AP diameter of the kidney) greater than 50% represents significant dilatation. Conversely, a measurement of 5–9 mm or a PD/KD ratio of less than 50% is usually physiological, and rarely progresses.

9.10. C True
Posterior urethral valves are the commonest cause of bladder outlet obstruction in males and may have variable sonographic features. These include a persistently dilated bladder with dilatation of the posterior urethra and thickening of the bladder wall (greater than 2 mm). Bilateral ureteric dilatation and pelvicaliceal dilatation may be seen in approximately 40% of cases, and therefore absence of hydronephrosis or unilateral dilatation does not exclude the diagnosis. Oligohydramnios is seen in about 50% of cases and urinoma is also a feature.

9.10. D False
Autosomal recessive, infantile polycystic kidney disease may have variable expression depending upon the amount of renal involvement. The degree of renal involvement is inversely correlated to the degree of hepatic involvement, and children with mild renal disease will present in later childhood with cirrhosis and portal hypertension. These cases may have normally appearing kidneys *in utero*. Conversely, severe cases manifest *in utero* as oligohydramnios and bilateral, enlarged, echogenic kidneys.

9.10. E True
PUJ obstruction may appear on antenatal ultrasound as dilatation of the renal pelvis and calices. The ureter is not seen as it is not dilated. Severe cases may manifest as a large abdominal cyst with little or no surrounding renal parenchyma. If the contralateral kidney is normal, the bladder and amniotic fluid volume may be normal, but both oligohydramnios and polyhydramnios can occur.

(Reference 23)

9.11. A False
Choroid plexus cysts are the commonest sonographically demonstrated intracranial abnormality. The vast majority identified initially in the second trimester are small, and typically resolve by 24 weeks. However, cysts that are large and expand the ventricular wall appear to carry a significant risk of underlying chromosomal abnormality, notably trisomy 18 (Edwards' syndrome).

9.11.   B False
Polyhydramnios is most severe when obstruction to the gastrointestinal tract is proximal, as occurs with oesophageal and duodenal atresia. If amniotic fluid can reach the small bowel, it will be reabsorbed. Consequently, in most distal obstructions the amniotic fluid volume is normal.

9.11.   C True
Gastroschisis is characterized by a relatively small defect (2–4 cm) involving all layers of the abdominal wall, and almost always located to the right of the umbilicus. There is no genetic association or increased risk of occurrence. Ultrasound may also demonstrate bowel dilatation, which suggests either obstruction and/or ischaemia. This may result in gangrene, perforation and meconium peritonitis, and is a major cause of mortality (although death from gastroschisis is very rare).

9.11.   D False
Omphalocele is characterized by a midline abdominal wall defect at the base of the umbilical cord that is covered by a membrane. The sac most commonly contains liver with or without bowel, or bowel alone. Omphalocele is commonly associated with other malformations, and 40% of babies will have chromosomal abnormalities (trisomy 18 and 13 most commonly but also trisomy 21, Turner's syndrome and triploidy).

9.11.   E False
The normal umbilical artery Doppler waveform exhibits continuous positive flow during diastole, reflecting the low resistance of the placental capillary network. With advancing gestational age, there is progressively lower resistance to flow, manifest as increased diastolic flow. Conversely, some pregnancies affected by disorders such as intrauterine growth retardation (IUGR) or chromosomal anomaly show abnormally increased resistance to blood flow. This results in a reduced diastolic component to the Doppler waveform, and absent to reversed diastolic flow in the umibilical artery is associated with perinatal death rates as high as 50–90%.

(References A, B, 23; C, D, E, 11)

9.12.   A False
The placenta averages 1 cm in thickness at 10 weeks gestation and 3 cm in thickness at term. The normal placenta is rarely thicker than 4.5 cm or thinner than 1 cm when it has a normal area of attachment to the interior of the uterus.

9.12.   B True
Other causes of a large placenta include maternal diabetes mellitus, maternal anaemia and Rhesus incompatibility. Small placental size is associated with pre-eclampsia and small-for-gestational-age babies.

9.12.   C True
Placenta membranacea is a rare placental anomaly in which the placenta covers most of the surface area of the uterine cavity. The placenta is generally thin but may show focal areas of thickening. Recognized complications include antepartum and intrapartum haemorrhage, premature delivery and IUGR.

9.12. D False

In cases of bilobate placenta, ultrasound demonstrates two lobes of similar size with a bridge of connecting tissue into which the cord inserts. With succenturiate placenta, an accessory lobe (the succenturiate lobe) is present, which is separated from that part of the placenta into which the cord inserts. The succenturiate lobe is connected to the main part of the placenta either by a membrane or by a bridge of tissue. Succenturiate placenta is a potential cause of postpartum haemorrhage.

9.12. E True

In the case of a succenturiate placenta with a membrane bridging the two lobes, velamentous insertion of the cord describes the insertion of the cord into this membrane, rather than into placental substance. If the membrane overlies the internal os, this situation results in vasa praevia. This condition has been demonstrated with transvaginal colour Doppler ultrasound.

(Reference 11)

9.13. A True

Scattered hypoechoic areas ranging from a few millimetres to several centimetres are often seen within the placenta and in most cases are due to intervillous thrombosis or perivillous fibrin deposition. In later pregnancy, decidual septal cysts may also result in hypoechoic areas. Subchorionic fibrin deposition may also result in hypoechoic regions under the chorion on the fetal side of the placenta. All of these conditions are of no clinical significance.

9.13. B False

In one large series, 86% of placental infarctions were isoechoic, the remainder appearing hypoechoic or anechoic due to associated fibrin deposition and haemorrhage. Occasional cases of placental infarction have been described that resulted in hyperechoic areas within the placenta.

9.13. C False

Chorioangiomas are usually single tumours and commonly appear as well-defined, hypoechoic masses on the fetal side of the placenta. Less commonly they appear as multiple masses within the placental substance. Approximately 30% are associated with polyhydramnios and 16% with stillbirths. Other uncommon complications include fetal anaemia, low birth weight and fetal hydrops. Other placental tumours include teratoma and metastasis.

9.13. D True

Elevation of the chorionic membrane is seen with the subchorionic type of placental abruption, which accounts for about 80% of cases of sonographically detected abruptions. Other ultrasound features of subchorionic haemorrhage include extension of the elevated membrane to the edge of the placenta. Also, if the gain settings are increased, the region between the membrane and the uterine wall will contain low-level echoes compared to the anechoic amniotic fluid, since it is filled with blood.

9.13. E True

In approximately 15% of cases, placental abruption appears as a retroplacental mass and, very rarely, as a preplacental mass. The echogenicity of the haematoma depends upon its age. Initially they may be hyperechoic, later becoming isoechoic, and manifest as placental thickening. Finally they become hypoechoic.

(Reference 11)

9.14. A False
Hydatidiform mole is the commonest and most benign form of gestational trophoblastic disease. In approximately 80% of cases, the disease follows a benign course with resolution after evacuation. In 12–15%, invasive mole develops and in 5–8% of cases metastatic choriocarcinoma results. In uncomplicated cases of second trimester molar pregnancy, ultrasound characteristically demonstrates a large, moderately echogenic mass filling the uterine cavity and containing numerous small cystic areas.

9.14. B False
Invasive mole is characterized pathologically by the presence of nests of trophoblastic cells as well as swollen chorionic villi within the myometrium, as opposed to the intracavitary location of hydatidiform mole. Ultrasound may be able to distinguish the two by the demonstration of haemorrhagic necrosis involving the myometrium and extending into the parametrium.

9.14. C False
Choriocarcinoma is the most malignant form of trophoblastic disease. Approximately 50% of cases follow a molar pregnancy (3–5% of molar pregnancies result in choriocarcinoma), 25% follow an abortion, 22% follow a normal pregnancy, and about 3% follow an ectopic pregnancy.

9.14. D True
Approximately 20–50% of cases of trophoblastic disease are associated with bilateral ovarian theca-lutein cysts, which are thought to be the result of the markedly elevated levels of human chorionic gonadotrophin (hCG) that accompany the disease. Ultrasound typically demonstrates multiseptate cysts that may not regress until 2–4 months following molar evacuation, and therefore cannot be used as evidence of persisting disease.

9.14. E False
Patients with metastatic choriocarcinoma may be grouped into low, inter-mediate and high-risk groups depending upon the duration of their disease, hCG levels and the sites of metastatic disease. Those with metastases to the lung or vagina are included in the low-risk group, whereas those with hepatic or CNS disease are in the high-risk group. Patients with other metastases are included in the intermediate group.

(Reference 25)

9.15. A False
Ovarian volume can be estimated using a simplified formula for a prolate ellipse (length times width times height divided by 2). At the menarche, ovarian volume is 4.18 ± 2.3 ml. In teenagers and young adults, the maximum

ovarian volume can be as high as 14 ml. For postmenopausal women, ovarian volumes should be 2.5 ml or less.

9.15. B False
In the postpubertal premenopausal period, multiple follicles can be seen developing during each menstrual cycle along the periphery of the ovarian cortex. These follicles do not exceed 25 mm in diameter and are round or ovoid, well-defined and anechoic. If a larger cyst is identified, a physiological cyst or other pathology needs to be considered.

9.15. C True
By transvaginal scanning, the central, more echogenic ovarian medulla can frequently be differentiated from the outer cortex. On transabdominal scanning the ovaries typically demonstrate a mid-level fine, homogeneous echotexture, with an echogenicity that is greater than that of the adjacent uterus and somewhat more homogeneous.

9.15. D False
Ovarian cysts have been identified in patients with cystic fibrosis, but are typically unilocular, unilateral and transitory. The ovaries in patients with the McCune–Albright syndrome may be enlarged and contain multiple cysts. Large ovarian cysts have also been reported in some patients with neurofibromatosis.

9.15. E False
The Stein–Leventhal syndrome consists of the combination of obesity, oligomenorrhea, hirsutism and polycystic ovarian disease (PCOD). The characteristic ultrasound appearance of the ovaries is one of bilaterally enlarged ovaries (mean volume 14 ml) containing multiple tiny cysts. The cysts are typically 5–8 mm in size and there are usually more than five in each ovary. These features are seen in 35–40% of cases. In approximately 30% of cases the ovarian volume is within normal limits. Ultrasound may also demonstrate hypoechoic ovaries or enlarged ovaries that are isoechoic with the uterus.

(Reference 25)

9.16. A True
Follicular cysts develop when a mature follicle fails to ovulate or involute. They may be 1–10 cm in size but can only be diagnosed when over 2.5 cm in diameter. They may cause symptoms due to pressure, haemorrhage or torsion but commonly regress spontaneously. When uncomplicated, ultrasound demonstates a unilocular, thin-walled, anechoic structure. Haemorrhage may result in development of internal echoes and occasionally septations.

9.16. B True
Ovarian hyperstimulation may occur in infertile women taking drugs such as clomiphene citrate to stimulate follicular development and ovulation. Ultrasound demonstrates enlarged ovaries (greater than 5 cm in length) containing multiple septate cystic structures. This condition is associated with increased likelihood of pregnancy, including multiple gestations. It may progress to the

ovarian hyperstimulation syndrome, which may be complicated by ovarian rupture.

9.16. C True
The ultrasound appearance of benign cystic teratomas (dermoid cysts) varies from completely cystic to inhomogeneously solid. The mass usually presents as a complex, predominantly solid lesion containing high-level echoes due to hair or calcifications arising within the cyst. The high echogenicity of the mass with distal shadowing may make it difficult to distinguish the mass from surrounding bowel.

9.16. D True
Three groups of malignancies tend to metastasize to the ovaries. These include metastases from other pelvic primaries, those from the gastrointestinal tract (especially stomach, colon, biliary tract and pancreas), and from breast carcinoma. Ultrasound usually demonstrates echogenic masses with varying anechoic spaces. However, they can also appear identical to cystadenocarcinomas.

9.16. E False
Ovarian fibromas account for approximately 5% of ovarian neoplasms. Multiple tumours are seen in 10% of cases, and ascites occurs in 40% of patients with tumours over 6 cm in size. Meigs' syndrome occurs in 1–3% of cases. Ultrasound characteristically demonstrates a large hypoechoic adnexal mass that causes marked acoustic attenuation.

(Reference 25)

9.17. A True
The thickness of the normal endometrial echo complex is approximately 2–4 mm in the proliferative phase of the menstrual cycle and 5–6 mm in the secretory (luteal) phase. In postmenopausal women in the absence of hormonal replacement therapy, the endometrium should measure 1–3 mm in thickness, and a measurement of above 5 mm should be considered abnormal.

9.17. B False
The ultrasound features of adenomyosis include diffuse uterine enlargement, with a normal central endometrial echo complex, normal echotexture of the myometrium and normal uterine contour. Occasionally, adenomyosis is focal and results in contour abnormality with preservation of the central echo complex. These cases may be difficult to distinguish from leiomyomas.

9.17. C True
Endometrial hyperplasia is the commonest cause of uterine bleeding and occurs predominantly in menstruating women but may also be seen following the menopause. It results from unopposed oestrogen stimulation. Ultrasound may demonstrate a prominent central uterine cavity echo complex.

9.17. D True
Endometrial carcinoma most commonly presents with postmenopausal bleeding, and if this occurs early in the course of the disease ultrasound may be normal or only demonstrate a thickened, irregular endometrial echo complex.

Occasionally, local invasion can be identified. More extensive disease appears as an enlarged uterus containing areas of low and high-level echoes. Ultrasound can distinguish carcinoma confined to the uterus (stages I and II) from that extending beyond the uterus (stages III and IV).

9.17. E True
Most endometrial polyps are asymptomatic but the commonest symptom is bleeding. Rarely, a polyp will have a pedicle long enough to allow it to protrude beyond the cervix or vagina. Ultrasound usually demonstrates thickening of the central echo complex. Occasionally, polyps may be seen as discrete mobile masses within the uterine cavity, and may even cause uterine enlargement.

(Reference 25)

9.18. A False
In one series, up to 28% of patients with PID had normal or near-normal pelvic ultrasound scans. Recognized sonographic features include free fluid, loss of the midline endometrial echo, fluid in the endometrial cavity, indistinct uterine margins, and adnexal masses. In severe cases, the whole of the pelvis may be filled by a heterogeneous mass which obscures the uterus.

9.18. B True
Tubo-ovarian abscess is a common complication of acute salpingitis, occurring after the purulent exudate spills out of the fimbriated end of the Fallopian tube into the pelvic cavity. Ultrasound typically shows a complex, solid-cystic, adnexal mass. An abscess that responds to therapy becomes more cystic and well-defined as it shrinks. Adnexal masses may also appear hypoechoic and solid, as in the case of acute pyosalpinx.

9.18. C True
An acute tubo-ovarian abscess may extend to involve the ureter, bowel, bladder or the opposite adnexa. Ultrasound may initially demonstrate a mild hydronephrosis, which usually resolves rapidly with treatment. If a chronic abscess develops, hydronephrosis can persist and worsen.

9.18. D True
Most significant uterine infections are related to the puerperium. Transient endometritis may also occur in association with PID. Ultrasound may demonstrate a dilated uterine cavity containing fluid, echogenic material or high-level echoes due to gas. Free pelvic fluid may also be identified. Ultrasound may also be normal. The ultrasound features are similar to those seen with retained products of conception.

9.18. E True
Pyometra or hydrometra occur when there is obstruction to drainage of the uterine cavity. Ultrasound demonstrates a symmetrically enlarged uterus with enlargement of the uterine cavity. The contents of the uterine cavity will appear cystic. Pyometra can be differentiated from hydrometra by the

presence of gas. A similar ultrasound picture can be seen in the case of a large degenerating or infected leiomyoma.

(Reference 24)

9.19.   A False
Leiomyomas are usually hypoechoic relative to the myometrium. They may also demonstrate varying degrees of decreased echotexture due to cystic change and internal haemorrhage. Increased echoes are related to calcification and fibrosis. Mixed echogenicity within a solitary lesion is not uncommon.

9.19.   B True
Ultrasound may also demonstrate uterine leiomyomas as deformity of the uterine contour, or markedly increased size of the uterus with a diffusely heterogeneous echotexture.

9.19.   C False
Uterine leiomyomas may undergo various changes during pregnancy. Small lesions have been noted to grow in the second and third trimesters and subsequently decrease in size in the third trimester. Larger lesions have demonstrated growth in the first trimester only with subsequent size reduction. Approximately 50% show no change in size. The finding of anechoic cystic spaces and heterogeneous echotexture has correlated clinically with episodes of abdominal pain and premature uterine contractions, presumably due to development of haemorrhagic infarction.

9.19.   D False
Leiomyosarcoma is a rare tumour, accounting for approximately 1–3% of uterine malignancies. Most are believed to arise from pre-existing leiomyomas. Ultrasound usually demonstrates a mass with areas of cystic degeneration. However, in the absence of local invasion or distant metastases, they are indistinguishable from leiomyomas.

9.19.   E True
Most leiomyosarcomas are intrumural and up to 75% present as solitary nodules within the uterus. Sarcomatous degeneration of a uterine leiomyoma should be suspected if it undergoes rapid enlargement. Other malignant tumours of the uterus include mixed Müllerian tumours, lymphoma and other sarcomas. All may have similar imaging features to leiomyosarcoma.

(References A, C, D, E, 24; B, 25)

9.20.   A False
In the normal adult, the long axis of the uterus measures about 7.5 cm, of which 5 cm represents the uterine body and 2.5 cm represents the cervix, resulting in a normal body : cervix ratio of 2 : 1. Infantile uterus is characterized by a shorter long axis measuring 3–4 cm and a reversal of body : cervix ratio (typically 1 : 2 to 1 : 4). Infantile uterus is seen in adults with hypogonadism during puberty. Congenital cervical incompetence is frequently associated.

9.20. B False
The internal cervical os can be identified as a normal narrowing of the cervical canal at its superior extent. On frontal views a definite angle can be identified between the cervical canal and the uterine cavity (the uterocervical angle) at the level of the internal os. Loss of this angle is a feature of cervical incompetence.

9.20. C True
In patients with cervical incompetence, the internal os diameter is greater than 8 mm. Such patients are at risk of late abortions and premature delivery. For accurate diagnosis, complete visualization of the cervix and uterine cavity is required.

9.20. D True
Mucosal folds (plicae palmitae) are frequently seen in the cervical canal, especially in nulliparas. Pseudodiverticula may be demonstrated at the level of the internal os following a low-segment Caesarean section. The cervical canal can also be distorted by polyps, leiomyomas or synechiae.

9.20. E True
The uterine cavity is normally triangular, with two slightly concave side walls and a slightly convex fundal dome. Fundal concavity (arcuate uterus) is a minor anomaly of no clinical significance. The lining of the endometrium changes its appearance during the phases of the menstrual cycle, appearing smooth in the follicular phase and irregular in the secretory phase due to endometrial growth and glandular secretions.

(Reference 24)

9.21. A True
Intrauterine synechiae (Asherman's syndrome) refers to the presence of adhesions that obliterate all or part of the endometrial and/or cervical canal. It most commonly follows curettage performed to terminate a pregnancy or evacuate retained products of conception and results in hypo- or amenorrhoea. Asherman's syndrome may be classified into cervicoisthmic and corporeal adhesions based on site, and various grades or types depending upon extent. Cervicoisthmic adhesions may affect the cervical canal alone, resulting in atresia, stenosis, multiple adhesions or a suprasisthmic diaphragm, or may be associated with concomitant corporeal adhesions.

9.21. B False
Three types of corporeal adhesions are described (mucous, fibrous and muscular). Hysterosalpingography cannot distinguish between these. Corporeal adhesions can result in intrauterine filling defects that range from a single small defect to almost complete obliteration and marked distortion of the uterine cavity. Venous and lymphatic intravasation is a frequent finding in patients with extensive adhesions.

9.21. C False
With adenomyosis, the endometrial glands within the myometrium may or may not be connected to the endometrial cavity. In the former case, hysterosalpingography demonstrates multiple diverticula orientated at right angles to the

uterine cavity. However, in a study of 150 cases of histologically proven adenomyosis, hysterosalpingography showed the characteristic features of the condition in only 38 patients. The extent of the disease may also be underestimated.

9.21. D True
The hysterographic features of endometrial hyperplasia depend upon the type of disease. Mucous hypertrophy results in a dense central lumen that may be surrounded by a partially opacified margin representing the hypertrophic endometrium. In the presence of cystic or adenomatous hyperplasia, the margins may be indented or undulating, and the cavity may appear grainy or even polypoid. The hysterographic findings are not diagnostic.

9.21. E False
Endometrial polyps may be single or multiple, sessile or pedunculated. They rarely exceed 1 cm in size in menstruating women. They appear as well-defined, round or oval filling defects. Pedunculated polyps may move during the investigation.

(Reference 24)

9.22. A True
Genital tuberculosis is usually secondary to haematogenous spread from a primary focus in the lungs, but may rarely be due to lymphatic spread from peritoneal, intestinal or urinary tract disease. The Fallopian tube is the initial site of genital involvement, with subsequent spread to the endometrium and ovaries in 50% and 30% of cases respectively. Tubal occlusion occurs in 50–70% of cases and may be proximal, midtubal or distal. Tubal abnormalities include rigidity, a 'club-shaped' end, alternating constrictions and dilatations, and diverticula, which may affect the distal or proximal portion of the tube.

9.22. B False
A normal hysterosalpingogram effectively excludes tuberculosis. Venous or lymphatic intravasation is a non-specific finding and may also be seen with other chronic infections and uterine tumours. Uterine tuberculosis may result in a shrunken, deformed uterine cavity with intrauterine adhesions. However, the uterus may also appear normal.

9.22. C True
Tubal endometriosis may affect any part of the tube. The intramural portion of the tube may be blocked by adenomyosis. Endometriosis of the extrauterine portion of the tube is less characteristic. It may result in multiple areas of diverticula formation, distortion of the ampullary cavity due to peritubal adhesions, and occlusion of the fimbrial end with hydrosalpinx formation.

9.22. D False
Salpingitis isthmica nodosa is a disease of unknown aetiology that most commonly affects the isthmic portion of the tube (unilaterally or bilaterally), but may involve the entire structure. It is associated with subfertility and an increased incidence of ectopic pregnancy. The typical hysterosalpingographic appearance is of multiple punctate 2–3 mm collections of contrast medium

adjacent to the true tubal lumen. These represent endosalpingeal diverticula. The tube is usually patent.

9.22. E True
Normal spill of contrast medium from the distal end of the Fallopian tube appears as curvilinear collections outlining small bowel loops. Pooling of contrast medium around the terminal end of the tube or ovaries suggests the presence of peritubal adhesions. Pelvic adhesions may also retract the uterus and cause stretching of the contralateral tube. Also, a loculated collection of contrast medium may occur and simulate a hydrosalpinx.

(Reference 24)

9.23. A False
On precontrast CT, the uterus appears as a triangular or oval soft-tissue density mass behind the urinary bladder. Occasionally, a low-density central area, possibly representing secretions in the uterine cavity, can be seen. On postcontrast scans, the myometrium shows marked enhancement, allowing better differentiation from the endometrial cavity.

9.23. B True
Adenocarcinoma of the endometrium is the commonest malignancy of the uterine body, affecting predominantly women in the fifth decade. On precontrast CT, focal or diffuse enlargement of the uterine body may be seen. On postcontrast scans, the tumour enhances to a lesser degree than the myometrium but becomes hyperdense to endometrial secretions. If the tumour occludes the internal cervical os, hydrometra, pyometra or haematometra may occur. CT will demonstrate a distended fluid-filled uterus.

9.23. C False
Lymphatic spread from tumours of the uterine body most commonly results in para-aortic, retroperitoneal and external iliac lymphadenopathy. Inguinal metastases occur via the round ligament from tumours of the fundus. Local spread to the broad ligament and adnexal structures, as well as metastases to the omentum and peritoneum, may occur. The incidence of distant metastases increases with increasing depth of myometrial invasion, which is best assessed on postcontrast scans.

9.23. D False
Extension of endometrial carcinoma into the cervix is manifest on CT as cervical enlargement and hypodense areas in the fibromuscular stroma of the cervix.

9.23. E False
Uterine leiomyomas are usually of soft-tissue density similar to the myometrium on precontrast CT. Necrosis or degeneration will result in low density areas. Contour deformity and calcification can also be seen. On postcontrast CT, leiomyomas show a similar degree of enhancement to that of the myometrium, a feature that may help to distinguish them from endometrial carcinoma.

(Reference 2)

9.24. A False
On CT, the normal cervix has a uniform rounded appearance. Carcinoma of the cervix may be recognized on precontrast CT as enlargement of the cervix with either regular or irregular borders. On postcontrast scans, the tumour appears as a soft-tissue mass enlarging the cervix, containing hypodense areas due to tumour necrosis and diminished enhancement relative to the normal cervical tissue.

9.24. B True
Cervical carcinoma may obstruct the endocervical canal, resulting in fluid within the uterine cavity and enlargement of the uterus. A similar CT appearance may be seen with endometrial carcinoma (see above), senile contraction of the cervix, radiation therapy or postsurgical scarring.

9.24. C True
CT features of parametrial spread of cervical carcinoma include irregularity or poor definition of the lateral cervical margins, prominent parametrial soft-tissue strands or eccentric soft-tissue mass, and obliteration of the periureteral fat plane. The latter two features are essential for the definitive diagnosis of parametrial spread. Poor definition of the cervical contour and linear increased densities in the pericervical fat may also occur with parametritis.

9.24. D True
Stage IIIb disease (FIGO classification) is indicated by extension to the pelvic side wall or hydronephrosis due to distal ureteric obstruction. CT may show soft-tissue extension to the obturator internus and piriformis muscles, which may also be enlarged.

9.24. E False
Lymphatic spread from cervical carcinoma initially involves the iliac nodes (external and internal), followed by the periaortic nodes. CT cannot demonstrate metastases in normal-sized lymph nodes, and some nodes will be enlarged due to inflammatory causes.

(Reference 2)

9.25. A False
CT may demonstrate the normal adult premenopausal ovaries as soft-tissue density masses, usually located posterolaterally to the uterus. Small cystic areas due to normal follicles may occasionally be seen. CT can also sometimes demonstrate the broad ligaments as linear soft-tissue bands extending from the lateral aspect of the uterus to the pelvic side wall. The lateral cervical and uterosacral ligaments are less commonly identified.

9.25. B False
Approximately 85% of ovarian cancers are of epithelial origin, the remainder being derived from germ or stromal cells. About 90% of epithelial carcinomas are either serous or mucinous cystadenocarcinoma, with the rest being endometrial or solid carcinomas. On CT, benign cystadenomas appear as well-defined, low-density masses with thick, irregular walls and multiple internal septa. Papillary projections may also be seen. Serous cystadenoma typically has central CT density near to that of water, whereas mucinous cystadenoma

may show attenuation values just below those of soft tissue. Calcification is an occasional finding with serous cystadenoma. CT cannot reliably distinguish cystadenoma from cystadenocarcinoma in the absence of metastases.

9.25. C True
Ovarian carcinomas usually spread by implanting widely on omental and peritoneal surfaces. Demonstration of an 'omental cake' is strongly suggestive of ovarian malignancy. An 'omental cake' appears on CT as an irregular sheet of nodular soft tissue deep to the anterior abdominal wall.

9.25. D True
Peritoneal implants from ovarian carcinomas appear on CT as soft-tissue nodules along the lateral peritoneal surfaces of the abdomen. Subdiaphragmatic metastases are most easily detected between the liver and the abdominal wall in the presence of ascites. Rarely, peritoneal and omental metastases may calcify. This may also occur following cisplatin therapy.

9.25. E False
Lymphatic metastases to the para-aortic and occasionally to the inguinal nodes occurs with ovarian carcinoma, particularly serous cystadenocarcinoma and other solid tumours. However, lymphadenopathy is not a major feature of metastatic ovarian carcinoma. Metastases to the liver are also uncommon, being seen in less than 10% of cases.

(Reference 2)

9.26. A True
On T2-WSE scans, the peripheral myometrium appears as a region of intermediate signal intensity and the endometrium is hyperintense. Between the two is a zone of low signal intensity (the junctional zone) which is thought to represent the more central, vascular region of the myometrium.

9.26. B False
The signal characteristics of the endometrium do not change on SE MRI sequences during the menstrual cycle but there is an increase in the endometrial thickness from approximately 1–3 mm in the proliferative phase to 5–7 mm in the mid-secretory phase of the cycle. The mid-secretory phase may also be associated with a slight increase in uterine volume, prominence of the junctional zone and increased signal intensity of the myometrium on T2-WSE scans.

9.26. C True
In women taking the oral contraceptive pill, the endometrium is less than 3 mm thick and is usually not visualized. Also, these patients show less variation in the cyclical changes mentioned above. The premenarchal and postmenopausal uterus typically appears smaller in size, with a decreased endometrial thickness and less prominence of the junctional zone.

9.26. D False
The cervix is relatively hypointense to the uterine body on all pulse sequences, presumably related to its increased fibrous tissue content. The central cervical canal appears hyperintense on T2-WSE scans due to its mucous content.

9.26. E True
The vaginal vault and body can both be demonstrated by MRI. On both T1- and T2-WSE sequences, the signal intensity of the vagina is similar to that of striated smooth muscle. High signal is occasionally seen in the vaginal vault on T2-WSE scans due to vaginal secretions.

(Reference 3)

9.27. A True
Uterine leiomyomas that are free of degenerative change appear as well-defined low-signal intensity masses. However, the presence of haemorrhage or cystic degeneration can result in variable signal intensities. Very low or absent signal is seen when there is extensive fibrosis or calcification.

9.27. B True
The MRI appearance of endometrial carcinoma is varied. It may not result in any abnormal signal intensity in the endometrium, and may then be indistinguishable from blood-clot or endometrial hyperplasia. It may also result in an area of reduced signal intensity in the endometrium on T2-WSE scans, while causing no change on T1-WSE scans. Other findings include hetero-geneous signal intensity on both pulse sequences, or simply an abnormally increased width of the endometrium.

9.27. C True
The presence of an intact hypointense junctional zone surrounding the endometrial cavity differentiates endometrial carcinoma that is confined to the cavity from that which has invaded the myometrium. However, this sign may be of limited value in postmenopausal women since the junctional zone may not be visible normally.

9.27. D True
On T1-WSE scans, the cervical carcinoma usually appears isointense to skeletal muscle. On T2-WSE scans, it usually appears as a hyperintense mass or less commonly as disruption of the shape of the normally hypointense cervix.

9.27. E False
Extension of cervical carcinoma into the parametrial fat is better demonstrated on T1-WSE scans, whereas extension into the body of the uterus is better demonstrated on sagittal T2-WSE scans. Sagittal scans are also useful for demonstrating tumour spread into the bladder or rectum, whereas axial or coronal scans demonstrate lateral extension and pelvic side wall involvement.

(References A, B, C, 24; D, E, 3)

9.28. A True
On T1-WSE sequences, the ovaries demonstrate intermediate signal intensity, with areas of lower or higher signal intensity due to the presence of follicular or haemorrhagic corpus luteum cysts. The signal intensity of the ovaries increases on T2-WSE scans and the ovarian follicles appear hyperintense, allowing easier recognition of the ovaries.

9.28. B True
Simple ovarian cysts show the MRI characteristics of cysts elsewhere, being uniformly hypointense on T1-WSE scans and hyperintense on T2-WSE scans. Polycystic ovaries may show characteristic appearances on T2-WSE scans, with a peripheral rim of small hyperintense cysts and a relatively hypointense central region representing the thecal stroma.

9.28. C False
The MRI appearances of ovarian teratoma depend upon its composition. Signal intensities corresponding to fat, fluid, blood or proteinaceous debris may be seen as well as signal void due to calcification, teeth, or ossification. The presence of chemical shift artefact at a fluid–fat interface, although a rare finding, may help in the diagnosis.

9.28. D True
Hyperintensity on both T1- and T2-WSE scans would be consistent with subacute haematoma with free methaemoglobin. Haemorrhagic pelvic masses may be due to pelvic inflammatory disease, ovarian carcinoma, or endometriosis. Their appearance will depend upon the age of the haemorrhage.

9.28. E False
The MRI appearance of the various ovarian neoplasms (with the exception of teratoma) is non-specific and variable due to the differing mucin and protein content of cyst fluid and whether the tumour is cystic or solid. MRI is therefore unable to differentiate reliably between the various types of ovarian tumours on the basis of signal intensities alone.

(Reference 3)

9.29. A True
The MRI diagnosis of endometriosis is suggested by the demonstration of multiple haemorrhagic cystic foci in the pelvis, commonly in the cul-de-sac, with or without free pelvic fluid. The cysts are commonly multiloculated and a low signal rim may be seen, related either to haemosiderin-laden macrophages in the cyst wall or a fibrous capsule. The signal intensity of the cyst contents depends upon the age of the contained blood.

9.29. B True
The loss of the clear outline to the uterine body is associated with diffuse pelvic involvement and adhesions. Adhesive tethering of the rectum may also be seen.

9.29. C False
Adenomyosis (internal endometriosis) is characterized by the presence of endometrial glandular tissue within the myometrium. It occurs in both focal and diffuse forms. Unlike endometriosis, haemorrhage is not a typical feature of adenomyosis, and therefore signal void due to haemosiderin is not seen.

9.29. D True
Irregular thickening, or a thickness of greater than 5 mm of the hypointense junctional zone of the uterus, is suggestive of diffuse adenomyosis. However, if the thickness of this zone is 3–5 mm (normal range), the diagnosis may still be

made by rescanning at a different stage of the menstrual cycle, since the normal junctional zone changes whereas adenomyosis will not change.

9.29.   E True
Focal adenomyosis has been demonstrated on T2-WSE scans as a poorly defined hypointense area containing tiny scattered regions of high signal intensity and located adjacent to the endometrium. It may be differentiated from leiomyoma by its indistinct border with the myometrium.

(References A, B, C, D, 3; E, 24)

9.30.   A False
The vessels of the breast are generally symmetrical in size and distribution. Increased vascularity may occur either due to distal obstruction or hyperaemia, and dilated vessels can accompany both inflammation and neoplasm. Also, differences in mammographic compression can result in apparent asymmetry.

9.30.   B True
A solitary dilated duct is a rare finding with carcinoma. If a single dilated duct contains a carcinoma, or more commonly an intraduct papilloma, the tumour may be proximal in the duct and not necessarily obstructing it distally near the nipple. A group of asymmetric ducts is a commoner finding and usually represents benign ectasia.

9.30.   C False
Intramammary lymph nodes are very common and histologically are found in all areas of the breast. On mammography, they are typically seen in the upper outer quadrant of the breast and rarely below the midline. They are virtually always at the edge of the mammary parenchyma and never deep within the breast tissue or in a subareolar location. They generally have a reniform shape with a fat-density hilar notch.

9.30.   D True
Radiolucent lesions contain encapsulated fat and are always benign. They appear as well-defined masses with a capsule which has a visible inner wall due to the fat content of the mass. Causes include lipoma, galactocele and post-traumatic oil cyst.

9.30.   E False
Metastases to the breast are most commonly from melanoma. Other sites include bronchus, stomach, ovary and sarcomas, as well as lymphoma. They may produce lesions that are similar to primary carcinomas but are more likely to be multiple and bilateral. They are usually round with fairly well-defined margins, although spiculations may rarely be seen.

(Reference 26)

9.31.   A True
The size of a carcinoma is directly related to the probability of axillary lymph node metastases, and therefore to prognosis. Carcinomas greater than 1 cm in size are twice as likely to have axillary nodal metastases than lesions under 1 cm in size.

9.31. B False
The 'halo' sign represents a thin, lucent band that may partially or completely surround a well-circumscribed mass. The lucent zone is an optical illusion (Mach effect) caused by an abrupt transition in density between the lesion and surrounding tissue. Although it has been considered to be pathognomonic of a benign lesion, sharply marginated malignant lesions can also produce this sign. These include lymphoma, papillary carcinoma and infiltrating carcinoma.

9.31. C False
In general, the flow of structures within the breast is directed towards the nipple along duct lines. Carcinomas can produce a cicatrization of tissues, pulling in the surrounding elements toward a point that is eccentric to the nipple, resulting in architectural distortion. This may be the only sign of a carcinoma but may also occur with benign disease such as postoperative scar, fat necrosis and radial scar.

9.31. D False
Metastatic spread of carcinoma from an intramammary tumour to an intramammary lymph node may result in enlargement and rounding of the node, with loss of its fat-density hilar notch. The node typically retains its well-defined outline. Metastatic spread to an intrammammary node carries the same diminished prognosis as axillary nodal involvement.

9.31. E True
The presence of spicules or a more diffuse stellate appearance is almost pathognomonic of carcinoma. The spiculation represents fibrosis interspersed with tumour cells extending into the tissue surrounding the cancer, and resulting in tissue distortion. The spicules may extend over several centimetres, or be only a few millimetres long. Rare causes of spiculated masses include extra-abdominal desmoid tumours, granular cell tumours and radial scars.

(Reference 26)

9.32. A True
Peripheral rim calcifications can also be seen with cysts and very rarely with circumscribed malignancies. However, the latter two conditions will be differentiated from fat necrosis by their internal density. When peripheral calcification is seen with malignant lesions, it tends to be relatively thick and continuous, whereas similar calcification in cyst walls tends to be patchy.

9.32. B False
Up to 50% of malignant masses have calcifications. Microcalcifications associated with malignancy are typically punctate, pointed and irregular and show heterogeneity of size and morphology, or fine, linear branching deposits that fill a narrowed duct lumen.

9.32. C False
The probability of malignancy increases with the increasing number of calcifications in the tissue volume. When the number of deposits is so great that they cannot be counted, the probability of malignancy is high. Some have shown that clusters containing at least five flecks of visible calcium in a 0.5 cm

× 0.5 cm area were all cancers. Others consider the presence of five microcalcifications within a square centimetre of the mammogram as being associated with a significant increase in the risk of malignancy.

9.32.   D False
Some carcinomas produce radiologically detectable calcifications early in their growth. Irregular, heterogeneous deposits grouped in isolation without an associated mass in a small volume of breast tissue and greater than four in number should arouse suspicion, since in approximately 20% of cases they represent cancer (typically an intraductal cancer or early infiltrating lesion).

9.32.   E True
Malignant microcalcifications may be up to 2 mm in diameter but are usually less than 0.5 mm. Calcifications as small as 0.2–0.3 mm can be observed on mammography, and the smaller the particles the more likely they are malignant in nature. In general, when all particles are greater than 2 mm in size, a benign process is likely. However, the significance of a cluster relates to the smallest calcifications in the group.

(Reference 26)

9.33.   A False
Round, hollow spherical calcifications with lucent centres are always benign. These occur in areas of fat necrosis, in association with benign calcifications in the ducts or periductal tissues that are thought to represent calcified debris, or secondary to inflammation caused by extrusion of ductal debris into the surrounding stroma. The latter types are typical of secretory calcifications and are usually only a few millimetres in size.

9.33.   B False
Extensive, thick, rod-shaped calcifications may represent secretory calcifications either within ducts or in the periductal tissues. These latter types may be associated with palpable thickening of the breast that has been called plasma cell mastitis, because it is accompanied by a plasma cell infiltrate. Those calcifications that form within the ducts are typically larger than 0.5 mm, do not branch and are orientated along duct lines. They may contain lucent centres.

9.33.   C False
This appearance is characteristically due to the benign precipitation of calcium (milk of calcium) within the small, cystically dilated acini of the lobules. On the craniocaudal view, the X-ray beam is perpendicular to the sediment of calcium, which therefore appears as a small, poorly defined dot. On the lateral view, the X-ray beam is parallel to the sedimented calcium which appears as a crescentic density.

9.33.   D True
Delicate, linear deposits under 0.5 mm in diameter that fill the narrowed duct lumen, may be branching, and that may be associated with punctate calcifications in a 'dot-dash' pattern, are likely to indicate a malignant process. These calcifications form a cast of the duct lumen that is irregularly narrowed by the heaped up malignant cells and are classical of ductal carcinoma.

9.33. E True
Extensive intraduct carcinoma can produce innumerable calcifications throughout large portions of the breast and, rarely, this can be bilateral. The microcalcifications are distinguished from benign conditions by their morphology (see above). The microcalcifications are relatively confined to a segment of the breast, rather than being diffuse and randomly distributed.

(Reference 26)

9.34. A True
Cooper's ligaments produce a fine trabecular pattern of curvilinear septations coursing through the breast. Fibrous extensions of Cooper's ligaments pass through the subcutaneous fat and insert in the skin as the retinacula cutis. Any condition that causes oedema of the breast can result in thickening of Cooper's ligaments. Benign causes include axillary node obstruction, SVC obstruction or congestive cardiac failure, and any inflammatory conditions. Tethering of the fibrous septations, or focal thickening, should raise the suspicion of cancer.

9.34 B False
Retraction or inversion of the nipple is usually the result of a benign idiopathic process, and may be bilateral or unilateral. Benign changes invariably occur over a long time period, and when nipple retraction occurs over a short period a malignant aetiology should be sought. Other benign causes that have been associated with nipple retraction include postsurgical scarring and fat necrosis.

9.34. C True
Skin thickening may be focal or diffuse. Infection or postsurgical scarring are the commonest cause of focal thickening. Generalized oedema may cause thickening of the dependent inferior skin of the breast. Diffuse skin (and trabecular) thickening are most commonly due to benign inflammation or to secondary oedema due to lymphatic obstruction from a tumour. Other causes of diffuse skin thickening include radiotherapy, psoriasis and congestive cardiac failure.

9.34. D True
Nipple calcifications are uncommon. Paget's disease of the nipple refers to the extension of intraductal carcinoma onto the surface of the nipple and can produce microcalcifications extending along the ductal network onto the nipple. Virtually all other causes of nipple calcification are benign and rare.

9.34. E True
The major lymphatic drainage of the breast is through the axillary nodes. Normal axillary nodes are less than 2 cm in size and have a typical hilar notch or central lucency. They may become extremely large when replaced by fat. Lymph nodes without central lucency that are larger than 1.5–2.0 cm should be considered abnormal. However, these are non-specific, and mammography cannot distinguish benign and malignant causes, unless microcalcification is present (very rare).

(Reference 26)

9.35. A False
Calcifications can only be demonstrated by breast ultrasound if they are several millimetres in size. This is one of the reasons why breast ultrasound is not a useful screening technique.

9.35. B True
A bright specular reflection is seen at the interface of the skin with the transducer. A second specular reflection is seen at the interface between the skin and the subcutaneous fat. Between these two, the skin is moderately echogenic, with a homogeneous echotexture. The skin is normally 0.5–2.0 mm in thickness. When thickened, there is an increase in separation between the two specular reflections which is visible by ultrasound, although the causes of skin thickening cannot be distinguished.

9.35. C True
Fat in the breast is relatively hypoechoic compared to the fibroglandular mammary parenchyma, which produces multiple reflective interfaces. Intra-parenchymal fat may mimic a hypoechoic mass, but generally, unlike masses that have a margin, collections of fat taper gradually and melt into the surrounding tissues. Subcutaneous fat appears as a relatively hypoechoic zone of variable thickness that surrounds the parenchymal cone.

9.35. D False
Cooper's ligaments are seen as echogenic reflecting surfaces in the normal breast and may cause difficulty with interpretation. They can cause scattering and refraction of sound, and may produce acoustic shadowing that is sometimes difficult to distinguish from the shadows produced by a scirrhous carcinoma. However, by angling the transducer so that the ultrasound beam is more perpendicular to the ligament, the tissues behind Cooper's ligaments can be assessed.

9.35. E True
There are many causes of a hypoechoic mass on breast ultrasound. Masses may be either well defined or poorly defined. Causes of well-defined lesions include fibroadenoma, infiltrating ductal carcinoma, lymphoma, metastasis, galactocele, oil cyst or abscess, as well as others. Ill-defined hypoechoic masses are more likely malignant. Benign causes of such an appearance include abscess, scars and fibroadenomas.

(Reference 26)

9.36. A False
Lipomas of the breast are hypoechoic relative to the breast parenchyma, having a similar echotexture to subcutaneous fat. They may be difficult to distinguish from normal lobules of fat within the breast, but this may be possible due to the specular reflection arising from the capsule of the lesion. Calcifications, when present, produce acoustic shadowing.

9.36. B True
Duct ectasia primarily affects the major ducts in the subareolar region, producing dilatation of one or more ducts that may be palpable or merely result in a mammographic abnormality. Mammography will show tubular,

serpiginous structures converging on the nipple, with or without the associated typical secretory calcifications. Ultrasound reveals anechoic tubular structures that may have visible branches, or more solid-appearing structures if the ducts are filled with debris.

9.36. C True
On ultrasound, galactoceles appear as well-defined masses containing low-level echoes, or as sonolucent lesions. Posterior acoustic enhancement is also a feature. A galactocele represents a milk-containing mass that develops during lactation or in the months following cessation of breast feeding. They may contain sufficient fat to appear radiolucent on mammography.

9.36. D True
Phyllodes tumour is an uncommon, usually benign tumour of the breast. Approximately 25% may recur locally if incompletely excised, and 10% can be expected to metastasize. The ultrasound appearances are similar to those of a fibroadenoma, although its relatively larger size may help as a distinguishing feature.

9.36. E False
Breast carcinomas are always hypoechoic. The classical appearance of a mass with an irregular anterior border and dense posterior acoustic shadowing is seen in approximately 35% of cases. About 25% of cancers appear well defined with a lobulated border, mimicking fibroadenoma. Although posterior shadowing is a frequent finding with carcinoma, it is not specific since it may occur with any lesion that contains a large amount of fibrosis.

(Reference 26)

9.37. A False
Fibroadenomas have no apparent increased risk of developing carcinoma, but since they contain epithelial elements carcinoma may develop in or immediately alongside a fibroadenoma. Such an occurrence may be suggested by the finding of spiculation, architectural distortion, or fine, irregular microcalcifications accompanying a mammographically obvious fibroadenoma.

9.37. B False
When calcifications begin within an involuting fibroadenoma, they may be very small and irregular, and indistinguishable from malignant microcalcifications. Generally, fibroadenomas calcify from the centre but peripheral calcification may occur. In the later stages, the calcifications are typically dense and large (popcorn), and when associated with a lobulated mass are diagnostic of an involuting fibroadenoma.

9.37. C True
Mammographically, a fibroadenoma is usually a sharply defined lesion. It may be surrounded by a thin, lucent 'halo', which is not specific. Fibroadenomas are often lobulated and may have relatively flattened contours, which helps to distinguish them from cysts. Occasionally, they may have a 'microlobulated' border which may also be seen with some carcinomas.

9.37. D True
Although statistically fibroadenomas are by far the commonest cause of a circumscribed hypoechoic lesion on breast ultrasound, a given lesion cannot be distinguished from other causes of a well-defined hypoechoic mass (see question 9.35E). Fibroadenomas are nearly always hypoechoic. Lateral wall reflective shadowing is also a feature in keeping with a mass that is round or oval.

9.37. E False
Posterior acoustic enhancement is usually seen with fibroadenomas, particularly if the lesion is cellular and round. However, a lesion containing abundant fibrosis may contain heterogeneous high-level echoes and be associated with posterior acoustic shadowing that is indistinguishable from the typical shadowing found with some cancers.

(Reference 26)

9.38. A True
Intraduct papilloma of the breast is one of the commonest causes of a serous or bloody nipple discharge. It represents a benign proliferation of ductal epithelium that projects into the lumen of the duct and is connected to it by a fibrovascular stalk. The majority are contained within a duct, extending longitudinally through its lumen, and are found only on microscopic section.

9.38. B True
Intraduct papillomas are usually located in the subareolar region within a major lactiferous duct. They may occasionally grow large enough to be imaged and are sometimes palpable. Solitary papillomas are always benign, and differ from multiple peripheral papillomas (papillomatosis), which are possibly associated with an increased incidence of malignancy.

9.38. C False
Most papillomas are not visible by mammography since they commonly do not expand the duct in which they arise. They may produce a well-defined, lobulated mass, which is almost invariably in the anterior portion of the breast. Solitary intraduct papillomas of the major ducts rarely calcify, and may occasionally produce a cluster of microcalcifications. Clustered microcalcifications are also an unusual feature of papillomatosis.

9.38. D False
Intraduct papillomas are indistinguishable from papillary carcinomas at ductography. Both may produce a filling defect in the contrast column, extravasation, or complete obstruction to retrograde flow.

9.38. E False
When large enough to be seen by ultrasound, intraduct papillomas produce a well-defined, hypoechoic mass which is usually lobulated. If they produce duct obstruction, they may appear as a frond-like mass within a cyst. However, they cannot be distinguished from an intracystic carcinoma, although the latter condition is very rare.

(Reference 26)

9.39. A True
In the immediate postoperative period, architectural distortion is a common finding and occasionally may persist for up to 6–12 months. Long-term fixed changes, however, are exceptional and the breast usually heals with no residual mammographic abnormality. When scarring does occur, it may be indistinguishable from a carcinoma, but it is rare for a carcinoma to develop in the region of a previous biopsy. If a mass does develop within 1 year of surgery, it is likely to be benign. However, progression of scarring beyond 1 year would be atypical for postsurgical change and biopsy is warranted.

9.39. B True
Radial scar (elastosis, indurative mastopathy or sclerosing duct hyperplasia) is one of the few benign lesions that can form a spiculated mass. Mammography demonstrates an area of architectural distortion with spiculations radiating from a central point. There may be associated microcalcifications. The centre of the lesion frequently contains lucent zones of fat, which helps to differentiate it from carcinoma, which typically has a central dense mass. Radial scar may also appear different on different projections.

9.39. C True
Benign calcifications may develop in approximately 33% of irradiated breasts, usually beginning 2–3 years after completion of therapy. These 'dystrophic' deposits are usually large and irregular in contour, with central lucencies. They always occur at the site of previous surgery.

9.39. D False
Breast cysts may be solitary but are usually multiple. Mammography typically demonstrates multiple well-defined round or oval densities, but they may have a lobulated margin. The well-defined margin may be obscured by overlying tissues or by pericystic fibrosis secondary to inflammation from leakage of cyst fluid into the surrounding stroma.

9.39. E True
At least 3% of women have asymmetric breast tissue that manifests as increased volume or density relative to the same projection of the contralateral breast. This usually represents normal asymmetric development.

(Reference 26)

9.40. A True
The fibroadenolipoma is a rare benign lesion of the breast that represents a proliferation of fibrous and adenomatous nodular elements in fat that is surrounded by a connective tissue capsule. Mammography characteristically demonstrates a lucent lesion containing lobulated densities surrounded by a thin capsule.

9.40. B False
Medullary carcinoma is a relatively uncommon form of breast cancer which is distinguished by its relatively large size at presentation. Mammography typically demonstrates a fairly well-defined, lobulated, smooth mass. Calcifications are not a particularly significant feature. Ultrasound usually demonstrates

a well-defined hypoechoic mass which, because of its uniform cellular composition, may have associated posterior acoustic enhancement.

9.40.   C False

Inflammatory carcinoma presents with a warm, erythematous breast that may have a classic peu d'orange appearance. Histologically, it represents diffuse early infiltration of the dermal lymphatics by an aggressive form of invasive carcinoma. Mammography demonstrates an increase in thickness of the skin which is non-specific, increased density of the breast and thickening of trabeculae. An associated tumour mass is only occasionally seen.

9.40.   D True

Phyllodes tumour appears on mammography as a well-defined mass with no particular distinguishing features. Spiculation does not occur and microcalcifications are not a feature. It generally presents as a rapidly enlarging mass that is several centimetres in diameter at the time of diagnosis.

9.40.   E True

Lymphoma of the breast accounts for approximately 0.1% of breast malignancy, and may be either primary (rare) or secondary. Mammographic features include multiple discrete nodules or a generalized increase in breast density. The nodules may be well or poorly defined but spiculation is not a feature. Enlarged axillary lymph nodes may also be seen.

(Reference 26)

10 Miscellaneous

10.1. Which of the following are true concerning fractures of the orbit?
A   Fractures of the orbital rim are most commonly isolated.
B   Fractures of the orbital walls most commonly affect the lateral wall.
C   Opacification of the ipsilateral ethmoid sinus is a feature of a blow-out fracture.
D   Soft-tissue herniation through the orbital floor in blow-out fractures indicates prolapse of the inferior rectus muscle.
E   Lateral wall fractures are associated with fractures of the orbital apex.

10.2. Concerning fractures of the zygoma and maxilla
A   in the tripod fracture, separation of the zygoma from the frontal bone typically occurs at the zygomaticofrontal suture.
B   a fracture of the inferior orbital rim is part of a tripod fracture.
C   reduced mandibular excursion is a recognized complication of zygomatic arch fractures.
D   isolated fractures of the maxilla usually involve the anterolateral wall.
E   LeFort fractures do not involve the pterygoid plates.

10.3. Which of the following are true of craniofacial fractures?
A   Pneumocephalus is most commonly associated with fractures of the ethmoid sinuses.
B   CSF rhinorrhoea is most commonly due to fracture of the frontal sinus.
C   Fractures of the mandible are typically unilateral.
D   A blow to the mandibular symphysis is a cause of bilateral condylar fractures.
E   Osteonecrosis of the mandibular condyle is a recognized complication of subcondylar fractures.

10.4. Concerning hyperflexion injuries to the cervical spine
A   flexion teardrop fractures are associated with posterior displacement of the involved vertebral body.
B   facet dislocation is a complication of flexion teardrop fracture.
C   hyperflexion sprain injuries are associated with reduced anterior disc space.
D   the hyperflexion compression fracture is invariably stable.
E   the clay-shoveller's fracture usually occurs between C6 and T1.

10.5. Which of the following are true of facet dislocations in the cervical spine?
A   Unilateral facet dislocation results in rotation of the spinous process toward the side of dislocation.
B   Radiculopathy is a recognized complication of unilateral facet dislocation.
C   Quadriplegia is an atypical finding with bilateral facet dislocation.
D   Interspinous widening is a feature of bilateral facet dislocation.
E   Rotation is not a feature of bilateral facet dislocation.

10.6. Concerning hyperflexion injuries to the cervical spine
A   Fractures of the atlas laminae are mechanically stable.
B   laminar fractures in the lower cervical spine are not associated with neurological signs.
C   the hyperextension dislocation (sprain) typically shows normal alignment on the lateral radiograph.
D   disc space abnormalities are seen in the majority of hyperextension sprains.
E   the extension teardrop fracture involves the anteroinferior corner of C2.

10.7. Which of the following are true of hangman's and pillar fractures?
A   Facet dislocation is a recognized feature of hangman's fracture.
B   Widening of the disc space is a feature of the hangman's fracture.
C   Vertebral artery injury is a recognized complication of the hangman's fracture.
D   The pillar fracture occurs most commonly at the C6–7 level.
E   Widening of the facet joint above the affected level is a feature of pillar fracture.

10.8. Concerning axial compression injuries to the cervical spine
A   unilateral fractures of the anterior and posterior arches of C2 are a recognized finding on CT.
B   atlantoaxial subluxation is not a recognized complication of the Jefferson fracture.
C   the Jefferson fracture is associated with anterior soft-tissue swelling.
D   the lower cervical burst fracture is mechanically stable.
E   disruption of the neurocentral joints is a feature of the lower cervical burst fracture.

10.9. Concerning fractures of the atlas and axis
A   the avulsion fracture from the anterior arch of the atlas runs in a vertical direction.
B   avulsion fractures of the anterior arch of the atlas are best assessed by CT.
C   the type I dens fracture is the commonest axis fracture.
D   non-union is a recognized complication of type II dens fracture.
E   type III dens fractures extend into the body of the axis.

10.10. Concerning injuries to the thoracolumbar spine
A   anterior vertebral body height less than posterior vertebral body height at T12 indicates a wedge fracture.
B   wedge fractures usually result in superior end-plate fractures.
C   the majority of wedge fractures are stable.
D   increased posterior vertebral body height is a feature of the 'seat-belt' injury.
E   interpedicular widening is a typical feature of the seat-belt injury.

10.11. Which of the following are true of injuries to the thoracolumbar spine?

A A horizontal fracture through the vertebral body is seen in flexion–rotation injuries.

B Flexion–rotation injury is a recognized cause of facet dislocation.

C Hyperextension injuries are characterized by an avulsion fracture from the anteroinferior corner of the vertebral body.

D Burst fractures are most common at L1.

E Burst fractures are associated with reduction of posterior vertebral body height.

10.12. Which of the following are true of fractures involving the pelvis?

A Avulsion of the anterior superior iliac spine is due to contraction of the sartorius muscle.

B Urinary incontinence is a complication of isolated sacral fractures.

C Ipsilateral fractures of both pubic rami may be caused by a direct lateral blow to the iliac crest.

D There is an association between unilateral pubic rami fractures and rupture of the ipsilateral iliolumbar ligament.

E There is an association between bilateral pubic rami fractures and sacroiliac joint diastasis.

10.13. Concerning fractures of the pelvis

A there is an association between bilateral pubic rami fractures and urethral tears.

B anterior compression on the anterior superior iliac spine (ASIS) is associated with disruption of the posterior sacroiliac ligament.

C the association of anterior compression on the ASIS with avulsion of the ischial spine increases the degree of pelvic instability.

D a Malgaigne fracture is associated with posterosuperior displacement of the injured hemipelvis.

E avulsion of the ipsilateral transverse process of L5 is a recognized finding with Malgaigne fracture.

10.14. Which of the following are true of soft-tissue injuries associated with pelvic fractures?

A Retroperitoneal haematoma is usually due to laceration of the common iliac artery.

B Arterial bleeding is best controlled by intra-arterial vasopressors.

C Retrograde urethrography does not demonstrate contrast leak in partial urethral tears.

D On retrograde urethrography, extravasation of contrast into the perineum indicates a complete tear of the bulbous urethra.

E Cystography cannot distinguish intraperitoneal and extraperitoneal bladder rupture.

10.15. Concerning fractures of the acetabulum
A a posterior column fracture involves the inferior pubic ramus.
B a pure transverse fracture is associated with posterior dislocation of the femoral head.
C an anterior column fracture results in disruption of the iliopectineal line.
D a combination of transverse acetabular and posterior rim fractures is a recognized occurrence.
E there is a recognized association between anterior and posterior column fractures.

10.16. Which of the following are true concerning injuries to the upper limb?
A Post-traumatic osteolysis following a shoulder injury affects only the distal clavicle.
B Dislocation of the sternoclavicular joint is usually in an anterior direction.
C A coracoclavicular distance greater than 2 cm is abnormal.
D The Hill–Sachs lesion of anterior shoulder dislocation is optimally seen on the AP radiograph with external rotation.
E There is an association between posterior dislocation of the shoulder and fracture of the lesser tuberosity.

10.17. Which of the following are true concerning imaging of the elbow joint?
A Visibility of the posterior fat pad indicates a haemarthrosis.
B On AP and lateral radiographs, a line drawn through the proximal radial shaft should pass through the capitellum.
C Supracondylar fractures in children may be associated with ventral displacement of the distal fracture fragment.
D Medial epicondyle fractures may be associated with widening of the humero-ulnar joint space.
E Avulsion of the medial epicondyle typically produces a joint effusion.

10.18. Concerning fractures about the elbow joint
A radial head fractures are invariably associated with displacement of the posterior fat pad.
B obliteration of the supinator fat stripe is a recognized finding with a radial head fracture.
C fractures of the capitellum are associated with proximal displacement of the fracture fragment.
D osteochondral fracture of the capitellum is a recognized cause of a displaced anterior fat pad.
E with dislocations of the elbow joint the radius and ulna are typically displaced medially.

10.19. Which of the following are true concerning injuries to the wrist and carpus?
A The Barton fracture involves the distal articular surface of the radius.
B Fractures of the triquetrum are best assessed on the lateral radiograph.
C Fractures of the capitate are typically isolated injuries.
D A fracture of the dorsal surface of the hamate is associated with dislocation of the fourth and fifth metacarpals.
E The risk of osteonecrosis of the scaphoid increases the more distal the fracture line.

10.20. Which of the following are true concerning carpal injuries?

A   A scapholunate distance of greater than 4 mm is a feature of scapholunate dissociation.

B   Foreshortening of the scaphoid on an AP radiograph is a feature of rotatory subluxation.

C   In dorsiflexion instability of the carpus, the scapholunate angle is equal to or greater than 70°.

D   In volarflexion instability, the scapholunate angle is less than 30°.

E   On wrist arthrography, communication between the radiocarpal and distal radio-ulnar joint is usually normal.

10.21. Which of the following are true concerning dislocations of the hip?

A   There is an association between posterior dislocation and fractured patella.

B   Persistent joint widening following reduction of a posterior dislocation is a recognized finding.

C   Osteonecrosis of the femoral head is a recognized complication of posterior dislocation.

D   Anterior dislocations invariably result in inferomedial displacement of the femoral head.

E   Anterior dislocations are not associated with osteonecrosis.

10.22. Concerning fractures and dislocations of the patella

A   patellar fractures are usually vertically orientated.

B   rupture of the infrapatellar tendon is associated with an avulsion fracture of the tibial tubercle.

C   dislocations of the patella are most commonly in a medial direction.

D   recurrent patellar subluxation is associated with patella alta (high patella).

E   patellar subluxation is best assessed on the axial view with 90° of knee flexion.

10.23. Concerning fractures about the knee joint

A   the majority of tibial plateau fractures involve the lateral side.

B   tibial plateau fractures are associated with knee instability.

C   lipohaemarthrosis is a feature of tibial plateau fracture.

D   superior dislocation of the proximal tibiofibular joint is usually an isolated injury.

E   there is an association between fracture of the fibular head and injury to the medial collateral ligament of the knee.

10.24. Which of the following are true of injuries to the ankle joint?

A   Supination–external rotation (inversion) injury is associated with an avulsion fracture of the anterior tibial tubercle.

B   Widening of the medial joint space is a finding in supination–external rotation injuries.

C   Supination–adduction injury is associated with a transverse lateral malleolar fracture.

D   Pronation–external rotation injury is associated with a transverse fracture of the medial malleolus.

E   There is an association between external rotation injuries and proximal fibular fractures.

10.25. Concerning trauma to the foot
A   avulsion fractures are the commonest talar fractures.
B   coronal fractures of the talar neck are associated with osteonecrosis of the distal fracture fragment.
C   the middle third of the lateral margin of the talar dome is a typical site of osteochondral fracture.
D   a Boehler's angle of 40° indicates a compression fracture of the calcaneus.
E   fracture of the cuboid is a recognized finding with tarsometatarsal fracture–dislocation.

10.26. Concerning ultrasound in the assessment of focal masses of the thyroid gland
A   a 'cystic' lesion with wall irregularity and internal echoes is usually malignant.
B   thyroid carcinomas are usually hypoechoic relative to normal thyroid tissue.
C   a hypoechoic mass containing hyperechoic foci is a recognized appearance of medullary carcinoma.
D   the presence of a hypoechoic 'halo' around a mass indicates a benign aetiology.
E   the demonstration of multiple masses indicates benign thyroid disease.

10.27. Concerning scintigraphy in the assessment of thyroid nodules
A   a colloid cyst is a cause of a focal defect (cold nodule).
B   thyroiditis is a recognized cause of a focal area of increased uptake (hot nodule).
C   a cold nodule has a 50% chance of being malignant.
D   suppression of activity with T3 is a typical feature of an autonomous non-toxic nodule.
E   progression of a non-toxic nodule to a toxic nodule can be identified.

10.28. Concerning the scintigraphic assessment of goitres
A   diffuse increased uptake of isotope is a feature of iodine deficiency.
B   diffuse reduction in isotope uptake is a recognized feature of lymphoma.
C   perchlorate discharge tests can only be performed with radioiodine.
D ^{99m}Tc-pertechnetate is the isotope of choice for demonstration of retrosternal extension.
E   Hashimoto's thyroiditis is a recognized cause of gallium-67 citrate uptake by the thyroid.

10.29. Concerning scintigraphy in the assessment of thyroid neoplasms
A   pure papillary carcinomas typically accumulate ^{131}I.
B   metastases from thyroid carcinomas only accumulate ^{131}I after ablation of the thyroid gland.
C   activity demonstrated in the pelvis on a ^{131}I whole-body scan indicates metastatic disease.
D   uptake of thallium-201 is a recognized feature of anaplastic carcinoma.
E   uptake of ^{131}I-*meta*-iodobenzylguanidine (MIBG) is a feature of medullary carcinoma.

10.30. Concerning ultrasound of the parathyroid glands
A normal sized glands can be demonstrated in the majority of patients.
B parathyroid adenomas are typically over 2 cm in greatest dimension at time of diagnosis.
C cystic change within a mass is a recognized finding with parathyroid adenoma.
D a superior parathyroid adenoma is typically located immediately cephalad to the upper pole of the thyroid gland.
E presence of a lobulated contour to the gland indicates carcinoma.

10.31. Concerning the parathyroid glands
A normal glands are visualized in 50% of cases on thallium/pertechnetate subtraction scintigraphy.
B ectopic parathyroid glands occur in less than 5% of cases.
C parathyroid adenomas result in increased uptake on the thallium-201 scan.
D a parathyroid adenoma is a cause of a 'cold' area on the pertechnetate scan.
E the majority of parathyroid carcinomas are functional.

10.32. Concerning the CT features of masses in the neck
A branchial cleft cysts typically cause posterolateral displacement of the sternocleidomastoid muscle
B rim enhancement is a feature of thyroglossal duct cyst on postcontrast scans
C schwannoma of the vagus nerve is a cause of a mass between the internal carotid artery and internal jugular vein
D on postcontrast CT, demonstration of a low-density mass with irregular peripheral enhancement is diagnostic of abscess.
E enhancement of the wall of the jugular vein is a CT feature of jugular vein thrombosis.

10.33. Concerning CT and MRI in the assessment of the paranasal sinuses
A bowing of a bony margin by a mass is indicative of benign disease.
B sclerosis of a sinus wall is a feature of chronic sinusitis.
C with an opaque sinus, tumour and secretions can be differentiated on postcontrast CT.
D enhancement of tissue within a sinus indicates inflamed mucosa rather than tumour.
E mucosal secretions and tumour are most clearly differentiated on T1-WSE MRI.

10.34. Concerning CT in the assessment of tumour spread from the maxillary sinuses
A spread to the infratemporal fossa is not a feature.
B the pterygopalatine fossa may be involved by perineural spread along the infraorbital nerve.
C effacement of fat below the inferior rectus muscle is a feature of intraorbital extension.
D extension into the ethmoid sinus occurs via the maxilloethmoid plate.
E lymphatic spread is not a feature.

10.35. Which of the following are true of neoplasms involving the paranasal sinuses and nasal cavity?

A   Squamous carcinoma is the commonest primary tumour.
B   Extension through the cribriform plate is a feature of aesthesioneuroblastoma.
C   Perineural spread is a feature of adenoid cystic carcinoma.
D   Bone destruction is a recognized feature of inverting papilloma.
E   Metastases to the middle cranial fossa are typically blood-borne.

10.36. Concerning CT in the assessment of benign conditions of the paranasal sinuses

A   bone destruction is a feature of mucormycosis.
B   aspergillosis is a recognized cause of an enhancing antral mass on postcontrast CT.
C   mucoceles are most commonly demonstrated in the maxillary antrum.
D   a mucocele is invariably associated with intact bony margins.
E   retention cysts characteristically have a convex upper margin on coronal scans.

10.37. Concerning CT in the assessment of the nasopharynx and parapharyngeal space

A   fluid in the mastoid air cells is a finding in nasopharyngeal carcinoma.
B   deep extension of nasopharyngeal carcinoma deforms the parapharyngeal space from the lateral aspect.
C   a mass confined to the prestyloid space is usually of parotid origin.
D   paraganglionomas typically arise in the poststyloid space.
E   a mass arising in the infratemporal fossa causes posteromedial displacement of the parapharyngeal fat.

10.38. Concerning CT and MRI in the assessment of the laryngeal anatomy

A   the false cords normally have soft-tissue density on CT.
B   the true cords lie in the plane of the vocal process of the arytenoids.
C   irregularity of calcification of the thyroid cartilage is a normal finding.
D   the pre-epiglottic space appears hyperintense to muscle on T1-WSE MRI.
E   the epiglottis is isointense to muscle on MRI.

10.39. Concerning CT in the assessment of laryngeal carcinoma

A   glottic tumours are not typically associated with cervical adenopathy at diagnosis.
B   supraglottic spread of glottic tumours may result in increased density of the false cord.
C   contralateral spread of glottic tumour may result in thickening of the anterior commissure.
D   supraglottic tumours are the commonest laryngeal carcinomas.
E   extension to the base of the tongue is a feature of carcinoma of the epiglottis.

10.40. Concerning CT in the assessment of cervical lymph nodes

A low density within a node on precontrast scans is always pathological.

B a jugulodigastric node measuring 10 mm in diameter should be considered abnormal.

C metastatic nodes from a squamous cell primary characteristically have a low-attenuation centre on postcontrast scans.

D tumours of the soft palate usually metastasize to the retropharyngeal nodes.

E tumours of the hypopharynx typically metastasize to the external jugular nodes.

Miscellaneous: Answers

10.1. A False
Orbital rim fractures may be isolated but are more commonly part of the tripod fracture complex. Isolated rim fractures result from a direct blow and are commonest inferolaterally. They are best assessed on the Waters' (OM 40°) view. The detached fracture fragment may rotate and result in the 'disappearing rim' sign. The fracture may also extend into the anterior orbital floor.

10.1. B False
Orbital wall fractures most commonly involve the posterior orbital floor medial to the infraorbital groove, where the floor is thinnest, or the lamina papyracea of the ethmoid bone. A 'blow-out' fracture, by definition, involves the orbital floor or medial wall with sparing of the orbital rim. However, orbital rim fractures may be associated.

10.1. C True
Blow-out fractures result from a direct blow to the globe. They are associated with fractures of the medial orbital wall in 20–40% of cases and result in opacification of the ipsilateral ethmoid sinus. Orbital emphysema is a frequent association. The fracture lines may only be visible by conventional or computed tomography. Entrapment of the medial rectus muscle is a very rare complication.

10.1. D False
The radiographic signs of blow-out fractures involving the floor include depression of bony fragments, soft-tissue prolapse through the orbital floor, and opacification of the maxillary antrum. Prolapsed soft tissue may consist of haematoma, orbital fat or herniated muscle. These can be differentiated with CT or MRI.

10.1. E True
Lateral orbital wall fractures are usually associated with ipsilateral zygomatic fractures and may extend to the orbital apex. The apex is best assessed by CT. Fractures of the orbital apex may cause blindness by damaging the optic nerve, or may produce the superior orbital fissure syndrome, consisting of ophthalmoplegia, ptosis, proptosis, pupillary dilatation, pain and optic neuralgia.

(Reference 34)

10.2. A True
The tripod (zygomaticomaxillary) fracture results in separation of the malar eminence of the zygoma from its frontal, temporal and maxillary attachments. Separation of the zygoma from the frontal bone may, less commonly, result from a fracture of the frontal process of the zygoma. Separation from the temporal bone usually occurs via a zygomatic arch fracture, or via separation of the zygomaticotemporal suture.

10.2. B True
Separation of the fracture fragment from the maxilla typically results from a fracture of the inferior orbital rim immediately adjacent to the zygomaticomaxillary suture. This is commonly associated with extension of the fracture into the anterior orbital floor and lateral maxillary wall.

10.2. C True
Isolated fractures of the zygomatic arch most commonly result in three fracture lines crossing the arch, with depression of the central two fragments and outer displacement of the zygomatic and temporal ends of the arch. The extent of depression is best assessed on an underpenetrated basal view. Severe depression of fracture fragments may result in impingement upon the coronoid process of the mandible and reduced mandibular excursion.

10.2. D True
Fractures involving the anterolateral wall may produce an abnormal linear density on the frontal view, due to fragments of displaced bone. There may be associated opacification of the sinus, but this may occur due to tearing of the mucosa, in the absence of bony injury. The walls of the maxilla are well demonstrated on the basal view.

10.2. E False
LeFort fractures are relatively uncommon and rarely follow the classic planes as described by LeFort in 1901. All three LeFort fractures result in separation of a large bony fragment, containing portions of the maxilla, from the remainder of the craniofacial skeleton. They involve the pterygoid plates, by definition, and are typically unstable.

(Reference 34)

10.3. A False
Pneumocephalus complicates approximately 8% of fractures of the paranasal sinuses, but is most commonly seen following depressed frontal sinus fractures. The fracture must involve the posterior wall of the sinus, allowing communication between the sinus and intracranial contents. It may be seen on the lateral view but CT is better in its detection. Opacification of the sinus is frequently seen.

10.3. B False
CSF rhinorrhoea may occur following a fracture involving the frontal sinus (posterior wall), but is commoner with an ethmoidal injury. The site of dural tear may require intrathecal contrast or radionuclide studies. CSF rhinorrhoea is noted in approximately 50% of patients with pneumocephalus, and meningitis or brain abscess develops in up to 25% of affected patients.

10.3. C False
The mandible behaves as a complete bony ring and, therefore fractures are bilateral in up to 50% of cases. The symphysis is the site of 10–20% of mandibular fractures. These fractures have a tendency to override, resulting in entrapment of adjacent soft tissues. The body and angle of the mandible are fractured in 50–70% of cases and the ramus in 3–9% of patients.

10.3.   D True
A blow to the mandibular symphysis may also result in bilateral body or angle fractures. The orientation of mandibular angle fractures is significant since fractures angulated posterosuperior to anteroinferior are impacted and stabilized by the masseter and pterygoid muscles, whereas fractures in the opposite direction are separated by the action of these muscles.

10.3.   E True
Condylar fractures result from a blow to the side of the mandible and are associated with a contralateral angle or body fracture. They account for 15–20% of fractures in adults, being commoner in children. Medial angulation of the condylar head is typical due to the action of the lateral pterygoid muscle. Significant complications include mandibular growth disturbance, osteonecrosis of the condyle and bony ankylosis of the temporomandibular joint.

(Reference 34)

10.4.   A True
The flexion teardrop fracture is the most severe form of flexion injury, with complete quadriplegia or the 'acute anterior cervical cord syndrome' being present in almost 90% of cases. The injury is usually at the C5–6 level. Radiographically, a triangular fragment of bone usually arising from the anteroinferior angle of the vertebral body is seen in 75% of cases. The fragment may arise from the upper margin.

10.4.   B True
On the lateral radiograph, there is flexion of the cervical spine above the fracture, interspinous widening, facet joint widening, or even dislocation. Flexion teardrop fractures do not cause extensive injury to the centre of the vertebral body, which helps distinguish them from compression fractures. There is disruption of all posterior and anterior ligaments as well as the intervertebral disc.

10.4.   C True
The features of a hyperflexion sprain include focal kyphosis, mild anterolisthesis, widened apophyseal and interspinous spaces, widening of the posterior disc space and narrowing of the anterior disc space. Facet subluxation and compression fractures often coexist. The injury is usually located at one level and late instability may occur in up to 50% of patients.

10.4.   D False
The hyperflexion compression feature is caused by compression between two vertebral end-plates. It usually results in anterior wedging of the superior end-plate and localized angulation. These injuries are usually stable, but if there is greater than 25% compression they may be unstable due to associated disruption of the posterior ligaments.

10.4.   E True
The clay-shoveller's fracture is most common at C7 and results when the head and cervical spine are in forced flexion, with rotation against the opposing interspinous ligaments. The oblique fracture is perpendicular to these

ligaments. The AP view may show a 'double spinous process' sign. The injury is stable, but may extend into the lamina.

(Reference 34)

10.5. A True
Unilateral facet dislocation results from a combination of flexion and rotation. It is usually observed at the C4–6 levels. Lateral radiographs show anterior subluxation, which is classically less than one-half the vertebral body width. The dislocated facet is anteriorly positioned, as are the articular masses above, which appear in an oblique position. There is interspinous widening, and on the AP view the spinous processes are rotated to the side of the dislocation.

10.5. B True
On oblique views, the dislocated facet may be seen lying in the intervertebral foramen if completely dislocated (causing radiculopathy) or perched on the lower articular mass, if incompletely dislocated. Fractures about the facets, particularly the inferior facet, are seen in 33% of cases and if large may result in instability.

10.5. C False
Bilateral facet dislocation is associated with disruption of the posterior ligamentous structures, posterior longitudinal ligament, intervertebral disc, and often the anterior longitudinal ligament. The injury is unstable and the incidence of quadriplegia is reported at 72%.

10.5. D True
In cases of bilateral facet joint dislocation, lateral radiographs show anterolisthesis of 50% or more of the vertebral body width, disc space narrowing, interspinous widening and facet dislocation. The inferior facets at the dislocation are located anterior to the facets below if the dislocation is complete, or are 'perched' if incomplete.

10.5. E True
Unlike unilateral facet dislocations, rotation is not a radiographic component of bilateral facet dislocations. Facet dislocations may be difficult to evaluate with CT because of slice orientation. CT may demonstrate the 'naked facet sign', in which there is lack of normal facet articulation.

(Reference 34)

10.6. A True
The majority of hyperextension fractures to C1 are due to compression between the occiput and the large posterior elements of the axis. They result in bilateral laminar fractures which may be seen on lateral or oblique radiographs and should not be confused with congenital clefts. There is no anterior soft-tissue swelling. Atlas fractures are stable and there is no neurological deficit.

10.6. B False
Extension forces can also result in laminar fractures in the lower cervical spine, most commonly at C5–7. These fractures usually occur in older patients with spondylosis. The fractures are best identified on lateral radiographs or CT.

They are mechanically stable but may produce neurological symptoms due to cord involvement by displaced bony fragments.

10.6.   C   True
The hyperextension sprain results from a direct force to the face, causing rupture of the anterior longitudinal ligament and disc, or avulsion of the inferior end-plate at the attachment of intact Sharpey's fibres. Continued force results in posterior subluxation of the vertebral body with impingement upon the spinal cord. Alignment is normal because the subluxation reduces spontaneously. However, neurological symptoms are usually present.

10.6.   D   False
The radiographic findings with hyperextension sprain include anterior soft-tissue swelling (which may be the only abnormality in one-third of patients), an avulsion fracture (seen in two-thirds of cases) which is longer horizontally than vertically, allowing differentiation from a flexion teardrop fracture, and widening of the disc space or a vacuum disc. Disc abnormalities are seen in only 15% of cases.

10.6.   E   True
The extension teardrop fracture is an avulsion from the attachment of the anterior longitudinal ligament at the anteroinferior corner of C2. The fragment's vertical height is equal to or greater than its horizontal dimension. The fracture usually occurs in older patients with osteoporotic spines and spondylosis. Soft-tissue swelling is usually present and the fracture is unstable in extension.

(Reference 34)

10.7.   A   True
The hangman's fracture refers to bilateral fractures of the pars interarticularis of C2 and represents a traumatic spondylolisthesis. These fractures have been classified into three subtypes: type I is a minimally displaced fracture; type II is associated with more displacement and involvement of the subjacent disc; type III includes bilateral facet dislocation with anterior displacement and angulation.

10.7.   B   True
The oblique pars fracture is best demonstrated on the lateral view. There may be anterolisthesis as well as posterior displacement of the spinolaminar line at this level. Widening of the disc space occurs with types II and III fractures and there may be an avulsion fracture from the anteroinferior aspect of the body of C2, at the attachment of the anterior longitudinal ligament.

10.7.   C   True
CT may demonstrate the extension of the hangman's fracture into the vertebral artery foramen. Other complications include associated fractures of C1 and upper thoracic injuries, which may be present in up to 10% of cases. Neurological signs are not prominent due to the 'autodecompression' produced by the bilateral posterior element fractures and the spacious spinal canal at this level.

10.7.   D   True
The pillar fracture results from combined hyperextension and rotation, causing compression of the articular mass between adjacent vertebrae on the side of the rotation. The fracture may extend to involve the lamina, pedicles or

transverse process. AP radiographs may demonstrate disruption of the lateral cortical margins, while lateral views may show focal loss of overlap of the posterior articular mass cortices.

10.7. E True
Pillar fractures are optimally shown on pillar or oblique views. These may demonstrate asymmetric foraminal narrowing and widening of the facet joint above the fracture. CT best identifies involvement of the foramen transversarium. The fracture is stable and spinal cord damage is unusual, although acute radiculopathy is not uncommon.

(Reference 34)

10.8. A True
In the Jefferson fracture, the axial compressive force drives the occipital condyles towards the weaker lateral masses of the atlas, which fracture and are displaced laterally. The fracture is well seen on CT, and although bilateral fractures of the anterior and posterior arches are usual, the fractures may be unilateral or asymmetrical.

10.8. B False
The 'open mouth' view is the best radiographic projection to show the Jefferson fracture and demonstrates lateral displacement of the atlas articular masses. Lateral displacement of the articular masses by more than 6.9 mm or an increase in the predental space of greater than 6 mm is associated with rupture of the transverse atlantal ligament, and atlantoaxial subluxation.

10.8. C True
Anterior soft-tissue swelling may be the only abnormality seen on the lateral radiograph in patients with a Jefferson fracture. This finding differentiates it from bilateral laminar fractures of the atlas. The retropharyngeal space at C2 in a patient without a nasogastric or endotracheal tube should not be greater than 7 mm and is usually less than 5 mm.

10.8. D True
The lower cervical 'burst' fracture is caused by a vertical force transmitted to the intervertebral disc, which drives the nucleus pulposus through the inferior vertebral end-plate. The vertebral body explodes from within, producing fragments with varying displacement. The fracture is neurologically unstable due to the posterior displacement of fracture fragments into the spinal canal and impingement upon the cord.

10.8. E True
In the burst fracture, the AP view demonstrates a vertical fracture through the vertebral body and disruption of the neurocentral joints. The lateral view shows compression of the vertebral body and posterior displacement of a bony fragment, frequently from the posterosuperior margin of the vertebral body. Anterior soft-tissue swelling is seen. Associated laminar fractures are best identified with CT.

(Reference 34)

10.9.   A False
Avulsion fractures from the anterior arch of C1 are rare extension injuries from the site of attachment of the anterior longitudinal ligament. They are commonly only identified on the lateral view and appear as a horizontal cleft in the mid- to inferior aspect of the anterior arch. They are associated with anterior soft-tissue swelling.

10.9.   B False
Avulsion fractures run in a horizontal plane. Axial CT may therefore be falsely negative. The fracture may also be confused with a normal variant, the ununited accessory ossification centre. However, the latter has a well-defined sclerotic rim and anterior soft-tissue swelling is not seen.

10.9.   C False
Odontoid fractures have been divided into three types. Type I injuries are rare and their existence has been questioned. The injury consists of an avulsion from the attachment of the alar ligament and indicates occipitoatlantal dissociation.

10.9.   D True
Type II dens fractures account for almost 60% of odontoid injuries. They are fractures of the odontoid base without extension into the body of the axis. The lateral view may only show anterior soft-tissue swelling but cortical disruption and displacement may also be seen. Non-union is reported in up to 72% of cases, and is more likely with posterior displacement. It results in atlantoaxial instability.

10.9.   E True
The type III dens fracture involves the body of C2, extending in a posteroinferior direction from the junction of the dens with the body. The lateral view shows disruption of the C2 'ring', which may be the only finding. However, soft-tissue swelling is usually present. The os odontoideum is distinguished from an acute dens fracture by the presence of a well-defined sclerotic rim.

(Reference 34)

10.10.   A False
Wedge fractures are the result of combined flexion and compression and are commonest at L1, L2 and T12 in adults. The impaction can be either anterior or lateral. Anterior compression results in loss of vertebral body height anteriorly. However, anterior vertebral body height may normally be 2 mm less than posterior body height and therefore a greater loss of height is necessary for a fracture to be diagnosed.

10.10.   B True
The radiographic features of wedge fractures include paraspinal soft-tissue swelling, increased kyphosis or scoliosis and loss of vertebral body height either anteriorly or laterally. Also, end-plate buckling, eburnation adjacent to the end-plate, impaction of the anterosuperior vertebral corner and disc space narrowing may be seen. The superior end-plate is most commonly involved, but both end-plates or only the inferior end-plate may be fractured.

10.10. C True
Flexion–compression fractures are usually stable since the middle and posterior columns (consisting of all structures posterior to the middle of the vertebral body) are intact. This manifests as normal posterior vertebral body height and lack of interpedicular widening. The injury may become unstable if the force is great enough to cause disruption of the posterior column by distraction.

10.10. D True
The seat-belt injury is a type of flexion–distraction injury and is commonest at L1–3. The injury may be purely osseous, ligamentous or combined. The Chance fracture refers to an osseous injury at a single level. Radiographs demonstrate horizontal fractures of the posterior arch and vertebral body, compression fracture, increased posterior vertebral body height and a widened posterior disc space.

10.10. E False
AP radiographs demonstrate horizontal clefts in the spinous process and pedicles, and lack of overlap between the vertebral body and the posterior elements (the empty vertebra sign) due to elevation of the posterior elements. Seat-belt injuries are associated with neurological involvement in about 20% of cases, but abdominal injuries are common, particularly bowel and visceral lacerations.

(Reference 34)

10.11. A True
Flexion–rotation injury is unusual but is one of the most unstable spinal fractures, with neurological deficit present in up to 66% of cases. There is disruption of the middle and posterior columns by tension and the anterior column by compression and rotation. The thoracolumbar junction is typically involved. The injury may extend through the disc or cause a horizontal fracture through the upper part of the vertebral body.

10.11. B True
The characteristic feature of a flexion–rotation injury is rotation between adjacent vertebrae. Frequent associations include fractures through the facets, which may be dislocated, and fractures of the lamina, transverse processes and ribs.

10.11. C False
The hyperextension injury is very rare in the thoracolumbar spine but may occur if a patient falls and hyperextends over an object. These fractures may occur in patients with ankylosing spondylitis with mild trauma. The lateral radiograph shows anterior disc space widening, posterior arch fractures, retrolisthesis, and a triangular avulsion fracture from the anterosuperior vertebral corner.

10.11. D True
Burst fractures occur when a vertical compression force causes the intervertebral disc to herniate through the vertebral end-plate, resulting in the vertebral body bursting from within. There is failure of the anterior and middle

columns. The fractures may occur from T4 to L5 but are commonest at L1. Nearly 50% of cases are associated with trauma at other levels.

10.11. E True
Radiographic findings in patients with burst fractures include widened interpedicular distance, a vertical vertebral body fracture, anterior wedging, vertical fracture at the spinolaminar junction, loss of posterior vertebral body height, and retropulsed fragments. Rotational components and compression of more than 50% are indicators of instability.

(Reference 34)

10.12. A True
Pelvic avulsion fractures typically occur in young individuals engaged in strenuous activity. Avulsion of the anterior inferior iliac spine may result from forceful contraction of rectus femoris, and contraction of the hamstrings may avulse a fragment from the ischial tuberosity.

10.12. B True
Isolated sacral fractures result from a direct posterior blow and may be missed on radiographs, requiring scintigraphy or CT for their detection. Injury to exiting nerve roots may result in incontinence, impotence, perineal paraesthesia or CSF leak. Rectal laceration is also recognized.

10.12. C True
Ipsilateral pubic rami fractures due to a blow against the iliac blade often run in a horizontal or coronal plane. This mechanism of injury may also result in crush fractures of the anterior articular surfaces of the sacrum and ilium. These may only be detectable by scintigraphy or CT. Posterior stability is maintained by intact anterior and posterior sacroiliac ligaments.

10.12. D True
A greater degree of lateral compression to the ilium may result in a variety of fracture combinations. Inward rotation of the hemipelvis may result in ipsilateral pubic rami fractures with disruption of the posterior sacroiliac and iliolumbar ligaments, making the injury unstable. Lateral compression combined with upward rotation results in contralateral pubic rami fractures and ipsilateral avulsion fractures from the posterior aspect of the sacroiliac joint.

10.12. E True
With the greatest degree of lateral compression, bilateral pubic rami fractures may occur. The posterior component of the injury consists of an iliac or sacral fracture or complete diastasis of the sacroiliac joint. It is an unstable injury association with vascular disruption.

(Reference 34)

10.13. A True
The so-called straddle fracture occurs due to a direct blow to the pubis. The separated pubic fragment may be displaced superiorly by contraction of the rectus abdominis muscle.

10.13. B False
AP compression forces on one or both anterior superior iliac spines results initially in diastasis of the symphysis pubis or a sagittally orientated fracture through the pubic rami near the symphysis. With greater force, opening of the symphysis to a width greater than 2.5 cm is associated with disruption of the anterior sacroiliac ligaments, or occasionally an avulsion fracture of the anterior margin of the ilium.

10.13. C True
AP compression associated with fracture of the ischial spine indicates avulsion by the sacrospinous ligament and a greater degree of pelvic instability, since there would be dissociation of the anterior and posterior aspects of the pelvis. In these cases, the posterior sacroiliac ligament is intact.

10.13. D True
The Malgaigne fracture is a vertical shear fracture characterized anteriorly by bilateral pubic rami fractures or symphysis diastasis with posterior disruption, either through the sacroiliac joint or the adjacent sacrum or ilium. Posterior displacement of the separated hemipelvis is best demonstrated on the inlet view, while the outlet view identifies the extent of superior displacement. A degree of medial displacement is also present.

10.13. E True
The ipsilateral transverse process fracture associated with vertical shear injury is due to avulsion by the iliolumbar ligament as the displaced hemipelvis moves cephalad. These injuries are always very unstable, with a high incidence of vascular disruption. The posterior displacement indicates disruption of the posterior sacroiliac ligament.

(Reference 34)

10.14. A False
Vascular injury and retroperitoneal haematoma are commoner when there is disruption of the posterior pelvis. Vascular injury may result from direct vascular trauma, laceration by a bone fragment or traction on the vessels. Also, the iliac bones are very vascular and their fracture may contribute significantly to blood loss. Retroperitoneal haemorrhage in pelvic fractures usually originates from venous plexuses, and blood loss from this route as well as from fractured bone is best treated by stabilization of the fracture.

10.14. B False
Vasopressors are of no value in the control of arterial bleeding outside the gastrointestinal tract. Arterial bleeding from pelvic fracture is best controlled by embolization.

10.14. C True
Urethral tears are particularly common with trauma to the pubic arch, or with severe, displaced pelvic fractures. Tears have been classified into three types. Type I tears are partial and urethrography shows irregularity of the urethra but no frank contrast extravasation.

10.14. D True
Type II tears are complete tears of the membranous urethra and will result in contrast extravasation above the urogenital diaphragm, seen as contrast above the pubic arch. Type III tears are complete tears of the bulbous urethra below the urogenital diaphragm. Extravasated contrast is seen in the perineum, below the pubic arch. Type III tears appear to be commoner than type II tears.

10.14. E False
Bladder rupture most commonly results from blunt trauma to a full bladder. Extraperitoneal bladder rupture is most commonly due to pelvic fracture, and cystography typically demonstrates streaky or stellate collections of contrast medium in the perivesical fat. In the case of intraperitoneal rupture, cystography demonstrates extravasated contrast medium outlining intraperitoneal organs.

(Reference 34)

10.15. A True
With a posterior column fracture, the fracture line runs across the posteroinferior aspect of the acetabular fossa and across the inferior pubic ramus, separating the posterior column from the rest of the innominate bone. Radiographs show breaks in the ilioischial line (representing the medial border of the posterior column) and posterior acetabular rim. The iliopectineal line, teardrop and anterior rim are intact. These fractures are associated with central hip dislocation.

10.15. B False
Pure transverse fractures are associated with central dislocations of the femoral head. The fracture line runs in an axial plane through the anterior and posterior columns, resulting in breaks through both the ilioischial and iliopectineal lines. Anterior and posterior rim fractures may also be seen, and the teardrop may be disrupted.

10.15. C True
The iliopectineal line represents the medial aspect of the anterior column. Anterior column fractures are associated with anterior femoral head dislocations (the rarest type). The majority of the anterior column is separated from the rest of the pelvis. Breaks through the iliopectineal line, anterior rim, obturator ring, and the teardrop are seen. The separated fragment is usually displaced medially.

10.15. D True
A combination of transverse and posterior wall fractures is one of the commonest acetabular fracture types. It is associated with posterior hip dislocation in 80% of cases, whereas in 20% the hip dislocates centrally.

10.15. E True
Combined anterior and posterior column fractures are also relatively common. At the point where the posterior column fracture reaches the acetabulum, a second fracture runs anteriorly and superiorly into the iliac crest, or more transversely into the anterior edge of the ilium. Other combinations include a

T-shaped fracture, anterior and hemitransverse fractures, and posterior column and posterior wall fractures.

(Reference 34)

10.16. A False
The clavicle may undergo focal osteolysis following an acute shoulder injury, or as a consequence of repetitive stress. The disorder may occur weeks to years after the trauma. Radiographs show osteopenia of the clavicle, subarticular cortical loss with resorption of up to 3 cm of distal clavicle. Similar, but usually less severe changes rarely occur in the acromion. Bone scintigraphy shows increased activity.

10.16. B True
Anterior sternoclavicular joint dislocation usually results from a blow to the shoulder. The lesion is commonly seen in patients under 25 years, and may then be a separation of the medial clavicular epiphysis rather than true joint dislocation. Posterior dislocation is more serious and may be associated with damage to major blood vessels, the trachea or oesophagus. These injuries are best assessed by CT.

10.16. C True
Acromioclavicular separation has been classified into three grades. Grade I is a ligamentous sprain and radiographs are normal. A grade II injury is characterized by rupture of the acromioclavicular ligament and joint capsule, resulting in joint space widening and elevation of the clavicle. A grade III injury indicates rupture of the coracoclavicular ligament with increased coracoclavicular distance (greater than 1.3 cm being considered abnormal).

10.16. D False
The Hill–Sachs lesion represents a fracture of the posterolateral humeral head due to impaction against the anterior glenoid rim at the time of anterior dislocation. It is best demonstrated on AP views with internal rotation, or on the Stryker view. An associated fracture of the glenoid rim is termed the Bankart lesion and is best seen on the West Point or apical oblique views.

10.16. E True
Injuries associated with posterior dislocation include fractures of the humeral head and posterior glenoid rim, with stretching of the posterior capsule. The radiographic signs on the AP view include an abnormally high or low position of the humeral head, widening of the space between the humeral head and anterior glenoid rim, and an impaction fracture of the medial humeral head (the trough sign).

(Reference 34)

10.17. A False
The anterior fat pad is normally seen on the lateral view of the elbow, but the posterior fat pad lies within the olecranon fossa, and is either not seen or appears as a thin (1 mm) lucency. Haemarthrosis from an intracapsular fracture causes anterior bulging of the anterior fat pad and posterior displacement of the posterior fat pad, rendering it visible on the lateral

radiograph. However, this may also occur in cases of synovial hypertrophy, pyarthrosis or simple effusion.

10.17.   B True
The radiocapitellar line may be useful in evaluating the Monteggia fracture–dislocation, in which case the line will pass through the distal humeral shaft rather than the capitellum, indicating ventral dislocation of the radial head in association with the fracture of the proximal ulnar shaft.

10.17.   C True
The anterior humeral line describes a line drawn along the anterior humeral shaft on the lateral view of the elbow. This line normally intersects the capitellum in the anterior portion of its middle third or at the junction of the anterior and middle thirds. In cases of supracondylar fracture, ventral displacement of the distal fracture fragment results in this line intersecting a more posterior portion of the capitellum, whereas the reverse is true if the distal fragment is displaced dorsally.

10.17.   D True
The avulsed medial epicondyle may become entrapped between the humerus and the ulna. This is manifest as focal widening of the humero-ulnar joint space on the AP view, and the displaced epicondyle can be seen on both AP and lateral views. There will also be localized soft-tissue swelling.

10.17.   E False
The medial epicondyle is largely extracapsular, and simple avulsion is not associated with displacement of fat pads. If a joint effusion or a small metaphyseal flake fracture is present, extension of the fracture into the trochlear ossification centre should be suspected. This is an unstable fracture.

(Reference 34)

10.18.   A False
Radial head fractures are the commonest elbow injuries in adults. Displaced fat pads are typical but not always seen and the fracture may be difficult to see on routine AP and lateral views since it may be minimally displaced. In these cases, a radial head–capitellum view may be of value.

10.18.   B True
The supinator fat stripe lies ventral to the radial head and neck and may be displaced or obliterated in the presence of a fracture around the elbow joint. This may be the only abnormality in patients with elbow trauma and requires further radiographic evaluation of the joint.

10.18.   C True
Displaced capitellar fractures are characterized on the lateral view by proximal displacement of the fracture fragment, which comes to lie above the radial head and coronoid process. It may not be identified on the AP view. The fragment may be rotated 90° so that the articular surface faces ventrally.

10.18. D True
Some capitellar injuries result in production of thin cartilaginous or osteocartilaginous fragments that may not be identified by routine radiography, being manifest only as displacement of fat pads. These osteocartilaginous fractures may be assessed by CT, CT arthrography or MRI.

10.18. E False
Elbow dislocations are associated with posterior or posterolateral displacement of the proximal radius and ulna in approximately 90% of cases. Associated injuries are frequent and include fractures of the medial humeral condyle or epicondyle, radial head and coronoid process. The presence of joint incongruity after reduction may indicate trapped intra-articular fragments, typically the medial epicondyle in children and the tip of the coronoid process in adults.

(Reference 34)

10.19. A True
The Barton fracture is an intra-articular fracture involving the dorsal or volar rim of the distal radius. The whole of the carpus is displaced with the fracture fragment, either in a dorsal or volar direction. Other distal radial fractures include Colles' fracture with dorsal angulation, and the Smith fracture with volar angulation.

10.19. B True
The triquetrum is the second commonest carpal bone to be fractured. The fracture may only be visible on the lateral view, which shows soft-tissue swelling over the dorsum of the wrist and a bony fragment over the proximal carpal row. The fracture is due to avulsion by the ulnotriquetral ligament.

10.19. C False
Fractures of the capitate usually occur in association with a scaphoid fracture or perilunate dislocation. The typical fracture is transverse and the proximal fracture fragment may be rotated 90° or 180°. The combination of a scaphoid fracture and 180° rotation of the proximal capitate fragment is termed the scaphocapitate syndrome and is associated with osteonecrosis of the proximal capitate fragment.

10.19. D True
Fractures of the hamate may involve the body (either isolated or as part of a perilunate fracture dislocation), the dorsal surface, or the hook. The latter injury is the result of a direct blow to the hypothenar eminence. On the AP view, the hook may appear sclerotic, have an ill-defined cortical outline, or be absent. CT is of value in assessing this injury.

10.19. E False
The scaphoid is the most commonly fractured carpal bone. Non-displaced fractures may only become radiographically apparent 7–10 days after the injury. Since the principal arterial blood supply to the scaphoid enters at the

waist, the risk of osteonecrosis and non-union increases the more proximal the fracture line.

(Reference 34)

10.20. A True
Two patterns of carpal dislocations/fracture–dislocations have been described. Lesser arc injuries are pure dislocations, whereas greater arc injuries are fractures or fracture–dislocations. Scapholunate dissociation is a lesser arc injury, resulting in an increase in the scapholunate space to greater than 2–3 mm on the AP view. This may only be seen on the AP view if the wrist is supinated or with the fist clenched.

10.20. B True
In rotatory subluxation, the axis of the scaphoid tilts in a palmar direction. On the AP view, the scaphoid appears foreshortened, and the distal pole is seen end-on, creating the 'ring sign' (NB: this may be a normal appearance with the wrist in radial deviation). Progression of a lesser arc injury results in dorsal dislocation of the capitate, volar subluxation of the lunate and, finally, lunate dislocation. The commonest greater arc injury is a trans-scaphoid perilunate dislocation.

10.20. C True
The two commonest patterns of carpal instability are dorsiflexion and volarflexion instability. They are best assessed on a true lateral radiograph, allowing measurement of the scapholunate angle, which is normally between 30° and 60°. In dorsiflexion instability, the axis of the scaphoid is tilted ventrally and the lunate is displaced in a ventral direction and tilted dorsally, increasing the angle.

10.20. D True
In volarflexion instability, both the scaphoid and lunate tilt in a ventral direction and the lunate is displaced dorsally. This results in a reduction of the scapholunate angle.

10.20. E False
Carpal instability may be assessed by fluoroscopy or by arthrography. Tears of the triangular fibrocartilage or interosseous ligaments can be indicated by abnormal intercompartmental communication. Contrast medium extending from the radiocarpal joint to the distal radio-ulnar joint is diagnostic of a triangular cartilage tear.

(Reference 34)

10.21. A True
Hip dislocations are classified as posterior, anterior or central, depending upon the relationship of the femoral head to the acetabulum. Posterior dislocation accounts for 80–85% of hip dislocations and commonly results from an anteriorly directed force to the flexed knee during a car accident. There is a

high incidence of associated femoral shaft and patellar fractures, as well as knee injuries.

10.21.   B True
Impaction of the posteriorly displaced femoral head may result in a fracture of the posterior acetabular rim, injury to the sciatic nerve or, less commonly, a shear fracture of the inferomedial aspect of the femoral head. Persistent joint widening following reduction suggests osseous or cartilaginous fragments within the joint, which can be demonstrated by CT.

10.21.   C True
Complications of posterior hip dislocation include secondary degenerative joint disease, periarticular ossification and femoral head osteonecrosis in up to 40% of cases. Osteonecrosis is commoner with increasing age and if there is a delay in reduction of more than 12 hours.

10.21.   D False
Anterior dislocation of the hip accounts for 10–15% of dislocations. It results from direct or indirect abduction forces. The femoral head may be displaced inferomedially (obturator dislocation), with the femur abducted and externally rotated, or anterosuperiorly (subspinous dislocation), with the femur extended and abducted.

10.21.   E False
Osteonecrosis complicates anterior hip dislocation in up to 8% of cases. Another association is a characteristic fracture of the femoral head resulting from impaction of the superolateral aspect of the head with the acetabulum. Recurrent dislocation is a rare complication.

(Reference 34)

10.22.   A False
A direct force to the patella may result in a vertical, stellate or comminuted fracture. However, 50–90% of patellar fractures are transverse and result from indirect forces from the quadriceps mechanism. These typically involve the midportion of the patella and, rarely, osteonecrosis of the proximal fragment may occur.

10.22.   B True
Less commonly, rupture of the infrapatellar tendon results in an avulsion fracture of the inferior pole of the patella. These fractures typically occur in active children. Radiographs show superior patellar displacement, infrapatellar soft-tissue swelling and increased density of the infrapatellar fat pad. Rupture of the quadriceps tendon may cause avulsion fractures of the superior pole of the patella. These occur most commonly in the elderly and have been associated with gout, renal failure, diabetes mellitus and hyperparathyroidism.

10.22.   C False
Acute traumatic dislocation of the patella is usually in a lateral direction. The dislocation is typically transient and not observed radiographically. Chondral or osteochondral fractures of the medial facet of the patella or lateral femoral

condyle often occur. These may be difficult to identify radiographically but can be demonstrated by arthrography, CT or MRI.

10.22. D True
Other factors predisposing to recurrent patellar subluxation include a previous dislocation, hypoplasia of the lateral femoral condyle, a shallow patellofemoral groove, a laterally located ligamentum patella and genu valgum. Patella alta may be diagnosed on a lateral view when the ratio of the length of the patellar tendon to the greatest length of the patella is more than 1.2.

10.22. E False
Patellar subluxation is best assessed on the axial view with the knee flexed to about 30°. Greater degrees of flexion may result in spontaneous reduction of the subluxed patella. Radiographic findings include a flat lateral femoral condyle, shallow patellofemoral groove, small medial facet, sclerosis of the lateral facet and lateral displacement of the patella.

(Reference 34)

10.23. A True
Approximately 80% of tibial plateau fractures involve the lateral tibial plateau and are caused by twisting or valgus forces. Various types include a vertical split of the joint margin, a local concave compression fracture due to impaction of the lateral femoral condyle and the split compression fracture, characterized by a peripheral vertical split fracture as well as compression of the medial surface of the lateral tibial plateau.

10.23. B True
Approximately 10–12% of tibial plateau fractures are associated with ligamentous injury, most commonly involving the medial collateral, anterior cruciate and lateral collateral ligaments. This may result in avulsion fractures from the medial and lateral femoral condyles, the intercondylar eminence, or proximal fibula.

10.23. C True
Lipohaemarthrosis is diagnostic of an intra-articular fracture and may be the only sign of a tibial plateau fracture, since the fracture line may be orientated obliquely to routine radiographic planes. These fractures may need to be investigated further by conventional or computed tomography to assess the extent of the fracture and the degree of depression.

10.23. D False
Dislocations of the proximal tibiofibular joint may be anterolateral (most commonly), anteromedial or superior in direction. Superolateral dislocation is associated with superior displacement of the fibular shaft and occurs with fractures of the tibial shaft or disruption of the distal tibiofibular joint.

10.23. E True
Fractures of the fibular head or neck usually occur in combination with ligamentous injuries or fractures about the knee or ankle. Fibular head fractures may result from a valgus force to the knee and are associated with medial collateral ligament injury or fracture of the lateral tibial plateau. Varus

stresses may cause an avulsion fracture of the proximal pole or styloid process of the fibula at the site of attachment of the lateral collateral ligament or biceps femoris muscle.

(Reference 34)

10.24. A True
Supination–external rotation (SER) injury is the commonest pattern of ankle injury. With the foot supinated, external rotation drives the talus against the fibula, resulting initially in rupture of the anteroinferior tibiofibular ligament, an avulsion fracture from the anterior tibial tubercle (Tilleaux fracture) or, rarely, a fracture of the anterior tip of the lateral malleolus.

10.24. B True
With greater degrees of SER, a short spiral fracture of the distal fibula may occur. This may be followed by an avulsion of the posterior lip of the tibia, fracture of the posterior malleolus, or rupture of the posteroinferior tibiofibular ligament. Finally, the deltoid ligament may rupture, causing widening of the medial joint space or there may be a transverse fracture of the medial malleolus.

10.24. C True
Supination–adduction (SAD) injury results in rupture of the lateral collateral ligament or a transverse fracture of the lateral malleolus. Further adduction drives the talus medially against the medial malleolus, producing an oblique or nearly vertical fracture of the medial malleolus.

10.24. D True
Pronation–external rotation (PER) initially results in rupture of the deltoid ligament or a transverse avulsion fracture of the medial malleolus. Further rotation forces the talus against the lateral malleolus, resulting in rupture of the anteroinferior tibiofibular ligament and distal interosseous membrane.

10.24. E True
External rotation injuries may produce a proximal fibular fracture termed the Maisonneuve fracture. This may be missed on the initial radiographic examination but should be suspected when an apparently isolated fracture of the posterior lip of the tibia is found with widening of the joint space, ankle instability, or other evidence of severe external rotation injury.

(Reference 34)

10.25. A True
Avulsion fractures may occur at the anterosuperior aspect of the talar neck (the site of attachment of the capsule), due to extreme plantar flexion. Eversion injury may result in a fracture from the medial attachment of the deltoid ligament. Severe dorsiflexion with external rotation can fracture the lateral process inferior to the lateral malleolus. Severe plantar flexion can fracture the posterior tubercle or the os trigonum between the calcaneus and posterior tibial lip.

10.25. B False
Coronal fractures of the talar neck are caused by a combination of dorsiflexion and a vertical force that drives the talus against the anterior tibial margin. This may result in posterior subluxation of the posterior fracture fragment. Osteonecrosis of the posterior fragment may occur in up to 80% of displaced fractures. Other complications include delayed or non-union, and secondary degenerative disease.

10.25. C True
The posterior third of the medial margin of the talar dome is also a characteristic site for osteochondral fracture. The lateral fracture results from inversion, while the medial fracture is due to plantar flexion with inversion and rotation. The osteochondral fragment may remain in situ, may rotate, or may be displaced.

10.25. D False
Boehler's angle normally measures 20–40°, and is decreased in the presence of a significant compression fracture of the calcaneus. These are the commonest calcaneal fractures and are due to a vertical force that drives the talus into the calcaneus. Fractures are usually intra-articular with involvement of the subtalar joint. About 10% are bilateral and injuries to the thoracolumbar spine and tibia are a recognized association.

10.25. E True
Tarsometatarsal fracture–dislocation (Lisfranc injury) results in lateral dislocation of the second, third and fourth metatarsals with dorsal or plantar displacement also. A second metatarsal base fracture is characteristic. The first metatarsal may dislocate laterally in the homolateral form of the injury, or medially in the divergent form.

(Reference 34)

10.26. A False
Any nodule that has a significantly cystic component on thyroid ultrasound is usually a benign colloid nodule or an adenomatous nodule that has undergone central haemorrhage or necrosis. Rarely, thyroid carcinoma, particularly the papillary type, may show varying degrees of cystic change.

10.26. B True
Although thyroid carcinomas are usually hypoechoic to normal thyroid tissue, this is a non-specific appearance. Since benign nodules are much commoner than malignant nodules, a solitary hypoechoic nodule is most likely to be benign. A solitary hyperechoic nodule is virtually always benign.

10.26. C True
Medullary thyroid carcinoma commonly exhibits echogenic foci (some of which may be associated with acoustic shadowing), either within the primary tumour, or within metastatically involved cervical lymph nodes. Pathologically, these represent foci of fibrosis or calcification around amyloid deposits. Punctate, diffuse calcifications are characteristic of papillary carcinoma.

10.26. D False
An anechoic halo surrounding a thyroid nodule (either complete or incomplete) has been reported in 60–70% of benign nodules and 15% of malignant nodules. In some cases, Doppler ultrasound has shown this halo to be due to vessels. Increased vascularity may be identified by Doppler ultrasound in both autonomously functioning adenomas and thyroid carcinomas.

10.26. E False
Benign thyroid nodules often coexist with malignant nodules. Also, at least 20% of cases of papillary carcinoma are multicentric. In one series, 64% of patients with thyroid carcinoma had at least one nodule detected by ultrasound in addition to the palpable dominant nodule.

(Reference 11)

10.27. A True
Recognized causes of a focal defect on scintigraphic examination of the thyroid gland include a cyst, a non-functioning adenoma, a colloid nodule, a malignant tumour, and focal thyroiditis. A cyst may commonly show very low background activity compared with a solid nodule, but this is an unreliable means of differentiating the two, and ultrasound is necessary. Presence of a non-functioning nodule should be considered if the isthmus is not well seen.

10.27. B True
Focal thyroiditis may appear as either a focal area of increased uptake or a region of reduced uptake. Also, scintigraphy may reveal a diffuse reduction or absence of activity. The scan appearances will become normal with resolution of symptoms. Other causes of a 'hot' nodule include a functioning adenoma.

10.27. C False
A solitary 'cold' nodule has approximately a 10% change of being malignant. Thyroid carcinomas tend to replace thyroid tissue rather than displace it. A functioning nodule is almost never malignant.

10.27. D False
Functioning thyroid nodules may or may not be autonomous. Functioning non-autonomous nodules are suppressible by T3 administration. Functioning autonomous nodules may appear as focal areas of increased uptake with partial suppression of the remainder of the gland. T3 administration will result in complete suppression of the normal gland, making the nodule more evident. Occasionally, the nodule will only be identified following administration of T3.

10.27. E True
A non-toxic nodule is associated with partial suppression of normal thyroid tissue. Development of a toxic nodule results in complete suppression of the remainder of the gland. The nodule will not be suppressible by T3 administration.

(Reference 5)

10.28. A True
Diffuse increase in uptake in a patient with a goitre may be due to Graves' disease, the early stages of Hashimoto's disease, iodine deficiency or organification defects. Diffuse normal uptake is seen in diffuse non-toxic (simple) goitres.

10.28. B True
Diffuse reduction in uptake in a patient with a goitre is seen with lymphoma, the later stages of Hashimoto's disease, iodine-induced goitre, and subacute (De Quervain's) thyroiditis. Multifocal areas of irregular uptake but with normal overall uptake is seen in cases of multinodular goitre, whereas irregular replacement of thyroid tissue is seen with diffuse carcinoma.

10.28. C True
The perchlorate discharge test is used to assess organification. The perchlorate molecule is similar to pertechnetate and competes with iodine. Oral or intravenous administration of perchlorate will block further uptake of iodine and displace any unbound iodine. In a normal test, the percentage uptake after perchlorate remains constant, while in a patient with inadequate organification the percentage uptake is decreased since the unbound iodine is displaced from the cell. Since pertechnetate is trapped but not bound, it cannot be used.

10.28. D False
An isotope of iodine (^{131}I or ^{123}I) is better than pertechnetate for demonstrating retrosternal extension of a goitre since there is higher uptake in the gland, better tissue-to-background ratio, and less bone absorption by the sternum. Neither technique will demonstrate non-functioning retrosternal tissue.

10.28. E True
Hashimoto's thyroiditis is associated with lymphocytic infiltration of the thyroid gland, resulting in uptake of gallium-67 citrate by the gland. Thyroid lymphoma will also cause increased uptake of gallium and is associated with Hashimoto's thyroiditis.

(Reference 5)

10.29. A False
Pure papillary carcinomas that have no colloid formation rarely take up significant amounts of ^{131}I. However, papillary carcinomas with colloid-producing follicular elements may accumulate significant amounts of ^{131}I. Uptake is a usual feature of follicular carcinoma but does not occur with anaplastic carcinomas, medullary carcinomas, or with lymphoma of the thyroid.

10.29. B True
At normal levels of TSH stimulation, metastases from differentiated thyroid carcinomas will not accumulate ^{131}I. Levels of TSH of at least 30 mU/litre must be shown to ensure that a functioning metastasis has not been missed.

Therefore, metastases from thyroid tumours will only accumulate ^{131}I after thyroid ablation, when TSH levels have risen.

10.29. C False
Following therapy (radioactive or surgical) for thyroid carcinoma, a whole-body ^{131}I scan can be performed to identify metastatic disease. However, using this technique, activity will normally be present in the nose, sinuses and mouth, the stomach and bowel and in the bladder. These must not be mistaken for sites of metastatic disease. Lung micrometastases may be demonstrated in the presence of a normal chest radiograph. Also, after a therapeutic dose of ^{131}I, residual uptake more than 6 months after treatment is likely to represent tumour.

10.29. D False
Anaplastic carcinoma does not accumulate any radioisotope. However, well-differentiated thyroid carcinoma may take up thallium-201 but does not give the same guide to ^{131}I treatment as a radioiodine scan.

10.29. E True
MIBG is taken up by a percentage of medullary carcinomas, both primary and metastatic. This tumour may also accumulate thallium-201 and ^{99m}Tc-labelled pentavalent DMSA, but appears as a 'cold' nodule on ^{131}I or ^{123}I scans. The tumour may be bilateral, corresponding to the distribution of 'C' cells.

(Reference 5)

10.30. A False
Normal parathyroid glands measure approximately $0.5 \times 0.3 \times 0.1$ cm in size and are rarely visualized sonographically. The smallest adenomas may not appear enlarged at surgery, but are found to be hypercellular at histological examination.

10.30. B False
At ultrasound, the typical parathyroid adenoma appears as a solid, oval, homogeneously hypoechoic mass measuring approximately 1 cm in greatest dimension. As the gland enlarges, it usually does so in a longitudinal fashion, and may become tubular in appearance. The largest adenomas may be up to 5 cm in size.

10.30. C True
The majority of parathyroid adenomas are uniformly hypoechoic relative to the thyroid gland, their homogeneous echotexture being related to their uniform hypercellularity. Cystic areas may be seen in about 2% of cases and calcification is a recognized but rare feature.

10.30. D False
Superior parathyroid adenomas are typically located deep to the middle portion of one of the lobes of the thyroid gland. The position of inferior adenomas is more variable but most are located either posterior to the caudal tip of a thyroid lobe or in the soft tissues, 1–2 cm below the thyroid.

10.30.   E False
Parathyroid carcinomas are frequently lobulated with heterogeneous echotexture and cystic areas. They are usually larger than adenomas, with an average diameter of 2 cm. However, these sonographic features can also be seen with larger adenomas, and in the absence of obvious local invasion or metastasis the two cannot be differentiated by ultrasound.

(Reference 11)

10.31.   A False
A parathyroid adenoma may be imaged using combined thallium-201 and pertechnetate subtraction scintigraphy. Thallium is taken up by both parathyroid and thyroid tissue, whereas pertechnetate is taken up only by the thyroid. Computer subtraction of the pertechnetate image from the thallium image identifies residual parathyroid uptake. Normal parathyroid glands and 50% of hyperplastic glands are not visualized by this technique.

10.31.   B False
The parathyroid glands are located in a normal site in approximately 75% of cases. The remaining glands are ectopic and may be found in a retropharyngeal, retrolaryngeal, intrathyroid, retro-oesophageal, cervical, and anterior or posterior mediastinal location, and also in relation to the carotid sheath. In approximately 5% of cases, more than four glands are present.

10.31.   C True
Whether a parathyroid gland will be visualized using the thallium/pertechnetate subtraction technique depends upon its size, with the majority of glands weighing more than 500 mg being identified (the normal gland weighs 20 mg). The technique has a sensitivity of about 70% for parathyroid adenoma detection.

10.31.   D True
If a parathyroid adenoma is large enough, its mass effect may result in a 'cold' defect on the pertechnetate scan. However, such a finding may also be due to a non-functioning thyroid mass such as a cyst or non-functioning solid nodule.

10.31.   E True
Primary hyperparathyroidism is due to an adenoma in approximately 83% of cases, hyperplasia in 15% and carcinoma in 2%. Thallium/pertechnetate subtraction imaging may be of value in identification of recurrent parathyroid carcinoma.

(Reference 5)

10.32.   A True
Branchial cleft cysts can arise anywhere from the tonsillar fossa to the supraclavicular area but most are located anteromedial to the sternocleidomastoid muscle. CT typically demonstrates a water-density mass that displaces the sternocleidomastoid muscle posteriorly, the carotid artery and jugular vein medially or posteromedially and the submandibular gland anteriorly.

10.32. B True
Thyroglossal duct cysts can occur anywhere from the foramen caecum in the tongue to the pyramidal lobe of the thyroid. Approximately 65% are infrahyoid, 20% suprahyoid, and 15% at the level of the hyoid (usually anterior but may be intrahyoid or posterior). On CT, the mass may be septated and show rim enhancement. Solid areas are rarely seen and may represent complicating papillary carcinoma.

10.32. C True
Neural tumours may occur anywhere along the course of the cranial or cervical nerve roots. Their attenuation on precontrast CT depends upon the relative proportions of neural and fibrous tissue and the presence of cystic change. They show a varying degree of enhancement on postcontrast scans.

10.32. D False
Such an appearance would be consistent with abscess, including tuberculous adenitis, but may also be seen with malignant lymph node involvement. Infected thyroglossal duct or branchial cysts also enter the differential diagnosis depending upon the site of the mass.

10.32. E True
CT findings in cases of jugular vein thrombosis include enlargement of the vein, a non-enhancing defect within the vessel lumen, enhancement of the vessel wall (presumably due to flow through the arterially supplied vasa vasorum), and demonstration of venous collaterals. Acute thrombus may be relatively hyperdense on precontrast CT and there may also be soft-tissue swelling and loss of normal fascial planes.

(Reference 2)

10.33. A False
Bowing of a bony wall implies slow growth and is classically seen with mucocele formation. It may also occur with benign neoplasms and rarely with malignant neoplasms. Bone destruction is also not diagnostic of malignancy, since it may be seen with other aggressive conditions such as granulomatous disease and mycotic infection.

10.33. B True
Reactive sclerosis of a sinus wall usually indicates a chronic lesion and is most commonly seen in association with inflammatory disease, such as chronic sinusitis. It may also occur in response to tumour, surgery or radiotherapy. Rarely, meningioma may be a cause of hyperostosis and sclerosis of a sinus wall, particularly the sphenoid sinus.

10.33. C True
Opacification of a sinus can be due to inflamed mucosa, cyst, tumour, retained secretions, blood, or occasionally as the result of abnormal development. In the context of neoplasm, opacification may be due to the tumour itself, or due to retained secretions within a sinus that is obstructed by tumour. Following a

rapid bolus of contrast medium, tumour may enhance allowing differentiation from non-enhancing secretions.

10.33. D False
When sinus mucosa becomes infected or inflamed, it appears enlarged and enhances on postcontrast scans. Tumour often enhances also and in this situation it may be difficult to distinguish between the two.

10.33. E False
On MRI, inflammatory tissue, retained secretions and tumour are most clearly differentiated on T2-WSE scans, in which most tumours have intermediate signal intensity, and inflamed mucosa and secretions have high signal intensity. However, some neoplasms such as neural tumours, salivary gland tumours and haemangiomas may be relatively hyperintense, making distinction less easy.

(Reference 35)

10.34. A False
Tumour spread through the posterolateral wall of the maxillary antrum will extend into the infratemporal fossa. Normally there is a well-defined fat plane behind the antral wall, separating it from the temporalis muscle. Effacement of this fat indicates tumour spread. Posteromedial spread may involve the pterygopalatine fossa, which also contains fat. Obliteration of this fat is a sign of tumour extension.

10.34. B True
The alveolar and palatine nerves can also carry tumour to the pterygopalatine fossa. Another route of perineural (transforaminal) spread includes that along the maxillary division of the trigeminal nerve, through the foramen rotundum, into the middle cranial fossa. Although enlargement of a foramen is good evidence of transforaminal spread, a normal sized foramen does not exclude tumour spread.

10.34. C True
Extension superiorly through the roof of the antrum into the orbit may be indicated by obliteration of the fat that is normally present below the inferior rectus muscle. Similarly, intraorbital extension of tumour from the ethmoid air cells may obliterate the fat normally seen medial to the medial rectus muscle.

10.34. D True
The maxilloethmoid plate is a bony septum that separates the maxillary antrum from the ethmoid air cells. Superomedial extension of tumour may destroy this, with resulting extension into the ethmoid sinus.

10.34. E False
Lymphatic spread from sinus and nasal cavity neoplasms is relatively uncommon. The nodal groups usually involved include the lateral pharyngeal nodes, the jugulodigastric nodes, and the deep cervical nodes. Tumours of the superior aspect of the nasal cavity can drain to the retropharyngeal nodes.

(Reference 35)

10.35. A True
Squamous carcinoma accounts for approximately 80% of malignant tumours of the sinuses and nasal cavity. Glandular tumours account for 10–14% and include adenoid cystic carcinoma, adenocarcinoma, mucoepidermoid and undifferentiated carcinomas, and pleomorphic adenomas. Rarer tumours include melanoma, plasmacytoma, lymphoma and sarcomas.

10.35. B True
Aesthesioneuroepithelioma (olfactory neuroblastoma) arises high in the nasal cavity and frequently extends through the cribriform plate into the anterior cranial fossa, and also into the ethmoid sinuses. The tumour enhances on postcontrast CT and has high signal intensity on T2-WSE MRI scans. It is commonest in young men.

10.35. C True
Perineural (transforaminal) spread is a characteristic feature of adenoid cystic carcinoma, resulting in intracranial spread. This tumour also appears to extend through bone without causing frank bone destruction.

10.35. D True
An inverting papilloma is a benign, locally invasive neoplasm. The lesion is almost always unilateral and typically arises from the lateral wall of the nasal cavity. Nasal polyps are inflammatory in origin and tend to cause bone remodelling rather than destruction. They are frequently bilateral, especially when related to allergy, and may or may not cause sinus obstruction.

10.35. E False
Haematogenous metastases are relatively uncommon from primary tumours of the sinuses and nasal cavity. Involvement of the middle cranial fossa may occur as a result of perineural spread from maxillary antrum tumours or as a consequence of direct lateral spread from a tumour in the sphenoid sinus, a relatively rare site for a primary tumour. This may manifest on postcontrast CT as lateral bowing of the cavernous sinus and extension into the temporal lobe.

(Reference 35)

10.36. A True
Mucormycosis and other fungal infections can destroy significant amounts of bone and have appearances that mimic malignant disease. Mucormycosis may extend through the orbital apex into the cavernous sinus, and occlusion of the ophthalmic and internal carotid arteries is a recognized complication. Wegener's granulomatosis and lethal midline granuloma are other benign conditions that may be associated with bone destruction.

10.36. B True
Aspergillosis may have a characteristic appearance on CT and MRI. On postcontrast CT, the mycetoma is hyperdense. On MRI, there is very little or no signal, mimicking an air-filled sinus. This is thought to be related to certain paramagnetic properties of the mycetoma itself.

10.36.   C False
A mucocele forms when a sinus is completely obstructed, usually due to inflammatory disease, but also due to neoplasm. Involvement of the frontal and ethmoid sinuses is most common, followed by the sphenoid sinus. Involvement of the antrum is relatively rare. In the case of the ethmoids, a single air cell may be obstructed and grossly dilated.

10.36.   D False
Mucocele typically causes expansion of the sinus with remodelling and thinning of the bony wall. However, defects in the bony cortex can occur, resulting in extension of the mucocele into the orbit or anterior cranial fossa. Usually, a mucocele has uniformly low soft-tissue attenuation on precontrast CT and does not enhance. Increased attenuation and enhancement may occur in the presence of infection.

10.36.   E True
A retention cyst usually presents in the base of the maxillary antrum. It has uniform low density, with a smooth convex upper margin. Unlike a mucocele, it typically does not fill the sinus.

(Reference 35)

10.37.   A True
Approximately 90% of nasopharyngeal malignancies are squamous cell carcinomas. Early lesions may efface the fossa of Rosenmüller, widen the deglutitional muscle ring or obstruct the Eustachian tube. This latter feature can be suspected when CT or MRI demonstrates fluid in an otherwise normally developed mastoid.

10.37.   B False
The parapharyngeal space appears on CT as an area of fat density just lateral to the deglutitional muscle (Passavant's) ring. It may be divided into an anterior prestyloid space and a posterior poststyloid space. Deep extension of nasopharyngeal masses deforms the parapharyngeal space from the medial aspect. Posterior or superior extension can result in erosion of the skull base. Intracranial spread may be by direct extension through bone or through cranial sutures.

10.37.   C True
The prestyloid space contains predominantly fat, but may also contain a small medial extension of the deep lobe of the parotid gland. A mass that is entirely within this space is most likely to be a pleomorphic adenoma. Masses in this site cause medial deviation of the parapharyngeal fat and may be continuous with the parotid gland through the stylomandibular canal, producing a dumbbell tumour. A mass that extends anteriorly into the masticator space or posteriorly into the poststyloid space is unlikely to be a benign parotid tumour, but a malignant lesion would have to be considered.

10.37.   D True
The poststyloid space contains the carotid artery, jugular vein, the IXth, Xth and XIIth cranial nerves, and the lateral pharyngeal lymph nodes. Therefore, primary masses are related to these structures and include paraganglionomas,

neuromas, and nodal masses. Paraganglionomas include glomus vagale tumours and inferior extensions of glomus jugulare tumours. On CT, they enhance dramatically, and on MRI they may have a 'salt and pepper' appearance due to multiple areas of signal void related to their vascularity.

10.37. E True
The infratemporal fossa (masticator space) contains the temporalis, pterygoid and masseter muscles, the mandible and the third division of the trigeminal nerve. Masses arising here are usually local extensions from primary tumours in the surrounding structures. Primary tumours of the infratemporal fossa include sarcomas, undifferentiated carcinomas, neural lesions, meningiomas, and tumours of the mandible.

(Reference 35)

10.38. A False
The false cords contain a relatively large amount of fat compared to the true cords and therefore have lower than soft-tissue density on CT. They are attached anteriorly to the inner surface of the thyroid cartilage, and at this site there may normally be some soft-tissue thickening which must not be mistaken for tumour spread. The false cords are separated from the true cords by the laryngeal ventricle.

10.38. B True
The true cords appear as bands of soft-tissue density that attach posteriorly to the vocal process of the arytenoid cartilage. Their anterior attachment is to the laryngeal prominence, forming the anterior commissure. The posterior commissure lies between the vocal processes on the anterior surface of the cricoid lamina. The true cords have soft-tissue density and taper anteriorly.

10.38. C True
There is great variation in the degree of calcification of the thyroid cartilage. The cartilage may be uniformly calcified or have a rim of calcification/ ossification around a low-density medullary cavity. The pattern of mineralization is frequently irregular, and segmental interruptions may simulate neoplastic invasion.

10.38. D True
The pre-epiglottic space is bordered superiorly by the valleculae and base of the tongue, posteriorly by the epiglottis, anteriorly by the hyoid bone, and laterally by the paralaryngeal space, thyrohyoid membrane and thyroid cartilage. It extends from the hyoid bone to just above the anterior commissure. It is predominantly fat containing, resulting in its low density on CT and high signal intensity on MRI.

10.38. E True
The epiglottis is composed of fibroelastic cartilage, and on MRI has intermediate signal intensity similar to skeletal muscle. The signal from the remainder of the laryngeal skeleton depends upon the degree of calcification/ ossification and also on the medullary fat content.

(Reference 2)

10.39.   A True
Approximately 50–60% of laryngeal carcinomas arise from the glottis and 75% of these originate on the anterior half of the true vocal cord. Cervical nodes at presentation are not typically seen due to the lack of lymph drainage from the cords. Most glottic tumours are localized and CT typically demonstrates focal or diffuse thickening of the true cord. A paralysed true cord may have an identical appearance.

10.39.   B True
Glottic tumours may spread superiorly into the laryngeal ventricle and false cord, replacing the normal fat density and enlarging the false cord. Subglottic extension results in the appearance of tumour between the cricoid ring and the airway when, in the normal situation, there is no soft tissue on the inner surface of the cricoid cartilage at a level just below the true cords.

10.39.   C True
The anterior commissure is normally less than 2 mm in AP diameter, and anterior extension may thicken this space. Tumour can then extend into the contralateral cord. Generally, more than 30% involvement of the contralateral cord precludes vertical hemilaryngectomy. Anterior spread may also involve the thyroid cartilage while posterior spread thickens the posterior commissure.

10.39.   D False
The supraglottis is bounded superiorly by the epiglottis and inferiorly by the laryngeal ventricle. Approximately 20–30% of primary laryngeal carcinomas arise in the supraglottic region, either from the epiglottis, the aryepiglottic fold, the laryngeal ventricle or the false cord. Pyriform sinus tumours account for 10–20%, while primary subglottic tumours account for less than 5% of laryngeal carcinomas.

10.39.   E True
Carcinoma of the epiglottis may appear on CT as a focal thickening of the epiglottis or as a bulky mass. Anterior spread is via the pre-epiglottic space, with superior extension into the valleculae and the base of the tongue. Inferiorly, tumour extension may thicken the anterior commissure, while lateral spread occurs via the paralaryngeal space. This space appears on CT as a low-density layer medial to the thyroid cartilage.

(Reference 2)

10.40.   A False
On precontrast CT, lymph nodes are usually of homogeneous soft-tissue density. Peripheral enhancement of a node is not normally seen. Fatty replacement of part of a node is a benign reactive process which is occasionally seen. CT demonstrates low density within the node which is characteristically in a peripheral, non-central location.

10.40.   B False
Normal cervical lymph nodes are usually 3–5 mm in diameter. However, they may normally be up to 1.5 cm in diameter in the upper neck (e.g. the jugulodigastric and submandibular nodes). The jugulodigastric node is the largest of the internal jugular group of deep cervical nodes, being seen just

above the hyoid bone near the junction of the internal jugular vein and the posterior belly of digastric muscle.

10.40.   C   True
Cervical nodes are considered abnormal if they measure greater than 1.5 cm in diameter. Also, a nodal mass with central low density and peripheral enhancement is considered abnormal, regardless of size. The presence of central necrosis is suggestive of malignant involvement rather than reactive hyperplasia.

10.40.   D   True
The lymph nodes of the head and neck may be divided into ten major groups. The first six (occipital, mastoid, parotid, submandibular, facial, and submental) are usually easily palpable. Sublingual and retropharyngeal nodes lie deep to these and are not easily palpable. The retropharyngeal nodes are usually located along the lateral edges of the longus capitis muscle at the C1–2 level. They are the major nodes involved from tumours of the nasal cavity, nasopharynx, and hard and soft palates.

10.40.   E   False
The remaining two nodal groups are the anterior and lateral cervical nodes. The latter are divided into superficial (external jugular) and deep (internal jugular, spinal accessory, and transverse cervical) nodes. The deep lateral cervical nodes are the major sites of metastases from laryngeal and hypopharyngeal carcinoma.

(Reference 2)

Bibliography

1. Husband, J. E. S. (ed.) (1989) *CT Review*, Churchill Livingstone, Edinburgh.
2. Lee, J. K. T., Sagel, S. S. and Stanley, R. K. (eds) (1989) *Computed Body Tomography with MRI Correlation*, Raven Press, New York.
3. Edelman, R. R. and Hesselink, J. R. (eds) (1990) *Clinical Magnetic Resonance Imaging*, Saunders, Philadelphia.
4. Armstrong, P., Wilson, A. G. and Dee, P. (eds) (1990) *Imaging of Diseases of the Chest*, Year Book Medical Publishers, Massachusetts.
5. Fogelman, I. and Maisey, M. (1988) *An Atlas of Clinical Nuclear Medicine*, Martin Dunitz, London.
6. Woodring, J. H. (ed.) (1990) *Lung Cancer*, Radiologic Clinics of North America, Vol. 28, No. 3, Saunders, Philadelphia.
7. Federle, M. P., Megibow, A. J. and Naidich, D. P. (eds) (1988) *The Radiology of AIDS*, Raven Press, New York.
8. Muller, N. L. (ed.) (1991) *High Resolution CT of the Chest*, Seminars in Roentgenology, Vol. 26, No. 2, Saunders, Philadelphia.
9. Haaga, J. R. and Alfidi, R. J. (eds) (1988) *Computed Tomography of the Whole Body*, Mosby, St Louis.
10. Kadir, S. (1986) *Diagnostic Angiography*, Saunders, Philadelphia.
11. Rifkin, M. D., Charboneau, J. W. and Laing, F. C. (eds) (1991) *Syllabus: Special Course. Ultrasound*, Presented at the 77th Scientific Assembly and Annual Meeting of the Radiological Society of North America, Chicago.
12. Raymond, H. W., Zwiebel, W. J. and Harnsberger, H. R.(eds) (1991) *Advances in Cardiac Imaging*, Seminars in Ultrasound, CT and MR, Vol. 12, No. 1, Saunders, Philadelphia.
13. Sacks, D. (ed.) (1992) *Noninvasive Vascular Imaging*, Seminars in Roentgenology, Vol. 27, No. 1, Saunders, Philadelphia.
14. Robinson, P. J. (ed.) (1986) *Nuclear Gastroenterology*, Churchill Livingstone, Edinburgh.
15. Amis, E. S. Jr (ed.) (1991) *Contemporary Uroradiology*, Radiologic Clinics of North America, Vol. 29, No. 3, Saunders, Philadelphia.
16. Landay, M. J. and Virolainen, H. (1991) 'Hyperdense aortic wall': A potential pitfall in screening for aortic dissection. *Journal of Computer Assisted Tomography*, **15**, 561–564.
17. Taylor, K. J. and Strandness, D. E. Jr (eds) (1990) *Duplex Doppler Ultrasound*, Clinics in Diagnostic Ultrasound, Vol. 26, Churchill Livingstone, Edinburgh.
18. Sharp, P. F., Gemmell, H. G. and Smith, F. W. (eds) (1989) *Practical Nuclear Medicine*, Oxford University Press, Oxford.
19. Bernardino, M. E. (ed.) (1991) *Imaging of the Liver and Biliary Tree*, Radiologic Clinics of North America, Vol. 29, No. 6, Saunders, Philadelphia.
20. Friedman, A. C. (ed.) (1987) *Radiology of the Liver, Biliary Tract, Pancreas and Spleen*, Golden's Diagnostic Radiology, Williams & Wilkins, Baltimore.

21. Freeny, P. C. (ed.) (1989) *Radiology of the Pancreas*, Radiologic Clinics of North America, Vol. 27, No. 1, Saunders, Philadelphia.
22. Rolfes, R. J. and Ros, P. R. (1990) *The Spleen: An Integrated Imaging Approach*, Critical Reviews in Diagnostic Imaging, Vol. 30, No. 1, pp. 41–83, CRC Press, Boca Raton, FL.
23. Arger, P. H. (ed.) (1990) *Obstetrical Ultrasound*, Seminars in Roentgenology, Vol. 25, No. 4, Saunders, Philadelphia.
24. Friedman, A., Radecki, P., Lev-Toaf, A. and Hulbert, P. (eds) (1990) *Clinical Pelvic Imaging*, Mosby, St Louis.
25. Callan, P. W. (1988) *Ultrasonography in Obstetrics and Gynaecology*, Saunders, Philadelphia.
26. Kopans, D. B. (1989) *Breast Imaging*, Lippincott, Philadelphia.
27. Atlas, S. W. (ed.) (1991) *Magnetic Resonance Imaging of the Brain and Spine*, Raven Press, New York.
28. Siegel, M. J. (1991) *Paediatric Sonography*, Raven Press, New York.
29. Babcock, D. S. (ed.) (1989) *Neonatal and Paediatric Ultrasonography*, Clinics in Diagnostic Ultrasound, Vol. 24, Churchill Livingstone, Edinburgh.
30. Cohen, M. D. (1992) *Imaging of Children with Cancer*, Mosby Year Book, St Louis.
31. Poznanski, A. K. and Kirkpatric, J. A. (eds) (1989) *Syllabus: A Categorical Course in Diagnostic Radiology, Paediatric Radiology*, Presented at the 75th Assembly and Annual Meeting of the Radiological Society of North America, Chicago.
32. Siegel, M. J. (ed.) (1988) *Paediatric Body CT*, Contemporary Issues in Computed Tomography, Vol. 10, Churchill Livingstone, Edinburgh.
33. Oh, K. S. and Bender, T. M. (eds) (1988) *Recent Advances in Practical Paediatric Radiology*, Radiologic Clinics of North America, Vol. 26, No. 2, Saunders, Philadelphia.
34. Sartoris, D. J. (ed.) (1989) *Musculoskeletal Trauma*, Radiologic Clinics of North America, Vol. 27, No. 5, Saunders, Philadelphia.
35. Latchaw, R. E. (ed.) (1991) *MR and CT Imaging of the Head, Neck, and Spine*, Mosby Year Book, St Louis.
36. Hall, C. M. and Shaw, D. G. (1991) *Non Accidental Injury*, Current Imaging, Vol. 3, No. 2, pp. 88–93, Churchill Livingstone, Edinburgh.
37. Mittelstaedt, C. A. (1987) *Abdominal Ultrasound*, Churchill Livingstone, Edinburgh.
38. Gore, R. M. (ed.) (1989) *CT of the Gastrointestinal Tract*, Radiologic Clinics of North America, Vol. 27, No. 4, Saunders, Philadelphia.
39. Whyte, A. M. (1991) *Imaging of the Salivary Glands*, Current Imaging, Vol. 3, No. 3, pp. 131–137, Churchill Livingstone, Edinburgh.
40. Johnson, R. J. and Carrington, B. M. (1992) *Pelvic Radiation Disease*, Clinical Radiology, Vol. 45, No. 1, pp. 4–12, Blackwell Scientific Publications, Oxford.
41. Scherrer, A., Reboul, F., Martin, D., Dupuy, J. C. and Menu, Y. (1990) CT of malignant anal canal tumours. *Radiographics*, **10**, 433–453.
42. Einstein, D. M., Singer, A. A., Chilcote, W. A. and Desai, R. K. (1991) Abdominal lymphadenopathy: Spectrum of CT findings. *Radiographics*, **11**, 457–472.
43. Lane, R. H., Stephens, D. H. and Reiman, H. M. (1989) Primary retroperitoneal neoplasms: CT findings in 90 cases with clinical and pathological correlation. *American Journal of Roentgenology*, **152**, 83–89.

44. Clements, R., Griffiths, G. J. and Peeling, W. B. (1991) 'State of the art' transrectal US imaging in the assessment of prostatic disease. *British Journal of Radiology*, **64**, 193–200.
45. Montana, M. A. and Richardson, M. L. (eds) (1988) *Ultrasonography of the Musculoskeletal System*, Radiologic Clinics of North America, Vol. 26, No. 1, Saunders, Philadelphia.
46. Modic, M. T. (ed.) (1991) *Imaging of the Spine*, Radiologic Clinics of North America, Vol. 29, No. 4, Saunders, Philadelphia.
47. Kaye, J. J. (ed.) (1990) *Imaging of Joints*, Radiologic Clinics of North America, Vol. 28, No. 5, Saunders, Philadelphia.
48. Cain, T. M., Fon, G. T. and Howie, D. W. (1990) *The Loose Hip Prosthesis*, Current Imaging, Vol. 2, No. 3, pp. 165–176, Churchill Livingstone, Edinburgh.
49. Resnick, D. and Niwayama, G. (eds) (1988) *Diagnosis of Bone and Joint Disorders*, 2nd edn, Vol. 1, Saunders, Philadelphia.
50. Boorstein, J. M., Kneeland, J. B., Dalinka, M. K., Iannotti, J. P. and Suh, J. (1992) *Magnetic Resonance Imaging of the Shoulder*, Current Problems in Diagnostic Radiology, Vol. 11, No. 1, Mosby Year Book, St Louis.
51. Kurtz, A. B. and Goldberg, B. B. (eds) (1988) *Gastrointestinal Ultrasonography*, Clinics in Diagnostic Ultrasound, Vol. 23, Churchill Livingstone, Edinburgh.